Studies in Logic

Mathematical Logic and Foundations
Volume 55

All about Proofs, Proofs for All

Studies in Logic Series Editor
Dov Gabbay   dov.gabbay@kcl.ac.uk

All about Proofs, Proofs for All

Edited by
Bruno Woltzenlogel Paleo
and
David Delahaye

ISBN 978-1-84890-166-7

College Publications
Scientific Director: Dov Gabbay
Managing Director: Jane Spurr

http://www.collegepublications.co.uk

Original cover design by Orchid Creative   www.orchidcreative.co.uk
Printed by Lightning Source, Milton Keynes, UK

Preface

The development of new and improved proof systems, proof formats and proof search methods is one of the most essential goals of Logic. But what is a proof? What makes a proof better than another? How can a proof be found efficiently? How can a proof be used? Logicians from different communities usually provide radically different answers to such questions. Their principles may be folklore within their own communities but are often unknown to outsiders.

The debate about the importance of proofs and what makes some proofs better than others has a long tradition that is probably as old as Logic itself. In the modern era we could, for example, quote Poincaré's opinion that "a collection of facts is no more a science than a heap of stones is a house"; cite Hilbert's unpublished 24th problem, which asks for criteria to judge the simplicity of proofs; and mention Kreisel's proof unwinding program and his question of "what more do we know if we have proved a theorem by restricted means other than if we merely know the theorem is true?". With the current multitude of proof-generating deduction tools, proof systems and practical applications of proofs, this debate is more pertinent and lively than ever. But it seems unfortunately fragmented across many communities.

The Vienna Summer of Logic presented a unique opportunity to promote a conversation about proofs among people from various Logic communities. To take advantage of this opportunity, we have organized a one-day workshop (on the 18th of July 2014), comprising 12 invited tutorials. Each tutorial was given by an invited expert from a different community, following a template of questions prepared by the editors with the aim of ensuring a broad coverage of topics. The chapters in this book reflect their tutorials.

After the workshop, we asked each invited speaker to indicate reviewers from his/her community, and we asked each indicated reviewer to review from 2 to 3 papers from other communities. The goal of this cross-community reviewing process was to ensure accessibility of every chapter to people from different Logic backgrounds.

We would like to thank all invited speakers and chapter authors for their invaluable contributions to the workshop and to this book. We are grateful to all the reviewers (listed in the next page), whose assistance was essential for achieving the goals of this book. The reviewing process ran smoothly and efficiently thanks to EasyChair. The organizers of the Vienna Summer of Logic, the largest logic conference ever, must be thanked for providing a great environment for our event. And finally, we are thankful to College Publications for agreeing to publish this book, and especially to Jane Spurr, for her guidance in the publication process.

We hope and wish that this book will become a useful resource for logicians interested in the current state-of-the-art with respect to proofs.

<table>
<tr><td>January 10, 2015</td><td>Bruno Woltzenlogel Paleo</td></tr>
<tr><td>Vienna and Paris</td><td>David Delahaye</td></tr>
</table>

Reviewers

Jesse Alama

Tomáš Balyo

Joshua Bax

David Cerna

Zakaria Chihani

Simon Cruanes

Etienne Duchesne

Andreas Fröhlich

Georg Hofferek

Chantal Keller

Giselle Reis

Andrew Reynolds

Martin Riener

Alexander Steen

Max Wisniewski

Table of Contents

Proofs for Satisfiability Problems

Marijn J.H. Heule[1] and Armin Biere[2]

[1] The University of Texas at Austin, United States
[2] Johannes Kepler University, Linz, Austria

1 Introduction

Satisfiability (SAT) solvers have become powerful tools to solve a wide range of applications. In case SAT problems are satisfiable, it is easy to validate a witness. However, if SAT problems have no solutions, a proof of unsatisfiability is required to validate that result. Apart from validation, proofs of unsatisfiability are useful in several applications, such as interpolation [64] and extracting a minimal unsatisfiable set (MUS) [49] and in tools that use SAT solvers such as theorem provers [4,65,66,67].

Since the beginning of validating the results of SAT solvers, proof logging of unsatisfiability claims was based on two approaches: *resolution proofs* and *clausal proofs*. Resolution proofs, discussed in zChaff in 2003 [69], require for learned clauses (lemmas) a list of antecedents. On the other hand, for clausal proofs, as described in Berkmin in 2003 [32], the proof checker needs to find the antecedents for lemmas. Consequently, resolution proofs are much larger than clausal proofs, while resolution proofs are easier and faster to validate than clausal proofs.

Both proof approaches are used in different settings. Resolution proofs are often required in applications like interpolation [47] or in advanced techniques for MUS extraction [50]. Clausal proofs are more popular in the context of validating results of SAT solvers, for example during the SAT Competitions or recently for the proof of Erdős Discrepancy Theorem [41]. Recent works also use clausal proofs for interpolation [33] and MUS extraction [11].

Proof logging support became widespread in state-of-the-art solvers, such as Lingeling [13], Glucose [7], and CryptoMiniSAT [57], since SAT Competition 2013 made unsatisfiability proofs mandatory for solvers participating in the unsatisfiability tracks. About half the solvers that participated in recent SAT Competitions can emit clausal proofs, including the strongest solvers around, for example the three solvers mentioned above. However, very few solvers support emitting resolution proofs.

The lack of support for resolution proofs is due to the difficulty to represent some techniques used in contemporary SAT solvers in terms of resolution. One such technique is conflict clause minimization [58], which requires several modifications of the solver in order to express it using resolution steps [62]. In contrast, emitting a clausal proof from SAT solvers such as MiniSAT [28] and Glucose requires only small changes to the code[3].

---

[3] A patch that adds clausal proof logging support to MiniSAT and Glucose is available on http://www.cs.utexas.edu/~marijn/drup/.

1

2 Proof Systems

2.1 Preliminaries and Notation

We briefly review necessary background concepts regarding the Boolean satisfiability (SAT) problem, one of the first problems that were proven to be NP-complete [21]. For a Boolean (or propositional) variable x, there are two *literals*, the positive literal, denoted by x, and the negative literal, denoted by $\bar{x}$. A *clause* is a finite disjunction of literals, and a CNF *formula* is a finite conjunction of clauses. When appropriate we also interpret a clause as a set of literals and a CNF formula as as set of clauses. A clause is a *tautology* if it contains both x and $\bar{x}$ for some variable x. The set of variable and literals occurring in a CNF formula F is denoted by $\mathsf{vars}(F)$ and $\mathsf{lits}(F)$, respectively. A *(truth) assignment* τ for a CNF formula F is a partial function that maps literals $l \in \mathsf{lits}(F)$ to $\{\mathbf{t}, \mathbf{f}\}$. If $\tau(l) = v$, then $\tau(\bar{l}) = \neg v$, where $\neg\mathbf{t} = \mathbf{f}$ and $\neg\mathbf{f} = \mathbf{t}$. An assignment can also be thought of as a conjunction of literals. Furthermore, given an assignment τ:

- A clause C is *satisfied* by τ if $\tau(l) = \mathbf{t}$ for some $l \in C$.
- A clause C is *falsified* by τ if $\tau(l) = \mathbf{f}$ for all $l \in C$.
- A formula F is *satisfied* by τ if $\tau(C) = \mathbf{t}$ for all $C \in F$.
- A formula F is *falsified* by τ if $\tau(C) = \mathbf{f}$ for some $C \in F$.

A CNF formula with no satisfying assignments is called *unsatisfiable*. A clause C is *logically implied* by CNF formula F if adding C to F does not change the set of satisfying assignments of F. The symbol ϵ refers to the unsatisfiable empty clause. Any CNF formula that contains ϵ is unsatisfiable. A proof of unsatisfiability shows why ϵ is redundant (i.e., its addition preserves satisfiability) with respect to a given CNF formula.

2.2 Resolution

The resolution rule [52] states that, given two clauses $C_1 = (x \vee a_1 \vee \ldots \vee a_n)$ and $C_2 = (\bar{x} \vee b_1 \vee \ldots \vee b_m)$ with a complementary pair of literals (in this case x and $\bar{x}$), the clause $C = (a_1 \vee \ldots \vee a_n \vee b_1 \vee \ldots \vee b_m)$, can be inferred by resolving on variable x. We say C is the *resolvent* of C_1 and C_2 and write $C = C_1 \diamond C_2$. C_1 and C_2 are called the *antecedents* of C. C is logically implied by any formula containing C_1 and C_2. A *resolution chain* is a sequence of resolution operations such that the result of the last operation is an antecedent of the next operation. Resolution chains are computed from left to right. Notice that the resolution operation is not associative. For example, $\big((a \vee c) \diamond (\bar{a} \vee b)\big) \diamond (\bar{a} \vee \bar{b}) = (\bar{a} \vee c)$, while $(a \vee c) \diamond \big((\bar{a} \vee b) \diamond (\bar{a} \vee \bar{b})\big) = (c)$.

Throughout this chapter we will use the following formula E as example to explain various concepts:

$$E := (\bar{b} \vee c) \wedge (a \vee c) \wedge (\bar{a} \vee b) \wedge (\bar{a} \vee \bar{b}) \wedge (a \vee \bar{b}) \wedge (b \vee \bar{c})$$

A *unit clause* is a clause of length one. A unit clause forces its only literal to be true. *Unit propagation* is an important technique used in SAT solvers and works as follows: Given a formula F, repeat the following until fixpoint: If F contains a unit clause (l), remove all clauses containing l and remove all literal occurrences of $\bar{l}$. If unit propagation on a formula F produces ϵ, denoted by $F \vdash_1 \epsilon$, F is unsatisfiable.

Let $C := (l_1 \vee l_2 \vee \cdots \vee l_k)$ be a clause. We denote with $\bar{C}$ the set of clauses $(\bar{l}_1) \wedge (\bar{l}_2) \wedge \cdots \wedge (\bar{l}_k)$. C is called a *reverse unit propagation* (RUP) clause with respect to F, if $F \wedge \bar{C} \vdash_1 \epsilon$ [61]. The prototypical RUP clauses are the learned clauses in CDCL solvers, the most common solvers, see [46] and Sect. 3. The conventional procedure to show that these learned clauses are implied by the formula applies unit propagations in the reverse order compared to deriving the clauses in the CDCL solver. This procedure gave rise to the name RUP.

A RUP clause C with respect to F is logically implied by F and one can construct a resolution chain for C using at most $|\text{vars}(F)|$ resolutions. For example, $E \wedge (\bar{c}) \vdash_1 (\bar{b}) \vdash_1 (a) \vdash_1 \epsilon$ uses the clauses $(\bar{b} \vee c)$, $(a \vee c)$, and $(\bar{a} \vee b)$. We can convert this in a resolution chain $(c) := (\bar{a} \vee b) \diamond (a \vee c) \diamond (\bar{b} \vee c)$.

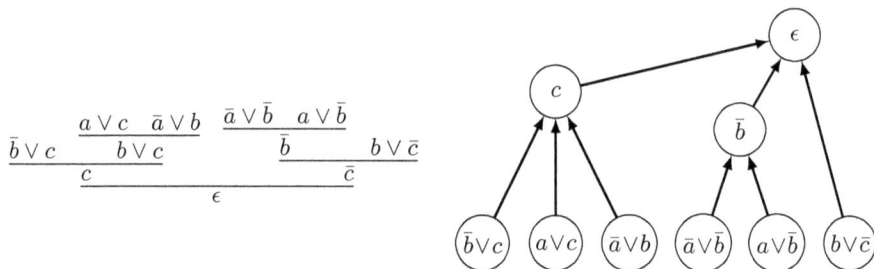

Fig. 1. A resolution derivation (left) and a resolution graph (right) for the example CNF formula E.

2.3 Extended Resolution and Its Generalizations

For a given CNF formula F, the *extension rule* [59] allows one to iteratively add definitions of the form $x := a \wedge b$ by adding clauses $(x \vee \bar{a} \vee \bar{b}) \wedge (\bar{x} \vee a) \wedge (\bar{x} \vee b)$ to F, where x is a new variable and a and b are literals in the current formula. *Extended Resolution* [59] is a proof system, whereby the *extension rule* is repeatedly applied to a CNF formula F, mixed with applications of the resolution rule. This proof system can even polynomially simulate extended Frege systems [22], which is considered to be one of the most powerful proof systems. For plain resolution this is not the case. Several generalizations of extended resolution have been proposed. Two important generalizations regarding proof systems are blocked clause addition [42] and resolution asymmetric clause addition [40].

Blocked Clauses Given a CNF formula F, a clause C, and a literal $l \in C$, the literal l *blocks* C with respect to F if (i) for each clause $D \in F$ with $\bar{l} \in D$, $C \diamond_l D$ is a tautology, or (ii) $\bar{l} \in C$, i.e., C is itself a tautology. Given a CNF formula F, a clause C is *blocked* with respect to F if there is a literal that blocks C with respect to F. Addition and removal of blocked clauses results in satisfiability-equivalent formulas [42].

Example 1. Recall the example formula E. Clause $(\bar{b} \vee c)$ is blocked on c with respect to E, because resolution on the only clause containing $\bar{c}$, results in a tautology, i.e., $(\bar{b} \vee c) \diamond (b \vee \bar{c}) = (\bar{b} \vee b)$. Since we know that E is unsatisfiable, $E \setminus \{(\bar{b} \vee c)\}$ must be unsatisfiable.

To see that blocked clause addition is a generalization of extended resolution, consider a formula containing variables a and b, but without variable x. The three clauses from the extension rule, i.e, $(x \vee \bar{a} \vee \bar{b})$, $(\bar{x} \vee a)$, and $(\bar{x} \vee b)$, are all blocked on x / $\bar{x}$ regardless of the order in which they are added. Hence blocked clause addition can add these three clauses, while preserving satisfiability.

In contrast to extended resolution, blocked clause addition can extend the formula with clauses that are not logically implied by the formula and do not contain a fresh variable. For example, consider the formula $F := (a \vee b)$. The clause $(\bar{a} \vee \bar{b})$ is blocked on $\bar{a}$ (and $\bar{b}$) with respect to F and can thus be added using blocked clause addition.

Resolution Asymmetric Tautologies Resolution asymmetric tautologies (or RAT clauses) [40] are a generalization of both RUP clauses and blocked clauses (and hence extended resolution). A clause C has RAT on l (referred to as the pivot literal) with respect to a formula F if for all $D \in F$ with $\bar{l} \in D$ holds that

$$F \wedge \bar{C} \wedge (\bar{D} \setminus \{(l)\}) \vdash_1 \epsilon.$$

Adding and removing RAT clauses results in a satisfiability-equivalent formula [40]. Given a formula F and a clause C that has RAT on $l \in C$ with respect to F. Let τ be an assignment that satisfies F and falsifies C. The assignment τ', which is a copy of τ with the exception that $\tau'(l) = \mathbf{t}$, satisfies $F \wedge C$. This observation can be used to reconstruct a satisfying assignment for the original formula in case it is satisfiable.

To see that RAT clauses are a generalization of blocked clauses and RUP clauses, observe the following. If a clause has RAT on some $l \in C$ with respect to a formula F, it also has RUP with respect to F because

$$F \wedge \bar{C} \vdash_1 \epsilon \implies F \wedge \bar{C} \wedge (\bar{D} \setminus \{(l)\}) \vdash_1 \epsilon.$$

Furthermore, if a clause C is blocked on l with respect to F, then for all $D \in F$ with $\bar{l} \in D$ holds that C contains a literal $k \neq l$ such that $\bar{k} \in D$. Now we have

$$F \wedge (k) \wedge (\bar{k}) \vdash_1 \epsilon \implies F \wedge \bar{C} \wedge (\bar{D} \setminus \{(l)\}) \vdash_1 \epsilon.$$

3 Proof Search

The leading paradigm to solve satisfiability problems is the conflict-driven clause learning (CDCL) approach [46]. In short, CDCL adds lemmas, typically referred to as conflict clauses, to a given input formula until either it finds a satisfying assignment or is able to learn (i.e., deduce) the empty clause (prove unsatisfiability). We refer to a survey on the CDCL paradigm for details [46].

An alternative approach to solve satisfiability problems is the lookahead approach [38]. Lookahead solvers solve a problem via a binary search-tree. In each node of the search-tree, the best splitting variable is selected using so-called lookahead techniques. Although it is possible to extract unsatisfiability proofs from lookahead solvers, it hardly happens in practice and hence we ignore lookahead solvers in the remainder of this chapter.

CDCL solvers typically use a range of preprocessing techniques, such as bounded variable elimination (also known as Davis-Putnam resolution) [25,26], blocked clause elimination [39], subsumption, and hyper binary resolution [8]. Preprocessing techniques are frequently crucial to solve large formulas efficiently. These preprocessing can also be used during the solving phase, which is known as inprocessing [40]. Most preprocessing techniques can be expressed using a few resolutions, such as bounded variable elimination and hyper binary resolution. Other techniques can be ignored in the context of unsatisfiability proofs, because they weaken the formula, such as blocked clause elimination and subsumption. A few techniques can only be expressed in extended resolution or its generalizations, such as bounded variable addition [45] and blocked clause addition [40].

Some CDCL solvers use preprocessing techniques which are hard to represent using existing proof formats. Examples of such techniques are Gaussian Elimination (GE), Cardinality Resolution (CR) [23] and Symmetry Breaking (SB) [1]. These techniques cannot be polynomially simulated using resolution: Certain formulas based on expander graphs are hard for resolution [60], i.e., resolution proofs are exponentially large, while GE can solve them efficiently. Similarly, formulas arising from the pigeon hole principle are hard for resolution [34], but they can be solved efficiently using either CR or SB. Consequently, resolution proofs of solvers that use these techniques may be exponentially large in the size of the solving time. At the moment, there is no solver that produces resolution proofs for these techniques.

Techniques such as GE, CR, and SB, can be simulated polynomially using extended resolution and its generalizations. However, it is not know how to simulate these techniques efficiently / elegantly using extended resolution. One method to translate GE into extended resolution proofs is to convert the GE steps into BDDs and afterwards translate the BDDs to extended resolution [55].

4 Proof Formats

Unsatisfiability proofs come in two flavors: resolution proofs and clausal proofs. A handful of formats have been designed for resolution proofs [69,28,12]. These formats differ in several details, such as whether the input formula is stored in the proof, whether resolution chains are allowed, and whether resolutions in the proofs must be ordered. This section focusses on the TraceCheck format which is the most widely used format for resolution proofs. The `tracecheck` [12] tool can be used to validate TraceCheck files.

For clausal proofs, there is essentially only one format, called RUP (reverse unit propagation) [61]. RUP can be extended with clause deletion information [36], and with a generalization of extended resolution [37]. The format with both extensions is known as DRAT [68] which is backward compatible with RUP. The `DRAT-trim` [68] tool can efficiently validate clausal proofs in the various formats.

4.1 Resolution Proofs

The proof checker TraceCheck can be used to check whether a trace represents a piecewise regular input resolution proof. A regular input resolution proof is also known as a trivial proof [10]. A trace is just a compact representation of general resolution proofs. The TraceCheck format is more compact than other resolution formats, because it uses resolution chains and the resulting resolvent does not need to be stated explicitly. The parts of the proof which are regular input resolution proofs are called chains in the following discussion. The whole trace consists of original clauses and chains.

Note that input clauses in chains can still be arbitrary derived clauses with respect to the overall proof and do not have to be original clauses. We distinguish between original clauses of the CNF, which are usually just called input clauses, and input clauses to the chains. Since a chain can be seen as new proof rule, we call its input clauses *antecedents* and the final resolvent just *resolvent*.

The motivation for using this format is that learned clauses in a CDCL solver can be derived by regular input resolution [10]. A unique feature of TraceCheck is that the chains do not have to be sorted, neither between chains (globally) nor their input clauses (locally). If possible the checker will sort them automatically. This allows a simplified implementation of the trace generation.

Chains are simply represented by the list of their antecedents and the resolvent. Intermediate resolvents can be omitted which saves quite some space if the proof generator can easily extract chains.

Chains can be used in the context of searched based CDCL to represent the derivation of learned clauses. It is even more difficult to extract a resolution proof directly, if more advanced learned clause optimizations are used. Examples are shrinking or minimization of learned clauses [58]. The difficult part is to order the antecedents correctly. The solver can leave this task to the trace checker, instead of changing the minimization algorithm [62].

Furthermore, this format allows a simple encoding of hyper resolution proofs. A hyper resolution step can be simulated by a chain. General resolution steps can also be encoded in this format easily by a trivial chain consisting of the two antecedents of the general resolution step. Finally, extended resolution proofs

6

can directly be encoded, since variables introduced in extended resolution can be treated in the same way as the original variables.

The syntax of a trace is as follows:

$$\langle\text{trace}\rangle = \{\langle\text{clause}\rangle\}$$
$$\langle\text{clause}\rangle = \langle\text{pos}\rangle\langle\text{literals}\rangle\langle\text{antecedents}\rangle$$
$$\langle\text{literals}\rangle = \text{``}*\text{''} \mid \{\langle\text{lit}\rangle\}\text{``0''}$$
$$\langle\text{antecedents}\rangle = \{\langle\text{pos}\rangle\}\text{``0''}$$
$$\langle\text{lit}\rangle = \langle\text{pos}\rangle \mid \langle\text{neg}\rangle$$
$$\langle\text{pos}\rangle = \text{``1''} \mid \text{``2''} \mid \cdots \mid \langle\text{maxidx}\rangle$$
$$\langle\text{neg}\rangle = \text{``} - \text{''}\langle\text{pos}\rangle$$

where | means choice, $\{\dots\}$ is equivalent to the Kleene star operation (that is a finite number of repetitions including 0) and $\langle\text{maxidx}\rangle$ is $2^{28} - 1$ (originally).

The interpretation is as follows. Original clauses have an empty list of antecedents and derived clauses have at least one antecedent. A clause definition starts with its index and a zero terminated list of its literals. This part is similar to the DIMACS format except that each clause is preceded by a unique positive number, the index of the clause. Another zero terminated list of positive indices of its antecedents is added, denoting the chain that is used to derive this clause as resolvent from the antecedents. The order of the clauses and the order of the literals and antecedents of a chain is arbitrary.

The list of antecedents of a clause should permit a regular input resolution proof of the clause with exactly the antecedents as input clauses. A proof is regular if variables are resolved at most once. It is an input resolution if each resolution step resolves at most one non input clause. Therefore it is also linear and has a degenerated graph structure of a binary tree, where each internal node has at least one leaf as child.

As example consider the following trace

```
1  -2   3 0 0
2   1   3 0 0
3  -1   2 0 0
4  -1  -2 0 0
5   1  -2 0 0
6   2  -3 0 0

7  -2   0 4 5 0
8   3   0 1 2 3 0
9   0   6 7 8 0
```

which consists of the six clauses from example CNF formula E. The corresponding DIMACS formula is shown in Fig. 2 (left).

input formula (DIMACS)	clausal proof (RUP)	resolution proof (TraceCheck)

```
   p cnf 3 6
   -2  3 0
    1  3 0
   -1  2 0
   -1 -2 0
    1 -2 0
    2 -3 0
```

```
   -2  0
    3  0
    0
```

```
1  -2   3  0  0
2   1   3  0  0
3  -1   2  0  0
4  -1  -2  0  0
5   1  -2  0  0
6   2  -3  0  0
7  -2   0  4 5 0
8   3   0  1 2 3 0
9   0   6 7 8 0
```

Fig. 2. An input formula (left) in the classical DIMACS format which is supported by most SAT solvers. A clausal proof for the input formula in RUP format (middle). In both the DIMACS and RUP formats, each line ending with a zero represents a clause, and each non-zero element represents a literal. Positive numbers represent positive literals, while negative numbers represent negative numbers. For example, -2 3 0 represents the clause $(\bar{x}_2 \lor x_3)$. A TraceCheck file (right) is a resolution graph that includes the formula and proof. Each line begins with a clause identifier (bold), then contains the literals of the original clause or lemma, and ends with a list of clause identifiers (bold).

The first derived clause with index **7** is the unary clause which consists of the literal -2 alone. It is obtained by resolving the original clause **4** against the original clause **5** on variable 1.

A chain for the last derived clause, which is the empty clause ϵ, can be obtained by resolving the antecedents **6**, **7** and **8**, first **6** with **7** to obtain the intermediate resolvent consisting of the literal -3 alone, which in turn can be resolved with clause **8** to obtain ϵ.

As discussed above, the order of the clauses, that is the order of the lines and the order of the antecedents indices is irrelevant. The checker will sort them automatically. The last two lines of the example can for instance be replaced by:

```
9  0  6 7 8 0
8  3  0  1 2 3 0
```

Note that the clauses **7** and **8** cannot be resolved together because they do not have a clashing literal. In this case the checker has to reorder the antecedents as in the original example.

The main motivation for having antecedents in the proof for each learned clause is to speed up checking the trace. While checking a learned clause, unit propagation can focus on the list of specified antecedents. It can further ignore all other clauses, particularly those that were already discarded at the point the solver learned the clause. An alternative is to include deletion information.

It might be convenient to skip the literal part for derived clauses by specifying a ⋆ instead of the literal list. The literals are then collected by the checker from

8

the antecedents. Since resolution is not associative, the checker assumes that the
antecedents are correctly sorted when $\star$ is used.

```
8  *  1  2  3  0
9  *  6  7  8  0
```

Furthermore, trivial clauses and clauses with multiple occurrences of the same
literal can not be resolved. The list of antecedents is not allowed to contain the
same index twice. All antecedents have to be used in the proof for the resolvent.

Beside these local restrictions the proof checker generates a global linear order
on the derived clauses making sure that there are no cyclic resolution steps. The
roots of the resulting DAG are the target resolvents.

4.2 Clausal Proofs

We appeal to the notion that *lemmas* are used to construct a proof of a theorem.
Here, lemmas represent the learned clauses and the theorem is the statement that
the formula is unsatisfiable. From now on, we will use the term clauses to refer
to input clauses, while lemmas will refer to added clauses.

$$\langle\text{proof}\rangle = \{\langle\text{lemma}\rangle\}$$
$$\langle\text{lemma}\rangle = \langle\text{delete}\rangle\{\langle\text{lit}\rangle\}\text{ "0"}$$
$$\langle\text{delete}\rangle = \text{ "" } | \text{ "d"}$$
$$\langle\text{lit}\rangle = \langle\text{pos}\rangle | \langle\text{neg}\rangle$$
$$\langle\text{pos}\rangle = \text{ "1" } | \text{ "2" } | \cdots | \langle\text{maxidx}\rangle$$
$$\langle\text{neg}\rangle = \text{ " } - \text{ "}\langle\text{pos}\rangle$$

There exist four proof formats for clausal proofs which have mostly the same
syntax and all of them can be expressed using the grammar above.

The most basic format is RUP (reverse unit propagation) [61]. A RUP proof is
a sequence of lemmas, with each lemma being a list of positive and negative integers to express positive and negative literals, respectively, which are terminated
with a zero.

Given a formula F, and a clausal proof $P := \{L_1, \ldots, L_m\}$. P is a valid RUP
proof for F if $L_m = \epsilon$ and for all L_i holds that

$$F \wedge L_1 \wedge \cdots \wedge L_{i-1} \wedge \bar{L}_i \vdash_1 \epsilon$$

Recall the example CNF formula E. The proof $P_E := \{(\bar{b}), (c), \epsilon\}$ is a valid
proof for E, because P_E terminates with ϵ and (with $\bar{\epsilon}$ being a tautlogy) we have

$$E \wedge (b) \qquad \vdash_1 \quad \epsilon$$
$$E \wedge (\bar{b}) \wedge (\bar{c}) \qquad \vdash_1 \quad \epsilon$$
$$E \wedge (\bar{b}) \wedge (c) \wedge \bar{\epsilon} \vdash_1 \quad \epsilon$$

9

The DRUP (delete reverse unit propagation) format [36] extends RUP by integrating clause deletion information into proofs. The main reason to add clause deletion information to a proof is to reduce the cost to validate a proof which will be discussed in Section 6.2. Clause deletion information is expressed using the prefix d.

4.3 Proofs with Extended Resolution

So far we only considered proof formats that validate techniques that can be simulated using resolution. Some SAT solver use techniques that cannot be simulated using resolution, such as blocked clause addition [42]. To validate these techniques, proof formats need to support a richer representation that includes extended resolution or one of its generalizations.

Resolution proofs, as the name suggests, can only be used to check techniques based on resolution. The TraceCheck format partially supports extended resolution in the sense that one can add the clauses from the extension rule using an empty list of antecedents. Hence these clauses are considered to be input clauses without actually validating them.

The RAT clausal proof format [37], which is syntactically the same as the RUP format, supports expressing techniques based on extended resolution and its generalizations. The difference between the RUP and RAT format is in the redundancy check that is computed in the checker for proofs in that format. A checker for RUP proofs validates whether a lemma is a RUP clause, while a checker of RAT proofs check whether each lemma is a RAT clause. The DRAT format [68] extends RAT by supporting clause deletion information.

Example 2. Consider the following CNF formula

$$G := (\bar{a} \vee \bar{b} \vee \bar{c}) \wedge (a \vee d) \wedge (a \vee e) \wedge (b \vee d) \wedge (b \vee e) \wedge (c \vee d) \wedge (c \vee e) \wedge (\bar{d} \vee \bar{e})$$

Fig. 3 shows G in the DIMACS format (left) using the conventional mapping from the alphabet to numbers, i.e., $(a\ 1)(\bar{a}\ \text{-}1)\ldots(e\ 5)(\bar{e}\ \text{-}5)$ and a DRAT clausal proof for G (middle). The proof for G uses a technique, called bounded variable addition (BVA) [45], that cannot be expressed using resolution steps. BVA can replace the first six binary clauses by five new binary clauses using a fresh variable f: $(f \vee a)$, $(f \vee b)$, $(f \vee c)$, $(\bar{f} \vee d)$, and $(\bar{f} \vee e)$. These new binary clauses are RAT clauses. Fig. 3 shows how easy it is to express BVA in the DRAT format: First add the new binary clauses, followed by deleting the old ones. After the replacement, the proof is short $\{(f), \epsilon\}$.

It is not clear how bounded variable addition can be expressed in a resolution-style format. Fig. 4 shows the main issue for the example formula G and the BVA based proof. The clauses $(f \vee a)$, $(f \vee b)$, and $(f \vee c)$ are trivially redundant with respect to G, because G does not contain any clause with variable f. Assigning f to t would satisfy these three clauses. However, $(\bar{f} \vee d)$ and $(\bar{f} \vee e)$ are not trivially redundant with respect to G after the addition of $(f \vee a)$, $(f \vee b)$, and $(f \vee c)$. There redundancy of $(\bar{f} \vee d)$ depends on the presence of $(a \vee d)$, $(b \vee d)$, and $(c \lor d)$. One option to express BVA is adding the dependency relationship to the proof, as suggested in recent work [68]. This results in a TraceCheck$^+$ proof for which each lemma has either a list of antecedents or a list of dependencies. However, there exists no procedure yet to validate such a TraceCheck$^+$ proof.

DIMACS formula

```
p cnf 5 8
-1 -2 -3 0
      1  4 0
      1  5 0
      2  4 0
      2  5 0
      3  4 0
      3  5 0
 -4 -5 0
```

DRAT clausal proof

```
        6  1 0
        6  2 0
        6  3 0
       -6  4 0
       -6  5 0
   d    1  4 0
   d    2  4 0
   d    3  4 0
   d    1  5 0
   d    2  5 0
   d    3  5 0
        6    0
             0
```

TraceCheck$^+$ resolution proof

1	1	4	0	**0**				
2	1	5	0	**0**				
3	2	4	0	**0**				
4	2	5	0	**0**				
5	3	4	0	**0**				
6	3	5	0	**0**				
7	-1	-2	-3	0	**0**			
8	-4	-5	0	**0**				
9	6	1	0	**0**				
10	6	2	0	**0**				
11	6	3	0	**0**				
12	-6	4	0	**1**	**3**	**5**	**0**	
13	-6	5	0	**2**	**4**	**6**	**0**	
14	6	0	**1**	**9**	**10**	**11**	**0**	
15	0	**8**	**12**	**13**	**14**	**0**		

Fig. 3. Example formula G in the classical DIMACS format (left). A clausal proof for the input formula in DRAT format (middle). A TraceCheck$^+$ file (right) is a dependency graph that includes the formula and proof. Each line begins with a clause identifier (bold), then contains the literals of the original clause or lemma, and ends with a list of clause identifiers (bold).

4.4 Open Issues and Challenges in Proof Formats

It is common practice to store proofs on disk and we discussed various formats for this purposes. However, in many applications where proofs have to be further processed and are used subsequently or even iteratively, disk I/O is considered a substantial overhead. There are only few publicly available SAT solvers, which keep proofs in memory. Beside the question, whether these proofs are stored as resolution or clausal proofs, and the technical challenge to reduce memory usage, partially addressed in two papers in 2008 [12,6], there is also no common understanding of what kind of API should be used to manipulate proofs.

Beside checking the proof online for testing and debugging, common operations might be, extracting a resolution proof from a clausal proof, generating interpolants, minimizing proofs, or to determine a clausal or a variable core. A generic API for traversing proof objects might also be useful. Last but not least it should be possible to dump these proofs to disk.

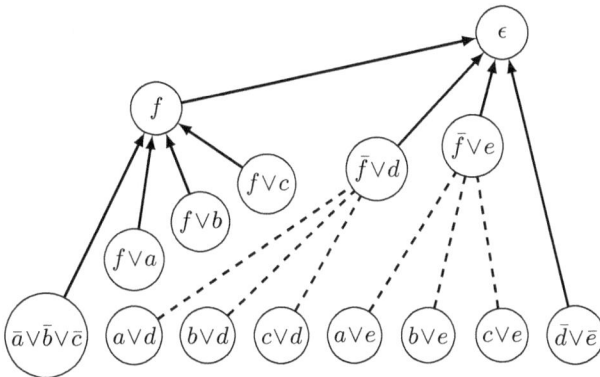

Fig. 4. A resolution-dependency graph illustrating a proof of example formula G. The clauses on bottom of the figure are the input clauses. The height of lemmas indicates the time that they were added to the proof: lower means earlier. Solid arrows represent resolution steps, while dashed lines represent a dependency relationship.

Related to reducing the memory usage of storing proofs is the question of a binary disk format for proofs, or specific compression techniques, as used in the version of MiniSAT with proof trace support or PicoSAT.

Finally, as SAT solving is at the core of state-of-the-art SMT solving and also used in theorem provers, producing and manipulating proofs for these more expressive logics will need to incorporate techniques discussed in this chapter. Interoperability, mixed formats, and APIs etc. are further open issues.

5 Proof Production

Proof logging of unsatisfiability results from SAT solvers started in 2003 of both resolution proofs [69] and clausal proofs [32]. Resolution proofs are typically hard to produce and tend to require lots of overhead in memory, which in turn slows down a SAT solver. Emitting clausal proofs is easy and requires hardly any overhead on memory. We will first describe how to produce resolution proofs and afterwards how to produce clausal proofs.

5.1 Resolution Proofs

The main motivation for adding proof support to PicoSAT was to make testing and debugging more efficient. In particular in combination with file based delta-debugging [20], proof trace generation allows to reduce discrepancies much more than without proof tracing.

The original idea was to use resolution traces. Originally it was however unclear how to extract resolution proofs during a-posteriori clause minimization [58]. This was the reason to use a trace format instead: clause minimization is obviously compatible with RUP, since required clause antecedents can easily be obtained. However, determining the right resolution order for a resolution proof is hard to generate directly and probably needs a RUP style algorithm anyhow. It was shown how clause minimization can be integrated into as DFS search for the first unique implication point clause [62], which at the same time can produce a resolution proof for the minimized learned clause. Currently it is unsolved how to further extend this approach to work with on-the-fly subsumption [35] as well. The solution is also not as easy to add to existing solvers as tracing added and deleted clauses.

5.2 Reducing Memory Consumption

As already discussed above, memory consumption of proofs stored in memory (or disk) can become a bottle neck. As pioneered by the (unpublished) disk format for proof traces for MiniSAT, and extended in the in-memory format for proof traces of PicoSAT, the antecedent clauses of a learned clause can be sorted, as well as literals of learned clauses. After sorting, the antecedent lists or literals can be stored as differences between literals or antecedent ids instead of absolute values. Then these differences can be encoded efficiently in byte sequences of variadic length. In practice this technique needs slightly more then one byte on average per antecedent (or literal) [12]. Experience shows that other classical compression techniques are also more effective after this byte-encoding, such that in combination a large reduction can be achieved.

Another option to reduce space requirements for storing proofs in memory, is to remove garbage clauses in the proof, which are not used anymore. This garbage collection was implemented with saturating reference counters in PicoSAT [12]. It has been shown that full reference counting can result in substantial reductions [6]. In principle it might also be possible to use a simple mark and sweep garbage collector, which should be faster, since it does not need to maintain and update reference counters.

5.3 Clausal Proofs

For all clausal proof formats, SAT solvers emit proofs directly to disk. Consequently, there is no memory overhead. In contrast to resolution proofs, it is easy to emit clausal proofs. For the most simple clausal proof format, RUP, one only needs to extend the proof with all added lemmas. This can be implemented by a handful lines of code. Below we discuss how to produce clausal proof formats that support deletion information and techniques based on generalized extended resolution.

Clausal proofs need deletion information for efficient validation, see Section 6.2 for details. Adding deletion information to a clausal proof is simple. As soon as a clause is removed from the solver, the proof is extended by the removed clause using a prefix expressing that it is a deletion step. If a solver removes a literal l from a clause C, then $C \setminus \{l\}$ is added as a lemma followed by deleting C.

Most techniques based on extended resolution or its generalizations can easily be expressed in clausal proofs. Similar to techniques based on resolution, for most techniques, one simply adds the lemmas to the proof. The RAT and DRAT formats only require that the pivot literal is the first literal of the lemma in the proof. However, as discussed in Section 3, there exist some techniques for which it is not known whether they can be elegantly be expressed in the DRAT format. Especially Gaussian elimination, cardinality resolution and symmetry breaking are hard to express in the current formats.

6 Proof Consumption

Although resolution proofs are harder to produce than clausal proofs, they are in principle easy to check and actually needed for some applications, like generating interpolants [47]. However, the large size of resolution proofs provides challenges, particular with respect to memory usage. See [63] for a discussion on checking large resolution proofs.

Clausal proofs are smaller, but validating clausal proofs is more complicated and more costly. Proofs are checked with dedicated tools, such as stc (short for simple tracecheck) for resolution proofs in TraceCheck format and DRAT-trim [68] for clausal proofs in DRAT format (and consequently in RUP, DRUP, and RAT formats due to backward compatibility).

6.1 Resolution Proofs

Resolution proofs can be checked in deterministic log space [61], a very low complexity class. The tool stc can efficient check proofs in the TraceCheck format. More details about the format and its use are discussed in Section 4.1.

Apart from validation, there exists a vast body of work on compression of resolution proofs [2,9,24,29,53,54]. One technique to make proofs more compact is *RecycleUnits* [9]: unit clauses in the proof are used to replace some clauses in the proof that are subsumed by the units. The replacement typically makes the proof illegal (i.e., some resolvents are no longer the result of resolving the antecedents). However, the proof can be fixed with a few simple steps. Figure 5 illustrates the RecycleUnits procedure on a derivation of the example CNF formula E. Notice in the example that replacing a clause by a unit may strengthen the resolvent.

Two other proof compression techniques are *LowerUnits* and *RecylcePivots*. LowerUnits uses the observation that one needs to resolve over a unit clause only a single time. In case a unit clause is occurs multiple times in a proof, it is lowered to ensure that it occurs only once. RecylcePivots reduces the irregularity in resolution proofs. A proof is irregular if it contains a path on which resolution is performed on the same pivot. Removing all irregularity in a proof may result in an exponential blow-up of the proof [31]. Hence, techniques such as RecylcePivots need to be restricted in the context of proof compression. The tool Skeptic [16] can be used to remove redundancy from resolution proofs and includes most of the compression techniques.

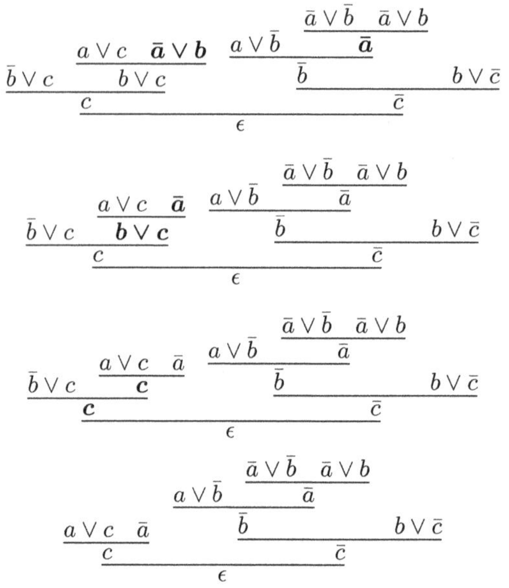

$$
\begin{array}{c}
\cfrac{}{}\\[-1ex]
\end{array}
$$

Derivation 1:

$$
\dfrac{\quad}{}
$$

$$
\begin{array}{ccccc}
 & a\vee c \quad \bar a\vee b & a\vee\bar b & \dfrac{\bar a\vee\bar b \quad \bar a\vee b}{\bar a} & \\
\dfrac{\bar b\vee c \qquad b\vee c}{c} & & \dfrac{}{\bar b} & & b\vee\bar c \\
 & & & \dfrac{}{\bar c} & \\
 & & \dfrac{}{\epsilon} & &
\end{array}
$$

Unit clause $\bar a$ can be recycled by replacing the non-antecedent $\bar a\vee b$

Derivation 2:

$$
\begin{array}{ccccc}
 & a\vee c \quad \bar a & a\vee\bar b & \dfrac{\bar a\vee\bar b \quad \bar a\vee b}{\bar a} & \\
\dfrac{\bar b\vee c \qquad b\vee c}{c} & & \dfrac{}{\bar b} & & b\vee\bar c \\
 & & & \dfrac{}{\bar c} & \\
 & & \dfrac{}{\epsilon} & &
\end{array}
$$

After the replacing $\bar a\vee b$ by $\bar a$, $b\vee c$ can no longer be produced using resolution. This is fixed by replacing $b\vee c$ with the actual resolvent c.

Derivation 3:

$$
\begin{array}{ccccc}
 & a\vee c \quad \bar a & a\vee\bar b & \dfrac{\bar a\vee\bar b \quad \bar a\vee b}{\bar a} & \\
\dfrac{\bar b\vee c \qquad c}{c} & & \dfrac{}{\bar b} & & b\vee\bar c \\
 & & & \dfrac{}{\bar c} & \\
 & & \dfrac{}{\epsilon} & &
\end{array}
$$

After the replacing $b\vee c$ by c, a similar problem occurs. Now the antecedent c and the "resolvent" are the same.

Derivation 4:

$$
\begin{array}{ccccc}
 & & a\vee\bar b & \dfrac{\bar a\vee\bar b \quad \bar a\vee b}{\bar a} & \\
\dfrac{a\vee c \qquad \bar a}{c} & & \dfrac{}{\bar b} & & b\vee\bar c \\
 & & & \dfrac{}{\bar c} & \\
 & & \dfrac{}{\epsilon} & &
\end{array}
$$

Removal of the redundant resolution step results in a correct and more compact derivation.

Fig. 5. An example of the proof compression technique RecycleUnits.

6.2 Clausal Proofs

Clausal proofs are checked using unit propagation. Recall that a clausal proof $\{L_1, \ldots, L_m\}$ is valid for formula F, if $L_m = \epsilon$ and for $i \in \{1, \ldots, m\}$ holds that

$$F \wedge L_1 \wedge \cdots \wedge L_{i-1} \wedge \bar L_i \vdash_1 \epsilon$$

The most simple, but very costly method to validate clausal proofs is to check for every $i \in \{1, \ldots, m\}$ the above equation holds.

One can reduce the cost to validate clausal proofs by checking them *backwards* [32]: Initially, only $L_m = \epsilon$ is marked. Now we loop over the lemmas in backwards order, i.e., $L_m, \ldots, L_1$. Before validating a lemma, we first check whether it is marked. If a lemma is not marked, it can be skipped, thereby reducing the computational costs. If a lemma is marked, we check whether the clause satisfies the above equation. If the check fails, the proof is invalid. Otherwise, we mark all clauses that were required to make the check succeed (using conflict analysis). For most proofs, over half the lemmas can be skipped during validation.

The main challenge regarding validating clausal proofs is efficiency. Validating a clausal proof is typically much more expensive than obtaining the proof using a SAT solver, even if the implementation uses backwards checking and the same data-structures as state-of-the-art solvers. Notice that for most other logics, such as first order logic, checking a proof is typically cheaper than finding a proof. There are two main reasons why checking unsatisfiability proofs is more expensive than solving.

First, SAT solvers aggressively delete clauses during solving, which reduces the cost of unit propagation. If the proof checker has no access to the clause deletion information, then unit propagation is much more expensive in the checker as compared to the solver. This was the main motivation why the proof formats DRUP and DRAT have been developed. These formats support expressing clause deletion information, thereby making the unit propagation costs between the solver and checker similar.

Second, SAT solvers reuse propagations in between conflicts, while a proof checker does not reuse propagations. Consider two consecutive lemmas L_i and L_{i+1} produced by a SAT solver. In the most extreme, but not unusual case, the branch that resulted in L_i and L_{i+1} may differ only in a single decision (out of many decisions). Hence most propagations will be reused in the solver. At the same time, it may be that the lemmas have no overlapping literals, i.e., $L_i \cap L_{i+1} = \emptyset$. Consequently, the checker would not be able to reuse any propagations. In case $L_i \cap L_{i+1}$ is nonempty, the checker could potentially reuse propagations, although no clausal proof checker implementation exploits this.

While checking a clausal proof, one can easily produce an unsatisfiable core and a resolution proof. The unsatisfiable core consists of the original clauses that were marked during backwards checking. For most unsatisfiable formulas that arise from applications, many clauses are redundant, i.e, are not marked and thus not in the unsatisfiable core. The resolution proof has for each marked lemma all the clauses that were required during its validation as antecedents.

The resolution proof that is produced by clausal proof checking might differ significantly from the resolution proof that would correspond to the actions of the SAT solver that emitted the clausal proof. For example, the resolution proof for the example formula E might be equal to the resolution graph shown in Fig. 1. On the other hand, the resolution proof produced by clausal proof checking might be equal to Fig. 6. Notice that the resolution graph of Fig. 6 (right) does not use all original clauses. Clause $(\bar{b} \vee c)$ is redundant and not part of the core of E.

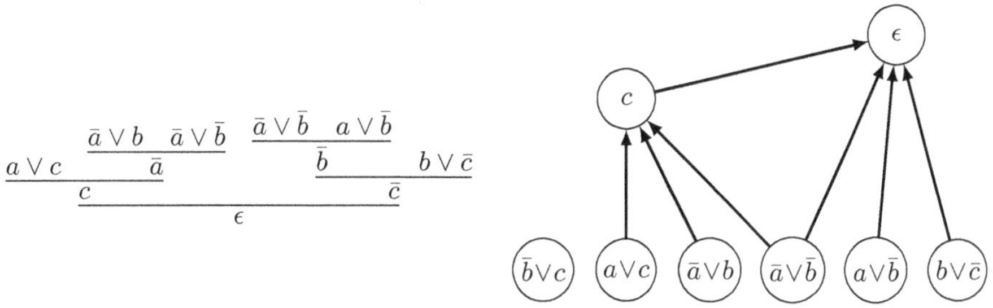

Fig. 6. A resolution derivation (left) and a resolution graph (right) for an example formula produce by checking a clausal proof.

7 Proof Applications

Proofs of unsatisfiability have been used to validate the results of SAT competitions[4]. Initially, during the SAT competition of 2007, 2009, and 2011, a special track was organized for which the unsatisfiability results were checked. For the SAT competitions of 2013 and 2014, proof logging became mandatory for tracks with only unsatisfiable benchmarks. The supported formats for SAT competition 2013 were TraceCheck and DRUP, but all solvers participating in these tracks opted for the DRUP format. For SAT competition 2014, the only supported format was DRAT, which is backwards compatible with DRUP.

As already mentioned above, one motivation for using proofs is to make testing and debugging of SAT solvers more effective. Checking learned clauses online with RUP allows to localize unsound implementation defects as soon they lead to clauses, which are not implied by reverse unit propagation.

Testing with forward proof checking is particularly effective in combination with fuzzing (generating easy formulas) and delta-debugging [19] (shrinking a formula that triggers a bug). Otherwise failures produced by unsound reasoning can only be observed if they turn a satisfiable instance into an unsatisfiable one. This situation is not only difficult to produce, but also tends to lead to much larger input files after delta-debugging.

However, model based testing [5] of the incremental API of a SAT solver is in our experience at least as effective as file based fuzzing and delta-debugging. More recently we added online proof checking capabilities to `Lingeling` [14], which allows to combine these two methods (model based testing and proof checking).

Probably the most important aspect of proof tracing is that it allows to generate a clausal (or variable) core (i.e., an unsatisfiable subset). These cores in turn can be used in many applications, including MUS extraction [50], MaxSAT [48], diagnosis [56,51], for abstraction refinement in model checking [27] or SMT [3,18]. Note that this list of references is very subjective and by far not complete. It should only be considered as a starting point for investigating related work on using cores.

Finally, extraction of interpolants is an important usage of resolution proofs, particularly in the context of interpolation based model checking [47]. Since resolution proofs are large and not easy to obtain, there has been several recent attempts to avoid proofs and obtain interpolants directly, see for instance [64]. Interpolation based model checking became the state-of-the-art until the invention of IC3 [17]. The IC3 algorithm is also based on SAT technology, and also uses cores, but usually in a much more light weight way. Typical implementations use assumption based core techniques as introduced in `MiniSAT` [28] (see also [44]) instead of proof based techniques.

---

[4] see http://www.satcompetition.org for details.

8 Conclusions

Unsatisfiability proofs are useful for several applications, such as computing interpolants and MUS extraction. These proofs can also be used to validate results of the SAT solvers that produced them and for tools that use SAT solvers, such as theorem provers.

There are two types of unsatisfiability proofs: resolution proofs and clausal proofs. Resolution proofs are used for most applications, but they are hard to produce. Therefore very few SAT solvers support resolution proof logging. Clausal proof logging is easy and therefore most state-of-the-art solvers support it. However, validating clausal proofs is costly, although recent advances significantly improved performance of checkers.

There are several challenges regarding unsatisfiability proofs. How can one store resolution proofs using much less space on disk and using much less memory overhead? Can the costs of validating clausal proofs be further be reduced? Last but not least, research is required to study how some techniques, such as Gaussian elimination, cardinality resolution, and symmetry breaking, can be expressed elegantly in unsatisfiability proofs.

References

1. Fadi A. Aloul, Karem A. Sakallah, and Igor L. Markov. Efficient symmetry breaking for Boolean satisfiability. *IEEE Trans. Computers*, 55(5):549–558, 2006.
2. Hasan Amjad. Compressing propositional refutations. *Electr. Notes Theor. Comput. Sci.*, 185:3–15, 2007.
3. Zaher S. Andraus, Mark H. Liffiton, and Karem A. Sakallah. Refinement strategies for verification methods based on datapath abstraction. In *Proc. ASP-DAC'06*, pages 19–24. IEEE, 2006.
4. Michaël Armand, Germain Faure, Benjamin Grégoire, Chantal Keller, Laurent Théry, and Benjamin Werner. Verifying SAT and SMT in Coq for a fully automated decision procedure. In *International Workshop on Proof-Search in Axiomatic Theories and Type Theories (PSATTT)*, 2011.
5. Cyrille Artho, Armin Biere, and Martina Seidl. Model-based testing for verification back-ends. In Margus Veanes and Luca Viganò, editors, *TAP*, volume 7942 of *Lecture Notes in Computer Science*, pages 39–55. Springer, 2013.
6. Roberto Asín, Robert Nieuwenhuis, Albert Oliveras, and Enric Rodríguez-Carbonell. Efficient generation of unsatisfiability proofs and cores in SAT. In Iliano Cervesato, Helmut Veith, and Andrei Voronkov, editors, *LPAR*, volume 5330 of *Lecture Notes in Computer Science*, pages 16–30. Springer, 2008.
7. Gilles Audemard and Laurent Simon. Glucose 2.3 in the SAT 2013 Competition. In Anton Belov, Marijn J. H. Heule, and Matti Järvisalo, editors, *Proceedings of SAT Competition 2013*, volume B-2013-1 of *Department of Computer Science Series of Publications B, University of Helsinki*, pages 42–43, 2013.
8. Fahiem Bacchus and Jonathan Winter. Effective preprocessing with hyper-resolution and equality reduction. In Giunchiglia and Tacchella [30], pages 341–355.
9. Omer Bar-Ilan, Oded Fuhrmann, Shlomo Hoory, Ohad Shacham, and Ofer Strichman. Linear-time reductions of resolution proofs. In Hana Chockler and Alan J. Hu, editors, *Haifa Verification Conference*, volume 5394 of *Lecture Notes in Computer Science*, pages 114–128. Springer, 2008.

10. Paul Beame, Henry Kautz, and Ashish Sabharwal. Towards understanding and harnessing the potential of clause learning. *JAIR*, 22:319–351, 2004.
11. Anton Belov, Marijn J. H. Heule, and Joao P. Marques-Silva. Mus extraction using clausal proofs. In Carsten Sinz and Uwe Egly, editors, *Theory and Applications of Satisfiability Testing SAT 2014*, volume 8561 of *Lecture Notes in Computer Science*, pages 48–57. Springer International Publishing, 2014.
12. Armin Biere. PicoSAT essentials. *Journal on Satisfiability, Boolean Modeling and Computation (JSAT)*, 4:75–97, 2008.
13. Armin Biere. Lingeling, Plingeling and Treengeling entering the SAT Competition 2013. In Anton Belov, Marijn J. H. Heule, and Matti Järvisalo, editors, *Proceedings of SAT Competition 2013*, volume B-2013-1 of *Department of Computer Science Series of Publications B, University of Helsinki*, pages 51–52, 2013.
14. Armin Biere. Yet another local search solver and Lingeling and friends entering the SAT Competition 2014. In Anton Belov, Marijn J. H. Heule, and Matti Järvisalo, editors, *SAT Competition 2014*, volume B-2014-2 of *Department of Computer Science Series of Publications B*, pages 39–40. University of Helsinki, 2014.
15. Roderick Bloem and Natasha Sharygina, editors. *Proceedings of 10th International Conference on Formal Methods in Computer-Aided Design, FMCAD 2010, Lugano, Switzerland, October 20-23*. IEEE, 2010.
16. Joseph Boudou, Andreas Fellner, and Bruno Woltzenlogel Paleo. Skeptik: A proof compression system. In Stéphane Demri, Deepak Kapur, and Christoph Weidenbach, editors, *Automated Reasoning - 7th International Joint Conference, IJCAR 2014, Held as Part of the Vienna Summer of Logic, VSL 2014, Vienna, Austria, July 19-22, 2014. Proceedings*, volume 8562 of *Lecture Notes in Computer Science*, pages 374–380. Springer, 2014.
17. Aaron R. Bradley. SAT-based model checking without unrolling. In Ranjit Jhala and David A. Schmidt, editors, *VMCAI*, volume 6538 of *Lecture Notes in Computer Science*, pages 70–87. Springer, 2011.
18. Robert Brummayer and Armin Biere. Effective bit-width and under-approximation. In *Proc. EUROCAST'09*, volume 5717 of *LNCS*, pages 304–311, 2009.
19. Robert Brummayer and Armin Biere. Fuzzing and delta-debugging SMT solvers. In *International Workshop on Satisfiability Modulo Theories (SMT)*, pages 1–5. ACM, 2009.
20. Robert Brummayer, Florian Lonsing, and Armin Biere. Automated testing and debugging of SAT and QBF solvers. In *Theory and Applications of Satisfiability Testing (SAT)*, volume 6175 of *LNCS*, pages 44–57. Springer, 2010.
21. Stephen A. Cook. The complexity of theorem-proving procedures. In *Proceedings of the Third Annual ACM Symposium on Theory of Computing*, STOC '71, pages 151–158, New York, NY, USA, 1971. ACM.
22. Stephen A. Cook and Robert A. Reckhow. The relative efficiency of propositional proof systems. *The Journal of Symbolic Logic*, 44(1):pp. 36–50, 1979.
23. W. Cook, C.R. Coullard, and Gy. Turán. On the complexity of cutting-plane proofs. *Discrete Applied Mathematics*, 18(1):25 – 38, 1987.
24. Scott Cotton. Two techniques for minimizing resolution proofs. In Ofer Strichman and Stefan Szeider, editors, *Theory and Applications of Satisfiability Testing SAT 2010*, volume 6175 of *Lecture Notes in Computer Science*, pages 306–312. Springer, 2010.
25. Martin Davis and Hilary Putnam. A computing procedure for quantification theory. *J. ACM*, 7(3):201–215, 1960.

26. Niklas Eén and Armin Biere. Effective preprocessing in SAT through variable and clause elimination. In Fahiem Bacchus and Toby Walsh, editors, *SAT*, volume 3569 of *Lecture Notes in Computer Science*, pages 61–75. Springer, 2005.

27. Niklas Eén, Alan Mishchenko, and Nina Amla. A single-instance incremental SAT formulation of proof- and counterexample-based abstraction. In Bloem and Sharygina [15], pages 181–188.

28. Niklas Eén and Niklas Sörensson. An extensible SAT-solver. In Giunchiglia and Tacchella [30], pages 502–518.

29. Pascal Fontaine, Stephan Merz, and Bruno Woltzenlogel Paleo. Compression of propositional resolution proofs via partial regularization. In Nikolaj Bjørner and Viorica Sofronie-Stokkermans, editors, *Automated Deduction - CADE-23 - 23rd International Conference on Automated Deduction, Wroclaw, Poland, July 31 - August 5, 2011. Proceedings*, volume 6803 of *Lecture Notes in Computer Science*, pages 237–251. Springer, 2011.

30. Enrico Giunchiglia and Armando Tacchella, editors. *Theory and Applications of Satisfiability Testing, 6th International Conference, SAT 2003. Santa Margherita Ligure, Italy, May 5-8, 2003 Selected Revised Papers*, volume 2919 of *Lecture Notes in Computer Science*. Springer, 2004.

31. Andreas Goerdt. Comparing the complexity of regular and unrestricted resolution. In Heinz Marburger, editor, *GWAI-90 14th German Workshop on Artificial Intelligence*, volume 251 of *Informatik-Fachberichte*, pages 181–185. Springer Berlin Heidelberg, 1990.

32. Evguenii I. Goldberg and Yakov Novikov. Verification of proofs of unsatisfiability for CNF formulas. In *Design, Automation and Test in Europe Conference and Exhibition (DATE)*, pages 10886–10891. IEEE, 2003.

33. Arie Gurfinkel and Yakir Vizel. Druping for interpolants. In *Proceedings of the 14th Conference on Formal Methods in Computer-Aided Design*, FMCAD '14, pages 19:99–19:106, Austin, TX, 2014. FMCAD Inc.

34. Armin Haken. The intractability of resolution. *Theor. Comput. Sci.*, 39:297–308, 1985.

35. HyoJung Han and Fabio Somenzi. On-the-fly clause improvement. In Kullmann [43], pages 209–222.

36. Marijn J. H. Heule, Warren A. Hunt, Jr., and Nathan Wetzler. Trimming while checking clausal proofs. In *Formal Methods in Computer-Aided Design (FMCAD)*, pages 181–188. IEEE, 2013.

37. Marijn J. H. Heule, Warren A. Hunt, Jr., and Nathan Wetzler. Verifying refutations with extended resolution. In *International Conference on Automated Deduction (CADE)*, volume 7898 of *LNAI*, pages 345–359. Springer, 2013.

38. Marijn J. H. Heule and Hans van Maaren. *Handbook of Satisfiability*, volume 185 of *Frontiers in Artificial Intelligence and Applications*, chapter Chapter 5, Look-Ahead Based SAT Solvers, pages 155–184. IOS Press, February 2009.

39. Matti Järvisalo, Armin Biere, and Marijn Heule. Simulating circuit-level simplifications on CNF. *J. Autom. Reasoning*, 49(4):583–619, 2012.

40. Matti Järvisalo, Marijn J. H. Heule, and Armin Biere. Inprocessing rules. In *International Joint Conference on Automated Reasoning (IJCAR)*, volume 7364 of *LNCS*, pages 355–370. Springer, 2012.

41. Boris Konev and Alexei Lisitsa. A SAT attack on the erdos discrepancy conjecture. 2014. Accepted for SAT 2014.

42. Oliver Kullmann. On a generalization of extended resolution. *Discrete Applied Mathematics*, 96-97:149–176, 1999.

43. Oliver Kullmann, editor. *Theory and Applications of Satisfiability Testing - SAT 2009, 12th International Conference, SAT 2009, Swansea, UK, June 30 - July 3, 2009. Proceedings*, volume 5584 of *Lecture Notes in Computer Science*. Springer, 2009.

44. Jean-Marie Lagniez and Armin Biere. Factoring out assumptions to speed up MUS extraction. In Matti Järvisalo and Allen Van Gelder, editors, *SAT*, volume 7962 of *Lecture Notes in Computer Science*, pages 276–292. Springer, 2013.

45. Norbert Manthey, Marijn J. H. Heule, and Armin Biere. Automated reencoding of Boolean formulas. In *Proceedings of Haifa Verification Conference (HVC)*, volume 6397 of *LNCS*, pages 102–117. Springer, 2012.

46. Joao P. Marques-Silva, Ines Lynce, and Sharad Malik. *Handbook of Satisfiability*, volume 185 of *Frontiers in Artificial Intelligence and Applications*, chapter Chapter 4, Conflict-Driven Clause Learning SAT Solvers, pages 131–153. IOS Press, February 2009.

47. Kenneth L. McMillan. Interpolation and SAT-based model checking. In Warren A. Hunt Jr. and Fabio Somenzi, editors, *CAV*, volume 2725 of *Lecture Notes in Computer Science*, pages 1–13. Springer, 2003.

48. António Morgado, Federico Heras, Mark H. Liffiton, Jordi Planes, and Joao Marques-Silva. Iterative and core-guided MaxSAT solving: A survey and assessment. *Constraints*, 18(4):478–534, 2013.

49. Alexander Nadel. Boosting minimal unsatisfiable core extraction. In Bloem and Sharygina [15], pages 221–229.

50. Alexander Nadel, Vadim Ryvchin, and Ofer Strichman. Efficient MUS extraction with resolution. In *FMCAD*, pages 197–200. IEEE, 2013.

51. Alexander Nöhrer, Armin Biere, and Alexander Egyed. Managing SAT inconsistencies with HUMUS. In *Proc. VaMoS'12*, pages 83–91. ACM, 2012.

52. J. A. Robinson. A machine-oriented logic based on the resolution principle. *J. ACM*, 12(1):23–41, January 1965.

53. Simone Fulvio Rollini, Roberto Bruttomesso, Natasha Sharygina, and Aliaksei Tsitovich. Resolution proof transformation for compression and interpolation. *Formal Methods in System Design*, 45(1):1–41, 2014.

54. Carsten Sinz. Compressing propositional proofs by common subproof extraction. In Roberto Moreno-Díaz, Franz Pichler, and Alexis Quesada-Arencibia, editors, *EUROCAST*, volume 4739 of *Lecture Notes in Computer Science*, pages 547–555. Springer, 2007.

55. Carsten Sinz and Armin Biere. Extended resolution proofs for conjoining bdds. In Dima Grigoriev, John Harrison, and Edward A. Hirsch, editors, *CSR*, volume 3967 of *Lecture Notes in Computer Science*, pages 600–611. Springer, 2006.

56. Carsten Sinz, Andreas Kaiser, and Wolfgang Küchlin. Formal methods for the validation of automotive product configuration data. *Artif. Intell. Eng. Des. Anal. Manuf.*, 17(1):75–97, January 2003.

57. Mate Soos. Strangenight. In A. Belov, M. Heule, and M. Järvisalo, editors, *Proceedings of SAT Competition 2013*, volume B-2013-1 of *Department of Computer Science Series of Publications B, University of Helsinki*, pages 89–90, 2013.

58. Niklas Sörensson and Armin Biere. Minimizing learned clauses. In Kullmann [43], pages 237–243.

59. Grigori S. Tseitin. On the complexity of derivation in propositional calculus. In *Automation of Reasoning 2*, pages 466–483. Springer, 1983.

60. Alasdair Urquhart. Hard examples for resolution. *J. ACM*, 34(1):209–219, 1987.

61. Allen Van Gelder. Verifying RUP proofs of propositional unsatisfiability. In *International Symposium on Artificial Intelligence and Mathematics (ISAIM)*, 2008.

62. Allen Van Gelder. Improved conflict-clause minimization leads to improved propositional proof traces. In Kullmann [43], pages 141–146.

63. Allen Van Gelder. Producing and verifying extremely large propositional refutations - have your cake and eat it too. *Ann. Math. Artif. Intell.*, 65(4):329–372, 2012.

64. Yakir Vizel, Vadim Ryvchin, and Alexander Nadel. Efficient generation of small interpolants in CNF. In Natasha Sharygina and Helmut Veith, editors, *CAV*, volume 8044 of *Lecture Notes in Computer Science*, pages 330–346. Springer, 2013.

65. Tjark Weber. Efficiently checking propositional resolution proofs in Isabelle/HOL. In *International Workshop on the Implementation of Logics (IWIL)*, volume 212, pages 44–62, 2006.

66. Tjark Weber and Hasan Amjad. Efficiently checking propositional refutations in HOL theorem provers. *Journal of Applied Logic*, 7(1):26–40, 2009.

67. Nathan Wetzler, Marijn J. H. Heule, and Warren A. Hunt, Jr. Mechanical verification of SAT refutations with extended resolution. In *Conference on Interactive Theorem Proving (ITP)*, volume 7998 of *LNCS*, pages 229–244. Springer, 2013.

68. Nathan Wetzler, Marijn J. H. Heule, and Warren A. Hunt, Jr. DRAT-trim: Efficient checking and trimming using expressive clausal proofs. In *Theory and Applications of Satisfiability Testing (SAT)*, 2014.

69. Lintao Zhang and Sharad Malik. Validating SAT solvers using an independent resolution-based checker: Practical implementations and other applications. In *Proceedings of the conference on Design, Automation and Test in Europe - Volume 1*, DATE '03, pages 10880–10885. IEEE Computer Society, 2003.

Proofs in Satisfiability Modulo Theories

Clark Barrett[1], Leonardo de Moura[2], and Pascal Fontaine[3]

[1] New York University
barrett@cs.nyu.edu
[2] Microsoft Research
leonardo@microsoft.com
[3] University of Lorraine and INRIA
pascal.fontaine@inria.fr

1 Introduction

Satisfiability Modulo Theories (SMT) solvers[4] check the satisfiability of first-order formulas written in a language containing interpreted predicates and functions. These interpreted symbols are defined either by first-order axioms (e.g. the axioms of equality, or array axioms for operators `read` and `write`,...) or by a structure (e.g. the integer numbers equipped with constants, addition, equality, and inequalities). Theories frequently implemented within SMT solvers include the empty theory (a.k.a. the theory of uninterpreted symbols with equality), linear arithmetic on integers and/or reals, bit-vectors, and the theory of arrays. A very small example of an input formula for an SMT solver is

$$a \leq b \wedge b \leq a + x \wedge x = 0 \wedge \left[f(a) \neq f(b) \vee (q(a) \wedge \neg q(b + x)) \right]. \tag{1}$$

The above formula uses atoms over a language of equality, linear arithmetic, and uninterpreted symbols (q and f) within some Boolean combination. The SMT-LIB language (currently in version 2.0 [7]) is a standard concrete input language for SMT solvers. Figure 1 presents the above example formula in this format.

SMT solvers were originally designed as decision procedures for decidable quantifier-free fragments, but many SMT solvers additionally tackle quantifiers, and some contain decision procedures for certain decidable quantified fragments (see e.g. [34]). For these solvers, refutational completeness for first-order logic (with equality but without further interpreted symbols) is an explicit goal. Also, some SMT solvers now deal with theories that are undecidable even in the quantifier-free case, for instance non-linear arithmetic on integers [16].

In some aspects, and also in their implementation, SMT solvers can be seen as extensions of propositional satisfiability (SAT) solvers to more expressive languages. They lift the efficiency of SAT solvers to richer logics: state-of-the-art SMT solvers are able to deal with very large formulas, containing thousands of atoms. Very schematically, an SMT solver abstracts its input to propositional logic by replacing every atom with a fresh proposition, e.g., for the above example,

$$p_{a \leq b} \wedge p_{b \leq a+x} \wedge p_{x=0} \wedge \left[\neg p_{f(a)=f(b)} \vee (p_{q(a)} \wedge \neg p_{q(b+x)}) \right].$$

---

[4] We refer to [6] for a survey on SMT.

```
(set-logic QF_UFLRA)
(set-info :source | Example formula in SMT-LIB 2.0 |)
(set-info :smt-lib-version 2.0)
(declare-fun f (Real) Real)
(declare-fun q (Real) Bool)
(declare-fun a () Real)
(declare-fun b () Real)
(declare-fun x () Real)
(assert (and (<= a b) (<= b (+ a x)) (= x 0)
            (or (not (= (f a) (f b))) (and (q a) (not (q (+ b x)))))))
(check-sat)
(exit)
```

Fig. 1. Formula 1, presented in the SMT-LIB 2.0 language

The underlying SAT solver is used to provide Boolean models for this abstraction, e.g.

$$\{p_{a\leq b}, p_{b\leq a+x}, p_{x=0}, \neg p_{f(a)=f(b)}\}$$

and the theory reasoner repeatedly refutes these models and refines the Boolean abstraction by adding new clauses (in this case $\neg p_{a\leq b} \lor \neg p_{b\leq a+x} \lor \neg p_{x=0} \lor p_{f(a)=f(b)}$) until either the theory reasoner agrees with the model found by the SAT solver, or the propositional abstraction is refined to an unsatisfiable formula. As a consequence, propositional reasoning and theory reasoning are quite well distinguished. Naturally, the interaction between the theory reasoning and the SAT reasoning is in practice much more subtle than the above naive description, but even when advanced techniques (e.g. online decision procedures and theory propagation, see again [6]) are used, propositional and theory reasoning are not very strongly mixed. SMT proofs are thus also an interleaving of SAT proofs and theory reasoning proofs. The SAT proofs are typically based on some form of resolution, whereas the theory reasoning is used to justify theory lemmas, i.e. the new clauses added from the theory reasoner. The entire proof ends up having the shape of a Boolean resolution tree, each of whose leaves is a clause. These clauses are either derived from the input formula or are theory lemmas, each of which must be justified by theory-specific reasoning. The main challenge of proof production is to collect and keep enough information to produce proofs, without hurting efficiency too much.

The paper is organized as follows. Section 2 discusses the specifics of what is required to implement proof-production in SMT along with several challenges. Next, Section 3 gives a historical overview of the work on producing proofs in SMT. This is followed by a discussion of proof formats in Section 4. We then look at (in Sections 5, 6, and 7) the particular approaches taken by CVC4, veriT, and Z3. Section 8 includes a discussion of the new *Lean* prover, which attempts to bridge the gap between SMT reasoning and proof assistants more directly by building a proof assistant with efficient and sophisticated built-in

SMT capabilities. Finally, Section 9 discusses current applications of SMT proofs, and then Section 10 concludes.

2 Implementing Proof-Production in SMT

Since the core of SMT solvers is mainly a SAT solver adapted for the needs of SMT, the core of the proof is also a SAT proof. We refer to the other article by Marijn Heule, in the same volume, for more details on SAT solving. Let us just recall that SAT solvers work on a conjunctive normal form, and build a (partial) assignment using Boolean propagations and decisions. If building this assignment fails (i.e. it falsifies a clause) because of some decisions, the algorithm analyses the conflict, learns a new clause that will serve as a guide to avoid repeating the same wrong decisions again, and backtracks. Only this learning phase is important for proofs, and it is easy to generate a resolution trace corresponding to this conflict analysis and production of the learned clause. Proof-producing SMT solvers rely on the underlying SAT solver to produce this resolution proof.

In the context of SAT solving, a conflict is always due to the assignment falsifying a clause. Another kind of conflict can occur in SMT: assume the SAT solver assigned $p_{a=b}$ (i.e. the propositional abstraction of the atom $a = b$) to true, and later $p_{f(a)=f(b)}$ to false. There may be no clause conflicting with such an assignment, but it is obviously inconsistent on the theory level. The theory reasoner embedded in the SMT solver checks assignments for theory consistency. In the present case, it would notify a conflict to the underlying SAT engine, and would produce the clause $\neg p_{a=b} \lor p_{f(a)=f(b)}$. This *theory lemma* refines the propositional abstraction with some knowledge from the theory. This knowledge can then be used like any other clause. If theory reasoners provide detailed proofs for theory lemmas, the SMT proof is just the interleaving of these theory proofs and the resolution proofs generated by the SAT solver.

2.1 Proofs of Theory Lemmas

Simple examples of theory reasoners include decision procedures for uninterpreted symbols and linear arithmetic. The satisfiability of sets of ground literals with equalities and uninterpreted symbols can be checked using congruence closure [45–47]. Efficient algorithms are non-trivial, but the idea of congruence-closure algorithms is simple. They build a partition of all relevant terms, according to the equalities, and check for inconsistencies with negated equalities. For instance, if the relevant terms are a, b, $f(a)$ and $f(b)$, the initial partition would be $\{\{a\}, \{b\}, \{f(a)\}, \{f(b)\}\}$. If an equality $a = b$ is asserted, the algorithm would merge the classes for a and b, because of this equality, and would also merge the classes for $f(a)$ and $f(b)$ because it merged the arguments of f and because equality is a congruence. The resulting partition would be $\{\{a, b\}, \{f(a), f(b)\}\}$. If the algorithm is aware that $f(a) \neq f(b)$ should hold, it detects a conflict as soon as $f(a)$ and $f(b)$ are put in the same congruence class. Producing proofs for congruence closure is thus simply a matter of keeping a trace of which (two) terms are merged and why (i.e. either because of a congruence, or an asserted equality). This information can be stored in an acyclic undirected[5] graph, where terms are nodes, and edges represent asserted equalities or congruences. If the equality of two terms can be deduced by transitivity using asserted equalities and congruences, then there is a (unique) path between the two terms in the graph. To prove a congruence (e.g. $g(a,c) = g(b,d)$), it suffices to first prove the equality of the corresponding arguments ($a = b$, $c = d$) and use an instance of the congruence axiom schema, e.g. $(a = b \wedge c = d) \Rightarrow g(a,c) = g(b,d)$. Proof steps (i.e. an equality being the consequence of transitivity, and congruence instances) are again simply connected using Boolean resolution.

Another very important theory supported by many SMT solvers is linear arithmetic on reals. Decision procedures for this theory are often based on a specialized simplex algorithm [26, 40]. The algorithm maintains (a) a set of linear equalities, (b) upper and lower bounds on the variables, and (c) an assignment. The bounds and the linear equalities come directly from asserted literals; new variables are introduced so that bounds only apply on variables and not on more complicated linear combinations. Starting from an assignment satisfying all equalities, the procedure tries to progressively satisfy all bounds on the variables. During the process, the linear equalities are combined linearly to produce new sets of equalities, always equivalent to the original set. If the right strategies are used, the procedure is guaranteed to terminate either with an assignment satisfying all equalities and bounds, or an assignment that causes a specific linear equality to be in contradiction with the bounds on its variables. It then suffices to collect the constraints corresponding to the equality and the bounds to derive a contradiction. In fact, the coefficients of the variables in the linear equality correspond to the coefficients of the linear combination related to Farkas' lemma (roughly, this lemma states that a set of linear inequalities is inconsistent if and only if there is a linear combination of those inequalities that reduces to $1 \leq 0$; see e.g. [13]).

The above procedure only applies to reals. Various techniques are used to adapt it for mixed real and integer linear arithmetic. One is branching: if, in the real algorithm, an integer variable x has a rational value (e.g. 1.5), the procedure may issue a new lemma stating $x \leq 1 \vee x \geq 2$. Notice that this lemma is a tautology of arithmetic. Another technique is to introduce cuts; cuts basically remove from the set of real-feasible solutions a subset with no integer solution. Very schematically, cuts are linear combinations of already available constraints, strengthened using the fact that variables are integer, e.g. $x + y \leq 1.5$ can be strengthened to $x + y \leq 1$ if x and y are integer variables.

Other theory reasoning engines found in modern solvers include modules for reasoning about arrays, inductive data types, bit-vectors, strings, and non-linear arithmetic. Their internals may vary a lot from one solver to another, and we will not consider them in more detail here. For some non-linear arithmetic techniques

---

[5] For efficiency reasons, modern congruence closure algorithms actually use a directed graph [47].

(see e.g. [13]) like cylindrical algebraic decomposition or virtual substitution, it is not even clear how to produce useful certificates.

The theory reasoner may be a decision procedure for a single theory, but it is often a combination of decision procedures for multiple theories. Combination frameworks [44, 61] decide sets of literals mixing interpreted symbols from several decidable languages (like the aforementioned linear arithmetic and uninterpreted symbols) by combining the decision procedures for the individual theories into one decision procedure for the union of languages. Combining the decision procedures is possible when the theories in the combination satisfy certain properties. A sufficient condition, for instance, is that the theories are *stably-infinite*[6] and disjoint; this is notably the case for the combination of linear arithmetic with uninterpreted symbols. Combining theories also involves guessing which shared terms are equal and which are not, or, equivalently, communicating disjunctions of equalities between decision procedures. In order for a formula to be *unsatisfiable*, each possible equivalence relation over the shared terms must be ruled out. This typically is done by producing many individual theory lemmas, which, taken together, rule out all of the equivalence relations. Proofs for combinations of theories are thus built from proofs of the theory lemmas in the individual theories, and the component proofs are simply combined using Boolean reasoning rules.

2.2 Challenging Aspects of Proof Production

Quantifier reasoning in SMT is typically done through Skolemization when possible, and heuristic instantiation otherwise [33, 25]. Proofs involving instantiations are easily obtained by augmenting ground proofs with applications of a simple instantiation rule. Formulas like $\forall \mathbf{x}\, \varphi(\mathbf{x}) \Rightarrow \varphi(\mathbf{t})$ — where $\mathbf{x}$ is a vector of variables and $\mathbf{t}$ is a vector of terms with the same length — are first-order tautologies and can be provided as lemmas for building proofs.

Skolemization, however, is not trivial. This is because Skolemization is typically implemented by introducing a new uninterpreted constant or function, but such a step is not locally sound. It is only when looking at a global view of the proof, noting that the conclusion does not involve any introduced symbols, that soundness can be shown. Proofs involving Skolemization must therefore track the introduced constants and ensure they are not used in the conclusion of the proof. Furthermore, for efficiency, the solver may resort to advanced techniques for Skolemization that are even less amenable to proofs.

Another challenge is *preprocessing*. SMT solvers typically include a module for preprocessing formulas to simplify them before running the main algorithm to search for a satisfying assignment. This module may perform anything from simple rewrites (such as rewriting $x = y$ to $y = x$) to complex global simplifications that significantly change the structure of the formula. At the end of preprocessing, the formula is also typically converted into conjunctive normal

---

[6] A theory is stably-infinite if every satisfiable set of literals in the theory has an infinite model.

form (CNF). The CNF clauses form leaves in the ultimate resolution proof. Thus, in order to justify one of these clauses, a proof must capture each preprocessing step that was used to derive it from the original formula. Such bookkeeping is tedious and error-prone. One strategy is to disable complicated preprocessing techniques during proof-production. Still, SMT solvers rely on some minimal preprocessing for correct functioning, so in order to generate complete proofs, at least this preprocessing has to be recorded as part of the proof.

3   A Short History of Proofs in SMT

The Cooperating Validity Checker (CVC) [56] was the first SMT solver to tackle the problem of proof-production. CVC was built by Aaron Stump and Clark Barrett, advised by David Dill, at Stanford University. CVC was designed to improve upon and replace the Stanford Validity Checker (SVC) [9] for use in the group's verification applications and also to serve as a research platform for studying SMT ideas and algorithms. CVC uses a proof format based on the Edinburgh Logical Framework (LF) [35] and its proofs can be checked using the flea proof-checker [57, 58], which was developed concurrently with CVC.

The motivation for producing proofs was primarily that it would provide a means of independently certifying a correct result, so that users would not have to rely on the correctness of a large and frequently-changing code base. Another motivation was the hope that proof-production could aid in finding and correcting nasty bugs in CVC. In retrospect, however, the most important and lasting contribution of CVC's proof infrastructure was that it enabled an efficient integration of a SAT solver for Boolean reasoning. Initial attempts to use a SAT solver within CVC suffered from poor performance because the theory-reasoning part of CVC was being used as a black box, simply flagging a particular branch of the SAT search as unsatisfiable. Without additional information (i.e. a conflict clause specifying a small reason for the unsatisfiability), the SAT solver could not benefit from clause learning or non-chronological back-jumping and frequently failed to terminate, even on relatively easy problems. During a conversation with Cormac Flanagan, the idea of using the proof infrastructure to compute conflict clauses emerged, and this was the key to finally making the integration with SAT efficient (see [10] for more details). The technique was implemented in CVC, as well as in Flanagan's solver called Verifun [29] and this laid the groundwork for the so-called DPLL(T) architecture [48] used in nearly every modern SMT solver.

The next important development in proof-producing SMT solvers was the early exploration of using proofs to communicate with skeptical proof assistants. The first work in this area [41] was a translator designed to import proofs from CVC Lite [5] (the successor to CVC) into HOL Lite [36], a proof assistant for higher-order logic with a small trusted core set of rules. The goals of this work were two-fold: to provide access to efficient decision procedures within HOL Lite and to enable the use of HOL Lite as a proof-checker for CVC Lite. Shortly thereafter [30, 38], a similar effort was made to integrate the haRVey SMT solver [28] with the Isabelle/HOL proof assistant [49]. These early efforts demonstrated the feasibility of such integrations.

In 2008, an effort was made to leverage this work to certify that the benchmarks in the SMT-LIB library were correctly labeled (benchmarks are labeled with a *status* that can be "sat", "unsat", or "unknown", indicating the expected result when solving the benchmark). The certification was done using CVC3 [8] (the successor to CVC Lite) and HOL Lite [32]. Many of the benchmarks in the library were certified and additionally, a bug in the library was found as two benchmarks that had been labeled satisfiable were certified as unsatisfiable. The same work reports briefly on an anecdote that further validates the value of proof-production. A latent bug in CVC3 was revealed during the 2007 SMT-COMP competition. Using HOL Light as a proof-checker, the cause of the bug was quickly detected as a faulty proof rule. The bug in the proof rule had persisted for years and would have been very difficult to detect without the aid of the proof-checker.

The same year also marked the appearance of proof-production capabilities in several additional SMT solvers. Fx7 was a solver written by Michał Moskal which emphasized quantified reasoning and fast proof-checking [43]. MathSAT4 used an internal proof engine to enable the generation of unsatisfiable cores and interpolants [20]. Z3 began supporting proof-production as well [23], though its use of a single rule for theory lemmas meant that checking or reconstructing Z3 proofs requires the external checker to have some automated reasoning capability (see Section 7). Finally, development on veriT, the successor to haRVey, began in earnest [19], with proof-production as a primary goal (see Section 6). Another important community development around this time was an attempt to converge on a standard proof format for SMT-LIB. We discuss this effort in Section 4 below.

At the time of this writing, the state of proofs in SMT is in flux. A few groups are making a serious effort to develop solvers (e.g. CVC4 and veriT) capable of producing self-contained, independently-checkable proofs, though these are in progress and no standard format has been agreed upon. Other solvers (like Z3) produce a trace of deduction steps, but reconstructing a full proof from these steps requires additional search in some cases. Still other solvers (e.g. MathSAT, SMTInterpol), use proof technology primarily to drive additional solver functionality, like computing explanations, interpolants, or unsatisfiable cores. In this paper, we focus on tools that produce proofs with the goal of having them checked or translated by an external tool.

4   Proof Formats for SMT

The SMT-LIB initiative has produced a series of documents standardizing formats for theories, logics, models, and benchmarks. An early goal of the initiative was to produce a standard format for proofs as well. However, this proved to be challenging, primarily because each solver implements its decision procedures in a different way, meaning the proof rules for each solver could vary significantly.

Another challenge is that it is difficult to formally express the wide variety of rules needed to cover the decision procedures used in SMT solvers using existing proof languages (such as the Edinburgh Logical Framework, LF [35]).

Probably the most ambitious effort to develop a standard proof format was the one led by Aaron Stump [50, 59] (at the time, a leader of the SMT-LIB initiative), who developed a language called LFSC, modeled after LF, but able to express more complicated rules by expressing side-conditions using pieces of trusted code in a small custom programming language. LFSC allows a user to define a set of arbitrary inference rules and then use these to construct a proof. LFSC was used as the proof language for the prototype proof-producing SMT solver CLSAT and was also integrated as an alternative proof-language in CVC3 [60]. LFSC is also the proof language used in CVC4 (see Section 5 for an example).

In 2011, the first Workshop on Proof Exchange for Theorem Proving (PxTP) was held, collocated with the 23rd Conference on Automated Deduction (CADE) in Wrocław, Poland. The topic of proof formats was central, with several papers on the subject. Besson, Fontaine, and Théry proposed a proof format for SMT based on the syntax of SMT-LIB [11]. This later formed the basis for the veriT proof format (see Section 6). Böhme and Weber made a plea from a user's perspective for proof formats to be clear, well-documented, and complete (i.e. not requiring any search to replay or check) [15].

Unfortunately, none of the proposed formats have caught on as a general proof language for SMT-LIB. One reason for this is that there are not nearly as many consumers of SMT proofs as there are of SMT technology generally, so the proof format has not been prioritized. Another is that there are differences of opinion on what features should be included in such a standard. And another is that the few tools that produce proofs have considerable momentum behind their own formats and shifting to a new format would require a lot of work. However, there are clear benefits to a standard format, especially for proof exchange between SMT solvers and other tools, so we expect this will continue to be a subject of debate and research.

5 The CVC Family of SMT Solvers

The original CVC tool used a proof system based on LF. Each deductive step in the tool had a parallel proof-production step. CVC had its own internal Boolean engine which recorded proofs. Proofs were not available when using a SAT solver for Boolean reasoning. In CVC Lite and CVC3, a different design decision was made: a special *Theorem* class was created to hold all derived facts. Theorem classes could only be constructed using special proof-rule functions. This allowed us to isolate all of the trusted code in the system in a few proof-rule files. This approach, while elegant, turned out not to be very efficient. In particular, many deductive steps were not needed in the final proof, but their proofs were recorded anyway. More significantly, there was no way to turn proof production off completely.

In CVC4, we opted for a new model in which proof production is done much more efficiently. First, all proof code is protected under a compile-time flag. By disabling this flag, CVC4 can be built with no proof code at all (for maximum efficiency). When proofs are compiled in, they are generated lazily: only a minimal amount of bookkeeping is done in the SAT solver and no bookkeeping at all is done in the theory solvers.[7] If the user asks for a proof, the SAT solver bookkeeping is leveraged to produce the resolution part of the proof. Proofs of the needed theory lemmas used in the resolution proof are then generated by *re-running* the theory decision procedures in proof-producing mode. In proof-producing mode, theory decision procedures log their deductive steps in a special *proof manager* object which then stitches all of the pieces together into a full proof.

As mentioned above, the proof format used by CVC4 is LFSC. A full description of LFSC is well beyond the scope of this paper, but a good introduction can be found in [60]. LFSC is versatile, and can be used to precisely represent proofs generated in various contexts. This makes it particularly suitable for SMT proofs, where parts of proofs may come from various, very different reasoners. The philosophy behind LFSC is quite similar to some other approaches, e.g. [21]. LFSC has a stand-alone efficient proof-checker, which can be used for any proof system expressible in LFSC. The input to the proof checker is an LFSC proof signature file (e.g. specifying the proof rules for a particular theory) and the proof, and the checker checks that the proof correctly uses the rules in the signature. Figure 2 shows an LFSC proof for the example from Section 1. The first section states what is being proved: given real numbers a, b, and x, function f and predicate q, together with four proofs each justifying one of the four conjuncts of Formula 1, it produces a proof of the empty clause (represented by `cln`). The next section assigns names to different formulas appearing in the proof (each formula gets both a propositional name, starting with v and a logical atom name, starting with a). Next, rules are given for converting Formula 1 into clauses in CNF, each of which is assigned a label starting with C. Next, a number of theory derivations are given, again each generating a clause starting with C. Finally, a propositional resolution tree is given (R and Q are resolution rules) showing how to derive the empty clause from the clauses introduced through CNF and through theory lemmas. The proof shown depends on several additional proof files (not shown) which define the rules for resolution, theory reasoning, CNF conversion etc.

At the time this article is being written, CVC4 supports proofs for uninterpreted functions and arrays. Support for the theory of bit-vectors is a current project, and support for arithmetic proofs (already present in CVC3) is expected to follow soon thereafter.

---

[7] This is not quite true for the bit-vector theory solver which includes a second internal SAT solver as part of its machinery. The same minimal bookkeeping is done in this internal SAT solver as is done in the main SAT solver.

```
(check
(% a var_real
(% b var_real
(% x var_real
(% f (term (arrow Real Real))
(% q (term (arrow Real Bool))
(% @F1 (th_holds (<=_Real (a_var_real a) (a_var_real b)))
(% @F2 (th_holds (<=_Real (a_var_real b) (+_Real (a_var_real a) (a_var_real x))))
(% @F3 (th_holds (= Real (a_var_real x) (a_real 0/1)))
(% @F4 (th_holds (or (not (= Real (apply _ _ f (a_var_real a)) (apply _ _ f (a_var_real b))))
                     (and (= Bool (apply _ _ q (a_var_real a)) btrue)
                          (= Bool (apply _ _ q (+_Real (a_var_real b) (a_var_real x)))
                             bfalse))))
(: (holds cln)
(decl_atom (<=_Real (a_var_real a) (a_var_real b)) (\ v1 (\ a1
(decl_atom (<=_Real (a_var_real b) (+_Real (a_var_real a) (a_var_real x))) (\ v2 (\ a2
(decl_atom (= Real (a_var_real x) (a_real 0/1)) (\ v3 (\ a3
(decl_atom (= Real (a_var_real a) (a_var_real b)) (\ v4 (\ a4
(decl_atom (= Real (apply _ _ f (a_var_real a)) (apply _ _ f (a_var_real b))) (\ v5 (\ a5
(decl_atom (= Bool (apply _ _ q (a_var_real a)) btrue) (\ v6 (\ a6
(decl_atom (= Bool (apply _ _ q (+_Real (a_var_real b) (a_var_real x))) bfalse) (\ v7 (\ a7
(decl_atom (<=_Real (a_var_real b) (a_var_real a)) (\ v8 (\ a8
(decl_atom (= Real (a_var_real a) (+_Real (a_var_real b) (a_var_real x))) (\ v9 (\ a9
(decl_atom (and (= Bool (apply _ _ q (a_var_real a)) btrue)
                (= Bool (apply _ _ q (+_Real (a_var_real b) (a_var_real x))) bfalse))
           (\ v10 (\ a10
; CNFication
(satlem _ _ (asf _ _ _ a1 (\ l1 (clausify_false (contra _ @F1 l1)))) (\ C1
(satlem _ _ (asf _ _ _ a2 (\ l2 (clausify_false (contra _ @F2 l2)))) (\ C2
(satlem _ _ (asf _ _ _ a3 (\ l3 (clausify_false (contra _ @F3 l3)))) (\ C3
(satlem _ _ (ast _ _ _ a5 (\ l5 (asf _ _ _ a6 (\ l6 (clausify_false (contra _
  (and_elim_1 _ _ (or_elim_1 _ _ (not_not_intro _ l5) @F4)) l6)))))) (\ C4
(satlem _ _ (ast _ _ _ a5 (\ l5 (asf _ _ _ a7 (\ l7 (clausify_false (contra _
  (and_elim_2 _ _ (or_elim_1 _ _ (not_not_intro _ l5) @F4)) l7)))))) (\ C5
; Theory lemmas
; ~a4 ^ a1 ^ a8 => false
(satlem _ _ (asf _ _ _ a4 (\ l4 (ast _ _ _ a1 (\ l1 (ast _ _ _ a8
  (\ l8 (clausify_false (contra _ l1 (or_elim_1 _ _ (not_not_intro _ (<=_to_>=_Real _ _ l8))
  (not_=_to_>=_=<_Real _ _ l4))))))))))) (\ C6
; a2 ^ a3 ^ ~a8 => false
(satlem _ _ (ast _ _ _ a2 (\ l2 (ast _ _ _ a3 (\ l3 (asf _ _ _ a8 (\ l8 (clausify_false
  (poly_norm_>= _ _ _ (<=_to_>=_Real _ _ l2) (pn_- _ _ _ _ _ (pn_+ _ _ _ _ _
  (pn_var a) (pn_var x)) (pn_var b)) (\ pn2
  (poly_norm_= _ _ _ (symm _ _ _ l3) (pn_- _ _ _ _ _ (pn_const 0/1) (pn_var x)) (\ pn3
  (poly_norm_> _ _ _ (not_<=_to_>_Real _ _ l8) (pn_- _ _ _ _ _ (pn_var b) (pn_var a)) (\ pn8
  (lra_contra_> _ (lra_add_>_>= _ _ _ pn8 (lra_add_=_>= _ _ _ pn3 pn2)))))))))))))))))))) (\ C7
; a4 ^ ~a5 => false
(satlem _ _ (ast _ _ _ a4 (\ l4 (asf _ _ _ a5 (\ l5 (clausify_false
  (contra _ (cong _ _ _ _ _ _ (refl _ f) l4) l5)))))) (\ C8
; a3 ^ a4 ^ ~a9 => false
(satlem _ _ (ast _ _ _ a3 (\ l3 (ast _ _ _ a4 (\ l4 (asf _ _ _ a9 (\ l9 (clausify_false
  (poly_norm_= _ _ _ (symm _ _ _ l3) (pn_- _ _ _ _ _ (pn_const 0/1) (pn_var x)) (\ pn3
  (poly_norm_= _ _ _ l4 (pn_- _ _ _ _ _ (pn_var a) (pn_var b)) (\ pn4
  (poly_norm_distinct _ _ _ l9 (pn_- _ _ _ _ _ (pn_+ _ _ _ _ _
  (pn_var b) (pn_var x)) (pn_var a)) (\ pn9
  (lra_contra_distinct _ (lra_add_=_distinct _ _ _
  (lra_add_=_= _ _ _ pn3 pn4) pn9))))))))))))))))))) (\ C9
; a9 ^ a6 ^ a7 => false
(satlem _ _ (ast _ _ _ a9 (\ l9 (ast _ _ _ a6 (\ l6 (ast _ _ _ a7 (\ l7 (clausify_false
  (contra _ (trans _ _ _ _ (trans _ _ _ _ (symm _ _ _ l6) (cong _ _ _ _ _ _
  (refl _ q) l9)) l7) b_true_not_false))))))))) (\ C10
; Resolution proof
(satlem_simplify _ _ _ (R _ _ (Q _ _ (Q _ _ C6 C1 v1) (Q _ _ (Q _ _ C7 C2 v2) C3 v3) v8)
(Q _ _ (Q _ _ (Q _ _ (Q _ _ (R _ _ C9 C10 v9) C3 v3) C4 v6) C5 v7) C8 v5) v4)
(\ x x))))))))))))))))))))))))))))))))))))))))))))))))))))))))))))))))))))))
```

Fig. 2. An LFSC proof for example Formula 1.

6 The veriT SMT Solver

The veriT SMT solver is developed jointly at Loria, Nancy (France) and UFRN, Natal (Brazil). It is open-source, under the BSD license. veriT is first a testing platform for techniques developed around SMT, but it is sufficiently stable to be used by third parties. Proofs in veriT are mainly resolution proofs interleaved with theory reasoning lemmas.

The proof trace language of veriT is inspired by the SMT-LIB 2.0 standard. A sample proof for our running example 1 is given in Figure 3. The language is quite coarse-grained and rather falls into the "proof trace" category rather than the category of full detailed proofs. It provides, however, a full account of the resolution proof, and equality reasoning is broken down into applications of the congruence and transitivity instances of the axiom schemas for equality. Symmetry of equality is silently used uniformly. Special proof rules are assigned to theory lemmas (from arithmetic), but veriT does not break down arithmetic reasoning to instances of e.g., Presburger axioms. The example here does not feature quantifier reasoning; proofs with quantifiers use quantifier instantiation rules.

The first line gives the input. Clauses c2 to c8 explain clausification of the input. Equality reasoning produces clauses labeled with eq_congruent(_pred) and eq_reflexive, whereas the linear arithmetic module produces all the clauses labeled la_disequality and la_generic. The rest of the proof is mainly resolutions, and ends with the empty clause.

In the future, the proof format of veriT will be further improved to even better stick to the SMT-LIB standard. In particular, when using shared terms (for simplicity, the given proof example does not use shared terms), veriT uses a notation to label repeated formulas which was inspired by the SVC custom input language, but which does not fit well with the spirit of the SMT-LIB language.

7 The Z3 SMT Solver

Z3 is a Satisfiability Modulo Theories (SMT) solver developed at Microsoft Research. Z3 source code is available online, and it is free for non-commercial purposes. Z3 is used in various software analysis and test-case generation projects at Microsoft Research and elsewhere. Proof generation is based on two simple ideas: (1) a notion of implicit quotation to avoid introducing auxiliary variables (this simplifies the creation of proof objects considerably); and (2) natural deduction style proofs to facilitate modular proof re-construction.

In Z3, proof objects are represented as terms. So a proof-tree is just a term where each inference rule is represented by a function symbol. For example, consider the proof-rule $mp(p, q, \varphi)$, where p is a proof for $\psi \rightarrow \varphi$ and q is a proof for ψ. Each proof-rule has a *consequent*, the consequent of $mp(p, q, \varphi)$ is φ.

A basic underlying principle for composing and building proofs in Z3 has been to support a modular architecture that works well with theory solvers that receive literal assignments from other solvers and produce contradictions or new

```
(set .c1 (input :conclusion ((and (<= a b) (<= b (+ a x)) (= x 0)
                                  (or (not (= (f b) (f a))) (and (q a) (not (q (+ b x)))))))))))
(set .c2 (and :clauses (.c1) :conclusion ((<= a b))))
(set .c3 (and :clauses (.c1) :conclusion ((<= b (+ a x)))))
(set .c4 (and :clauses (.c1) :conclusion ((= x 0))))
(set .c5 (and :clauses (.c1) :conclusion
              ((or (not (= (f b) (f a))) (and (q a) (not (q (+ b x))))))))
(set .c6 (and_pos :conclusion ((not (and (q a) (not (q (+ b x))))) (q a))))
(set .c7 (and_pos :conclusion ((not (and (q a) (not (q (+ b x))))) (not (q (+ b x))))))
(set .c8 (or :clauses (.c5) :conclusion
              ((not (= (f b) (f a))) (and (q a) (not (q (+ b x)))))))
(set .c9 (eq_congruent :conclusion ((not (= a b)) (= (f b) (f a)))))
(set .c10 (la_disequality :conclusion ((or (= a b) (not (<= a b)) (not (<= b a))))))
(set .c11 (or :clauses (.c10) :conclusion ((= a b) (not (<= a b)) (not (<= b a)))))
(set .c12 (resolution :clauses (.c11 .c2) :conclusion ((= a b) (not (<= b a)))))
(set .c13 (la_generic :conclusion ((not (<= b (+ a x))) (<= b a) (not (= x 0)))))
(set .c14 (resolution :clauses (.c13 .c3 .c4) :conclusion ((<= b a))))
(set .c15 (resolution :clauses (.c12 .c14) :conclusion ((= a b))))
(set .c16 (resolution :clauses (.c9 .c15) :conclusion ((= (f b) (f a)))))
(set .c17 (resolution :clauses (.c8 .c16) :conclusion ((and (q a) (not (q (+ b x)))))))
(set .c18 (resolution :clauses (.c6 .c17) :conclusion ((q a))))
(set .c19 (resolution :clauses (.c7 .c17) :conclusion ((not (q (+ b x))))))
(set .c20 (eq_congruent_pred :conclusion ((not (= a (+ b x))) (not (q a)) (q (+ b x)))))
(set .c21 (resolution :clauses (.c20 .c18 .c19) :conclusion ((not (= a (+ b x))))))
(set .c22 (la_disequality :conclusion
              ((or (= a (+ b x)) (not (<= a (+ b x))) (not (<= (+ b x) a))))))
(set .c23 (or :clauses (.c22) :conclusion
              ((= a (+ b x)) (not (<= a (+ b x))) (not (<= (+ b x) a)))))
(set .c24 (resolution :clauses (.c23 .c21) :conclusion
              ((not (<= a (+ b x))) (not (<= (+ b x) a)))))
(set .c25 (eq_congruent_pred :conclusion
              ((not (= a b)) (not (= (+ a x) (+ b x))) (<= a (+ b x)) (not (<= b (+ a x))))))
(set .c26 (eq_congruent :conclusion ((not (= a b)) (not (= x x)) (= (+ a x) (+ b x)))))
(set .c27 (eq_reflexive :conclusion ((= x x))))
(set .c28 (resolution :clauses (.c26 .c27) :conclusion ((not (= a b)) (= (+ a x) (+ b x)))))
(set .c29 (resolution :clauses (.c25 .c28) :conclusion
              ((not (= a b)) (<= a (+ b x)) (not (<= b (+ a x))))))
(set .c30 (resolution :clauses (.c29 .c3 .c15) :conclusion ((<= a (+ b x)))))
(set .c31 (resolution :clauses (.c24 .c30) :conclusion ((not (<= (+ b x) a)))))
(set .c32 (la_generic :conclusion ((<= (+ b x) a) (not (= a b)) (not (= x 0)))))
(set .c33 (resolution :clauses (.c32 .c4 .c15 .c31) :conclusion ()))
```

Fig. 3. The proof output by veriT for example Formula 1. The output has been slightly edited to cut and indent long lines.

literal assignments. The theory solvers should be able to produce independent and opaque explanations for their decisions. Conceptually, each solver acts upon a set of hypotheses and produces a consequent. The basic proof-rules that support such an architecture can be summarized as: *hypothesis*, which introduces assumptions, *lemma*, which eliminates hypotheses, and *unit resolution*, which handles basic propagation. We say that a proof-term is *closed* when every path that ends with a hypothesis contains an application of rule lemma. If a proof-term is not closed, it is open.

The main propositional inference engine in Z3 is based on a DPLL(T) architecture. The DPLL(T) proof search method lends itself naturally to producing resolution style proofs. Systems, such as zChaff and a version of MiniSAT, produce proof logs based on logging the unit propagation steps as well as the conflict resolution steps. The resulting log suffices to produce a propositional resolution proof.

The approach taken in Z3 bypasses logging, and instead builds proof objects during conflict resolution. With each clause we attach a proof. Clauses that were produced as part of the input have proofs that were produced from the previous steps. This approach does not require logging resolution steps for every unit-propagation, but delays the analysis of which unit propagation steps are useful until conflict resolution. The approach also does not produce a resolution proof directly. It produces a natural deduction style proof with hypotheses.

The theory of equality can be captured by axioms for reflexivity, symmetry, transitivity, and substitutivity of equality. In Z3, these axioms are inference rules, and these inference rules apply for any binary relation that is reflexive, symmetric, transitive, and/or reflexive-monotone.

In the DPLL(T) architecture, decision procedures for a theory T identify sets of asserted T-inconsistent literals. Dually, the disjunction of the negated literals are T-tautologies. Consequently, proof terms created by theories can be summarized using a single form, here called *theory lemmas*. Some theory lemmas are annotated with hints to help proof checkers and reconstruction. For example, the theory of linear arithmetic produces theory lemmas based on Farkas' lemma. For example, suppose p is a proof for $x > 0$, and q is a proof for $2x + 1 < 0$, then $farkas(1, p, -1/2, q, \neg(x > 0) \vee \neg(2x + 1 < 0))$ is a theory lemma where the coefficients 1 and $-1/2$ are hints.

The Z3 simplifier applies standard simplification rules for the supported theories. For example, terms using the arithmetical operations, whether for integer, real, or bit-vector arithmetic, are normalized into sums of monomials. A single proof rule called *rewrite* is used to record the simplification steps. For example, $rewrite(x + x = 2x)$ is a proof for $x + x = 2x$.

Notice that Z3 does not axiomatize the legal rewrites. Instead, to check the rewrite steps, one must rely on a proof checker to be able to apply similar inferences for the set of built-in theories. Thus, Z3 proofs are coarse-grained and also fall into the "proof trace" category, similarly to veriT.

Since Z3 proofs are terms, they can be traversed using the Z3 API. Proofs can also be output in the SMT-LIB 2.0 language. A sample proof for our running example 1 is given in Figure 4.

```
(let (($x82 (q b)) (?x49 (* (- 1.0) b)) (?x50 (+ a ?x49))
      ($x51 (<= ?x50 0.0)) (?x35 (f b)) (?x34 (f a))
      ($x36 (= ?x34 ?x35)) ($x37 (not $x36))
      ($x43 (or $x37 (and (q a) (not (q (+ b x))))))
      ($x33 (= x 0.0)) (?x57 (+ a ?x49 x)) ($x56 (>= ?x57 0.0))
      ($x44 (and (<= a b) (<= b (+ a x)) $x33 $x43))
      (@x60 (monotonicity (rewrite (= (<= a b) $x51))
                         (rewrite (= (<= b (+ a x)) $x56))
                         (= $x44 (and $x51 $x56 $x33 $x43))))
      (@x61 (mp (asserted $x44) @x60 (and $x51 $x56 $x33 $x43)))
      (@x62 (and-elim @x61 $x51)) ($x71 (>= ?x50 0.0)))
(let ((@x70 (trans (monotonicity (and-elim @x61 $x33) (= ?x57 (+ a ?x49 0.0)))
                  (rewrite (= (+ a ?x49 0.0) ?x50)) (= ?x57 ?x50))))
(let ((@x74 (mp (and-elim @x61 $x56) (monotonicity @x70 (= $x56 $x71)) $x71)))
(let ((@x121 (monotonicity (symm ((_ th-lemma arith eq-propagate 1 1) @x74 @x62 (= a b))
                                (= b a))
                          (= $x82 (q a)))))
(let (($x38 (q a)) ($x96 (or (not $x38) $x82)) ($x97 (not $x96)))
(let ((@x115 (monotonicity (symm ((_ th-lemma arith eq-propagate 1 1) @x74 @x62 (= a b))
                                (= b a))
                          (= ?x35 ?x34))))
(let (($x100 (or $x37 $x97)))
(let ((@x102 (monotonicity (rewrite (= (and $x38 (not $x82)) $x97))
                          (= (or $x37 (and $x38 (not $x82))) $x100))))
(let (($x85 (not $x82)))
(let (($x88 (and $x38 $x85)))
(let (($x91 (or $x37 $x88)))
(let ((@x81 (trans (monotonicity (and-elim @x61 $x33) (= (+ b x) (+ b 0.0)))
                  (rewrite (= (+ b 0.0) b)) (= (+ b x) b))))
(let ((@x87 (monotonicity (monotonicity @x81 (= (q (+ b x)) $x82)) (= (not (q (+ b x)))
                                                                      $x85))))
(let ((@x93 (monotonicity (monotonicity @x87 (= (and $x38 (not (q (+ b x)))) $x88))
                          (= $x43 $x91))))
(let ((@x103 (mp (mp (and-elim @x61 $x43) @x93 $x91) @x102 $x100)))
(let ((@x119 (unit-resolution (def-axiom (or $x96 $x38))
                             (unit-resolution @x103 (symm @x115 $x36) $x97) $x38)))
(let ((@x118 (unit-resolution (def-axiom (or $x96 $x85))
                             (unit-resolution @x103 (symm @x115 $x36) $x97) $x85)))
(unit-resolution @x118 (mp @x119 (symm @x121 (= $x38 $x82)) $x82) false)))))))))))))))))
```

Fig. 4. The proof output by Z3 for example Formula 1. The output has been slightly edited to cut and indent long lines.

8   The Lean Prover

Lean is a new open source theorem prover being developed by Leonardo de Moura and Soonho Kong [24]. Lean is not a standard SMT solver; it can be used as an automatic prover like SMT solvers, but it can also be used as a proof assistant. The Lean kernel is based on dependent type theory, and is implemented in two layers. The first layer contains the type checker and APIs for creating and manipulating the terms, the declarations, and the environment. The first layer has several configuration options. For example, developers may instantiate the kernel with or without an impredicative Prop sort. They may also select whether Prop is proof-irrelevant or not. The first layer consists of 5k lines of C++ code. The second layer provides additional components such as inductive families (500 additional lines of code). When the kernel is instantiated, one selects which of these components should be used. The current components are already sufficient for producing an implementation of the Calculus of Inductive Constructions (CIC). The main difference is that Lean does not have universe cumulativity; it instead provides universe polymorphism. The Lean CIC-based kernel is treated as the standard kernel. Another design goal is to support the new Homotopy Type System (HTS) proposed by Vladimir Voevodsky. HTS is going to be implemented as another kernel instantiation.

Lean is meant to be used as a standalone system and as a software library. It provides an extensive API and can be easily embedded in other systems. SMT solvers can use the Lean API to create proof terms that can be independently checked. The API can also be used to export Lean proofs to other systems based on CIC (e.g., Coq and Matita).

Having a more expressive language for encoding proofs provides several advantages. First, we can easily add new "proof rules" without modifying the proof checker (i.e., type checker). Proof rules such as *mp* and *monotonicity* used in Z3 are just theorems in Lean. When a new decision procedure (or any other form of automation) is implemented, the developer must first prove the theorems that are needed to justify the results produced by the automatic procedure. For example, suppose a developer is implementing a procedure for Presburger arithmetic, she will probably use a theorem such as:

```
theorem add_comm (n m:nat) : n + m = m + n
:=
  induction_on m
    (trans (add_zero_right _) (symm (add_zero_left _)))
    (take k IH,
      calc
        n + succ k = succ (n+k) : add_succ_right _ _
        ... = succ (k + n) : {IH}
        ... = succ k + n : symm (add_succ_left _ _))
```

Preprocessing steps such as Skolemization can be supported in a similar way. Whenever a preprocessing procedure applies a Skolemization step, it uses the following theorem to justify it.

```
theorem skolem_th {A : Type} {B : A -> Type} {P : forall x : A, B x -> Bool} :
        (forall x, exists y, P x y) = (exists f, (forall x, P x (f x)))
:= iff_intro
        (assume H : (forall x, exists y, P x y), @axiom_of_choice _ _ P H)
        (assume H : (exists f, (forall x, P x (f x))),
                take x, obtain (fw : forall x, B x) (Hw : forall x, P x (fw x)), from H,
                    exists_intro (fw x) (Hw x))
```

Most SMT solvers make extensive use of preprocessing steps, and as pointed out before, it is not easy to be proof-producing for every single rewriting step that can occur in the solver. In Lean, this issue is addressed by providing a generic rewriting engine that can use any previously proved theorems. The engine accepts two kinds of theorems: congruence theorems and (conditional) equations. For example, the following two theorems can be used to distribute universal quantifiers over disjunctions when the left (right) hand side does not reference the bound variable.

```
theorem forall_or_distributel {A : Type} (p : Bool) (q : A -> Bool)
        : (forall x, q x \/ p) = ((forall x, q x) \/ p)
theorem forall_or_distributer {A : Type} (p : Bool) (q : A -> Bool)
        : (forall x, p \/ q x) = (p \/ forall x, q x)
```

9 Applications

While the ability to use a proof to independently check the result of an SMT solver is generally seen as a valuable objective, other applications have thus far been more effective in driving the development of proof-producing SMT solvers, particularly the desire to use SMT solvers in skeptical proof assistants and the need for interpolation. Below, we discuss some of the specific applications of the solvers discussed in this paper.

9.1 CVC

Early applications of proofs in CVC and CVC Lite were covered in Section 3. Later, CVC3 was instrumented to produce LFSC-proofs (instead proofs in its *ad hoc* native format) both as a platform for experimenting with proof systems [51] and as a way to produce certified interpolants [53].

Proofs in CVC4 are still at an early stage, but a current project[8] has as its objective the completion of proof-production for the theories of uninterpreted functions, arrays, and bit-vectors in CVC4 by August 2014 and a facility for translating those proofs to Coq by August 2015.

9.2 veriT

Proofs were first implemented in veriT to allow proof reconstruction of SMT proofs within proof assistants. This feature has been successfully used with Isabelle [30], and later with Coq within the SMTCoq tool [2,39].

Georg Hofferek et al. [37] used the proof producing capability of veriT in the context of controller synthesis for pipelined processors. The method basically builds several interpolants from a unique proof. In the Rodin plugin to SMT [27], proofs from veriT are used as a means to quickly extract unsatisfiable cores.

9.3 Z3

The two main applications for Z3 proof certificates are: proof reconstruction in interactive proof assistants such as Isabelle [14]; and interpolation generation [42].

---

[8] funded by the DARPA HACMS program

In Isabelle/HOL, Z3 proofs are reconstructed in a completely different system based on a secure proof kernel. Proof reconstruction is quite an involved process, because proof rules such as *theory lemmas* require several steps of reasoning in Isabelle/HOL.

The interpolation prover iZ3 [42] is implemented on top of Z3. It uses Z3 proofs to guide the construction of proofs by a secondary, less efficient, interpolating prover. It translates Z3 proofs into a proof calculus that does admit feasible interpolation, with "gaps", or lemmas, that must be discharged by the secondary prover.

10 Conclusions

Building a proof-producing SMT solver appears easy on the surface: the underlying SAT solver generates Boolean resolution proofs, and, as long as the theories in the combination also produce certificates, these can also be integrated using Boolean resolution. There are, however, challenges both small and large related to SMT proofs. First, it is necessary to collect and store all the necessary information to produce the final proof; this is mostly a technical problem, but it can be a bottleneck during proof search. Indeed, this may involve an enormous amount of bookkeeping, and, if using main memory, may quickly exhaust the available memory. Second, SMT often relies on many preprocessing techniques, some being necessary for the soundness of tool, some being useful for efficiency. The attitude followed by most SMT developers is to provide some kind of high level trace for essential preprocessing. Non-essential preprocessing is turned off, with consequences for efficiency. It may also be non-trivial to generate good proofs for some preprocessing techniques, e.g. for Skolemization, or symmetry breaking. Finally, SMT solvers may use external tools as specialized reasoners for some theories. Producing a proof in SMT may thus, in some cases, require certificates from those external reasoners as well.

Producing short proofs is an active research area, not only relevant in the context of SMT solving. Since SMT proofs often contain an important part of Boolean reasoning, the efforts for improving SMT proofs mainly target the propositional part of the proofs. Besides trivial post-processing methods like pruning irrelevant proof steps, some advanced techniques provide significant proof compression ratios [1, 55, 3, 22, 54, 4, 31, 18]. The standalone tool Skeptic [17] is dedicated to post-processing proofs from SAT and SMT solvers, and implements many of these techniques.

An important challenge for the SMT community is to provide a proof format which is sufficiently flexible to accommodate all needs, sufficiently compact to be practical, and sufficiently elegant (which is very subjective) to be accepted by most. Some formats have been proposed [12, 23, 52], but it seems it is still too early for a consensus on how to represent SMT proofs.

Acknowledgments: Clark Barrett would like to thank Andrew Reynolds for providing the LFSC proof example in Figure 2. Pascal Fontaine would like to thank David Déharbe, who jointly designed veriT with him. The development of veriT is partially supported by the project ANR-13-IS02-0001 of the Agence Nationale de la Recherche. Proof-production in CVC4 is funded in part by DARPA award FA8750-13-2-0241.

We are grateful to the anonymous reviewers for their careful read and their comments.

References

1. Hasan Amjad. Compressing propositional refutations. *Electr. Notes Theor. Comput. Sci.*, 185:3–15, 2007.
2. Michaël Armand, Germain Faure, Benjamin Grégoire, Chantal Keller, Laurent Thery, and Benjamin Werner. A Modular Integration of SAT/SMT Solvers to Coq through Proof Witnesses. In Jean-Pierre Jouannaud and Zhong Shao, editors, *CPP - Certified Programs and Proofs - First International Conference - 2011*, volume 7086 of *Lecture notes in computer science - LNCS*, pages 135–150, Kenting, Taïwan, Province De Chine, December 2011. Springer.
3. Omer Bar-Ilan, Oded Fuhrmann, Shlomo Hoory, Ohad Shacham, and Ofer Strichman. Linear-time reductions of resolution proofs. In Hana Chockler and Alan J. Hu, editors, *Haifa Verification Conference*, volume 5394 of *Lecture Notes in Computer Science*, pages 114–128. Springer, 2009.
4. Omer Bar-Ilan, Oded Fuhrmann, Shlomo Hoory, Ohad Shacham, and Ofer Strichman. Reducing the size of resolution proofs in linear time. *STTT*, 13(3):263–272, 2011.
5. Clark Barrett and Sergey Berezin. CVC Lite: A new implementation of the cooperating validity checker. In Rajeev Alur and Doron A. Peled, editors, *Proceedings of the 16th International Conference on Computer Aided Verification (CAV '04)*, volume 3114 of *Lecture Notes in Computer Science*, pages 515–518. Springer-Verlag, July 2004. Boston, Massachusetts.
6. Clark Barrett, Roberto Sebastiani, Sanjit A. Seshia, and Cesare Tinelli. Satisfiability modulo theories. In Armin Biere, Marijn J. H. Heule, Hans van Maaren, and Toby Walsh, editors, *Handbook of Satisfiability*, volume 185 of *Frontiers in Artificial Intelligence and Applications*, chapter 26, pages 825–885. IOS Press, February 2009.
7. Clark Barrett, Aaron Stump, and Cesare Tinelli. The SMT-LIB standard : Version 2.0, December 2010.
8. Clark Barrett and Cesare Tinelli. CVC3. In Werner Damm and Holger Hermanns, editors, *Proceedings of the 19th International Conference on Computer Aided Verification (CAV '07)*, volume 4590 of *Lecture Notes in Computer Science*, pages 298–302. Springer-Verlag, July 2007. Berlin, Germany.
9. Clark W. Barrett, David L. Dill, and Jeremy R. Levitt. Validity checking for combinations of theories with equality. In Mandayam Srivas and Albert Camilleri, editors, *Proceedings of the 1st International Conference on Formal Methods In Computer-Aided Design (FMCAD '96)*, volume 1166 of *Lecture Notes in Computer Science*, pages 187–201. Springer-Verlag, November 1996. Palo Alto, California.
10. Clark W. Barrett, David L. Dill, and Aaron Stump. Checking satisfiability of first-order formulas by incremental translation to SAT. In Ed Brinksma and Kim Guldstrand Larsen, editors, *Proceedings of the 14th International Conference on Computer Aided Verification (CAV '02)*, volume 2404 of *Lecture Notes in Computer Science*, pages 236–249. Springer-Verlag, July 2002. Copenhagen, Denmark.

11. Frédéric Besson, Pascal Fontaine, and Laurent Théry. A flexible proof format for SMT: a proposal. In *First International Workshop on Proof eXchange for Theorem Proving - PxTP 2011*, Wroclaw, Pologne, August 2011.

12. Frédéric Besson, Pascal Fontaine, and Laurent Théry. A flexible proof format for smt: a proposal. In *Workshop on Proof eXchange for Theorem Proving*, 2011.

13. Alexander Bockmayr and V. Weispfenning. Solving numerical constraints. In John Alan Robinson and Andrei Voronkov, editors, *Handbook of Automated Reasoning*, volume I, chapter 12, pages 751–842. Elsevier Science B.V., 2001.

14. Sascha Böhme. Proof reconstruction for Z3 in Isabelle/HOL. In *7th International Workshop on Satisfiability Modulo Theories (SMT '09)*, 2009.

15. Sascha Böhme and Tjark Weber. Designing proof formats: A user's perspective — experience report. In *First International Workshop on Proof eXchange for Theorem Proving - PxTP 2011*, Wroclaw, Pologne, August 2011.

16. Cristina Borralleras, Salvador Lucas, Rafael Navarro-Marset, Enric Rodríguez-Carbonell, and Albert Rubio. Solving non-linear polynomial arithmetic via sat modulo linear arithmetic. In Renate A. Schmidt, editor, *CADE*, volume 5663 of *Lecture Notes in Computer Science*, pages 294–305. Springer, 2009.

17. Joseph Boudou, Andreas Fellner, and Bruno Woltzenlogel Paleo. Skeptik: A proof compression system. In Stéphane Demri, Deepak Kapur, and Christoph Weidenbach, editors, *Automated Reasoning - 7th International Joint Conference, IJCAR 2014, Held as Part of the Vienna Summer of Logic, VSL 2014, Vienna, Austria, July 19-22, 2014. Proceedings*, volume 8562 of *Lecture Notes in Computer Science*, pages 374–380. Springer, 2014.

18. Joseph Boudou and Bruno Woltzenlogel Paleo. Compression of propositional resolution proofs by lowering subproofs. In Didier Galmiche and Dominique Larchey-Wendling, editors, *Automated Reasoning with Analytic Tableaux and Related Methods - 22th International Conference, TABLEAUX 2013, Nancy, France, September 16-19, 2013. Proceedings*, volume 8123 of *Lecture Notes in Computer Science*, pages 59–73. Springer, 2013.

19. Thomas Bouton, Diego B. Caminha de Oliveira, David Déharbe, and Pascal Fontaine. veriT: An open, trustable and efficient SMT-solver. In Renate A Schmidt, editor, *Automated Deduction – CADE-22*, volume 5663 of *Lecture Notes in Computer Science*, pages 151–156. Springer Berlin Heidelberg, 2009.

20. Roberto Bruttomesso, Alessandro Cimatti, Anders Franzén, Alberto Griggio, and Roberto Sebastiani. The MathSAT 4 SMT solver. In Aarti Gupta and Sharad Malik, editors, *Computer Aided Verification*, volume 5123 of *Lecture Notes in Computer Science*, chapter 28, pages 299–303. Springer Berlin Heidelberg, Berlin, Heidelberg, 2008.

21. Guillaume Burel. A shallow embedding of resolution and superposition proofs into the λΠ-calculus modulo. In *Third International Workshop on Proof eXchange for Theorem Proving - PxTP 2013*, Lake Placid, USA, june 2013.

22. Scott Cotton. Two techniques for minimizing resolution proofs. In Ofer Strichman and Stefan Szeider, editors, *Theory and Applications of Satisfiability Testing – SAT 2010*, volume 6175 of *Lecture Notes in Computer Science*, pages 306–312. Springer, 2010.

23. Leonardo de Moura and Nikolaj Bjørner. Proofs and refutations, and Z3. In *Proceedings of the LPAR 2008 Workshops, Knowledge Exchange: Automated Provers and Proof Assistants, and the 7th International Workshop on the Implementation of Logics, Doha, Qatar, November 22, 2008*, volume 418 of *CEUR Workshop Proceedings*. CEUR-WS.org, 2008.

41

24. Leonardo de Moura and Soonho Kong. Lean theorem prover: `http://github.com/leanprover`, 2014.

25. Leonardo Mendonça de Moura and Nikolaj Bjørner. Efficient E-matching for SMT solvers. In Frank Pfenning, editor, *Proc. Conference on Automated Deduction (CADE)*, volume 4603 of *Lecture Notes in Computer Science*, pages 183–198. Springer, 2007.

26. Bruno Dutertre and Leonardo de Moura. A Fast Linear-Arithmetic Solver for DPLL(T). In *Computer Aided Verification (CAV)*, volume 4144 of *Lecture Notes in Computer Science*, pages 81–94. Springer-Verlag, 2006.

27. David Déharbe, Pascal Fontaine, Yoann Guyot, and Laurent Voisin. Smt solvers for rodin. In John Derrick, John Fitzgerald, Stefania Gnesi, Sarfraz Khurshid, Michael Leuschel, Steve Reeves, and Elvinia Riccobene, editors, *Abstract State Machines, Alloy, B, VDM, and Z*, volume 7316 of *Lecture Notes in Computer Science*, pages 194–207. Springer Berlin Heidelberg, 2012.

28. David Déharbe and Silvio Ranise. Light-weight theorem proving for debugging and verifying units of code. In *First International Conference on Software Engineering and Formal Methods, 2003.*, pages 220–228. IEEE, 2003.

29. Cormac Flanagan, Rajeev Joshi, Xinming Ou, and James B. Saxe. Theorem proving using lazy proof explication. In Warren A. Hunt and Fabio Somenzi, editors, *Computer Aided Verification*, volume 2725 of *Lecture Notes in Computer Science*, pages 355–367. Springer Berlin Heidelberg, 2003.

30. Pascal Fontaine, Jean-Yves Marion, Stephan Merz, Leonor Prensa Nieto, and Alwen Tiu. Expressiveness + automation + soundness: Towards combining SMT solvers and interactive proof assistants. In Holger Hermanns and Jens Palsberg, editors, *Tools and Algorithms for the Construction and Analysis of Systems*, volume 3920 of *Lecture Notes in Computer Science*, pages 167–181. Springer Berlin Heidelberg, 2006.

31. Pascal Fontaine, Stephan Merz, and Bruno Woltzenlogel Paleo. Compression of propositional resolution proofs via partial regularization. In Nikolaj Bjørner and Viorica Sofronie-Stokkermans, editors, *Automated Deduction - CADE-23 - 23rd International Conference on Automated Deduction, Wroclaw, Poland, July 31 - August 5, 2011. Proceedings*, volume 6803 of *Lecture Notes in Computer Science*, pages 237–251. Springer, 2011.

32. Yeting Ge and Clark Barrett. Proof translation and SMT-LIB benchmark certification: A preliminary report. In *Proceedings of the 6th International Workshop on Satisfiability Modulo Theories (SMT '08)*, 2008.

33. Yeting Ge, Clark Barrett, and Cesare Tinelli. Solving quantified verification conditions using satisfiability modulo theories. *Annals of Mathematics and Artificial Intelligence*, 55(1-2):101–122, February 2009.

34. Yeting Ge and Leonardo Mendonça de Moura. Complete instantiation for quantified formulas in satisfiabiliby modulo theories. In Ahmed Bouajjani and Oded Maler, editors, *CAV*, volume 5643 of *Lecture Notes in Computer Science*, pages 306–320. Springer, 2009.

35. Robert Harper, Furio Honsell, and Gordon Plotkin. A framework for defining logics. *J. ACM*, 40(1):143–184, January 1993.

36. John Harrison. Hol light: A tutorial introduction. In Mandayam Srivas and Albert Camilleri, editors, *Formal Methods in Computer-Aided Design*, volume 1166 of *Lecture Notes in Computer Science*, pages 265–269. Springer Berlin Heidelberg, 1996.

37. Georg Hofferek, Ashutosh Gupta, Bettina Könighofer, Jie-Hong Roland Jiang, and Roderick Paul Bloem. Synthesizing multiple boolean functions using interpolation on a single proof. In *FMCAD 2013 - Formal Methods in Computer-Aided Design*, pages 77 – 84. IEEE, 2013.

38. Clément Hurlin, Amine Chaib, Pascal Fontaine, Stephan Merz, and Tjark Weber. Practical proof reconstruction for first-order logic and set-theoretical constructions. In Moa Johansson Lucas Dixon, editor, *The Isabelle Workshop 2007 - Isabelle'07*, pages 2–13, Bremen, Germany, 2007.

39. C. Keller. *A Matter of Trust: Skeptical Communication Between Coq and External Provers*. PhD thesis, École Polytechnique, June 2013.

40. Timothy King, Clark Barrett, and Bruno Dutertre. Simplex with sum of infeasibilities for SMT. In *Proceedings of the 13th International Conference on Formal Methods In Computer-Aided Design (FMCAD '13)*, pages 189–196. FMCAD Inc., October 2013. Portland, Oregon.

41. Sean McLaughlin, Clark Barrett, and Yeting Ge. Cooperating theorem provers: A case study combining HOL-Light and CVC Lite. In Alessandro Armando and Alessandro Cimatti, editors, *Proceedings of the 3rd Workshop on Pragmatics of Decision Procedures in Automated Reasoning (PDPAR '05)*, volume 144(2) of *Electronic Notes in Theoretical Computer Science*, pages 43–51. Elsevier, January 2006. Edinburgh, Scotland.

42. Kenneth L. McMillan. Interpolants from z3 proofs. In *Proceedings of the International Conference on Formal Methods in Computer-Aided Design*, FMCAD '11, pages 19–27, 2011.

43. Michał Moskal. Rocket-Fast proof checking for SMT solvers. In C R Ramakrishnan and Jakob Rehof, editors, *Tools and Algorithms for the Construction and Analysis of Systems*, volume 4963 of *Lecture Notes in Computer Science*, pages 486–500. Springer Berlin Heidelberg, 2008.

44. Greg Nelson and Derek C. Oppen. Simplifications by cooperating decision procedures. *ACM Transactions on Programming Languages and Systems*, 1(2):245–257, October 1979.

45. Greg Nelson and Derek C. Oppen. Fast decision procedures based on congruence closure. *Journal of the ACM*, 27(2):356–364, April 1980.

46. Robert Nieuwenhuis and Albert Oliveras. Union-find and congruence closure algorithms that produce proofs. In Cesare Tinelli and Silvio Ranise, editors, *Pragmatics of Decision Procedures in Automated Reasoning (PDPAR)*, 2004.

47. Robert Nieuwenhuis and Albert Oliveras. Fast congruence closure and extensions. *Information and Computation*, 205(4):557–580, 2007.

48. Robert Nieuwenhuis, Albert Oliveras, and Cesare Tinelli. Solving SAT and SAT Modulo Theories: from an Abstract Davis-Putnam-Logemann-Loveland Procedure to DPLL(T). *Journal of the ACM*, 53(6):937–977, November 2006.

49. Tobias Nipkow, Lawrence C Paulson, and Markus Wenzel. *Isabelle/HOL: a proof assistant for higher-order logic*, volume 2283. Springer, 2002.

50. Duckki Oe, Andrew Reynolds, and Aaron Stump. Fast and flexible proof checking for SMT. In *Proceedings of the 7th International Workshop on Satisfiability Modulo Theories - SMT '09*, pages 6–13, New York, New York, USA, August 2009. ACM Press.

51. Andrew Reynolds, Liana Hadarean, Cesare Tinelli, Yeting Ge, Aaron Stump, and Clark Barrett. Comparing proof systems for linear real arithmetic with LFSC. In *Proceedings of the 8th International Workshop on Satisfiability Modulo Theories (SMT '10)*, July 2010. Edinburgh, Scotland.

52. Andrew Reynolds, Liana Haderean, Cesare Tinelli, Yeting Ge, Aaron Stump, and Clark Barrett. Comparing proof systems for linear real arithmetic with LFSC. In *International Workshop on Satisfiability Modulo Theories (SMT)*, 2010.

53. Andrew Reynolds, Cesare Tinelli, and Liana Hadarean. Certified interpolant generation for EUF. In S. Lahiri and S. Seshia, editors, *Proceedings of the 9th International Workshop on Satisfiability Modulo Theories (Snowbird, USA)*, 2011.

54. Simone Fulvio Rollini, Roberto Bruttomesso, and Natasha Sharygina. An efficient and flexible approach to resolution proof reduction. In Sharon Barner, Ian Harris, Daniel Kroening, and Orna Raz, editors, *Hardware and Software: Verification and Testing*, volume 6504 of *Lecture Notes in Computer Science*, pages 182–196. Springer, 2011.

55. Carsten Sinz. Compressing propositional proofs by common subproof extraction. In Roberto Moreno-Díaz, Franz Pichler, and Alexis Quesada-Arencibia, editors, *EUROCAST*, volume 4739 of *Lecture Notes in Computer Science*, pages 547–555. Springer, 2007.

56. Aaron Stump, Clark W. Barrett, and David L. Dill. CVC: A cooperating validity checker. In Ed Brinksma and Kim Guldstrand Larsen, editors, *Proceedings of the 14th International Conference on Computer Aided Verification (CAV '02)*, volume 2404 of *Lecture Notes in Computer Science*, pages 500–504. Springer-Verlag, July 2002. Copenhagen, Denmark.

57. Aaron Stump, Clark W. Barrett, and David L. Dill. Producing proofs from an arithmetic decision procedure in elliptical LF. In Frank Pfenning, editor, *Proceedings of the 3rd International Workshop on Logical Frameworks and Meta-Languages (LFM '02)*, volume 70(2) of *Electronic Notes in Theoretical Computer Science*, pages 29–41. Elsevier, July 2002. Copenhagen, Denmark.

58. Aaron Stump and David L. Dill. Faster proof checking in the edinburgh logical framework. In Andrei Voronkov, editor, *Proc. Conference on Automated Deduction (CADE)*, volume 2392 of *Lecture Notes in Computer Science*, pages 392–407. Springer Berlin Heidelberg, 2002.

59. Aaron Stump and Duckki Oe. Towards an SMT proof format. In *Proceedings of the Joint Workshops of the 6th International Workshop on Satisfiability Modulo Theories and 1st International Workshop on Bit-Precise Reasoning - SMT '08/BPR '08*, page 27, New York, New York, USA, July 2008. ACM Press.

60. Aaron Stump, Duckki Oe, Andrew Reynolds, Liana Hadarean, and Cesare Tinelli. SMT proof checking using a logical framework. *Formal Methods in System Design*, 42(1):91–118, July 2013.

61. Cesare Tinelli and Mehdi T. Harandi. A new correctness proof of the Nelson–Oppen combination procedure. In F. Baader and Klaus U. Schulz, editors, *Frontiers of Combining Systems (FroCoS)*, Applied Logic, pages 103–120. Kluwer Academic Publishers, March 1996.

Proof Generation for Saturating First-Order Theorem Provers

Stephan Schulz[1] and Geoff Sutcliffe[2]

[1] DHBW Stuttgart
schulz@eprover.org
[2] University of Miami
geoff@cs.miami.edu

1 Introduction

First-order Automated Theorem Proving (ATP) is one of the oldest and most developed areas of automated reasoning. Today, the most widely used first-order provers are fully automatic and process first-order logic with equality. Many state-of-the-art ATP systems consist of a clausifier, translating a full first-order problem specification into clause normal form, and a saturation procedure that tries to derive the empty clause to complete a proof by contradiction. Saturation procedures are typically based on variants of the superposition calculus, often combining restricted forms of paramodulation with resolution and strong redundancy elimination techniques, in particular simplification via rewriting and subsumption. The first widely used ATP system in this mold was Bill McCune's Otter [38], now succeeded by Prover9 [36]. Other major examples include Vampire [69, 49], SPASS [70, 72], and E [55, 57].

Proof production was not a primary concern early on, and provers offered different levels of support for explicit proof objects. Information was output in a variety of formats. Nowadays, proof object output is supported by most major provers, and many systems support the syntax used in the Thousands of Problems for Theorem Provers (TPTP) project [65] for proof output. In this syntax, proofs are represented as directed acyclic graphs (DAGs), where each node is annotated with a clause or formula used in the proof. The original axioms, assumptions and goals are nodes with in-degree 0, i.e. they correspond to leaves if the DAG is unfolded into a proof tree. Inner nodes represent derived clauses and formulas, linked to the premises used in their derivation via incoming edges. The final node of the proof graph (i.e. the root in a proof tree) is the empty clause, concluding the proof by contradiction. Nodes are annotated with the inference(s) used to produce them.

The main difficulty in obtaining proof objects is the very high rate of inferences and simplifications during proof search. Most of these inferences do not contribute to the final proof, but the actual proof steps can only be identified a-posteriori. Hence careful book-keeping is necessary. If done naively, the amount of data quickly becomes unmanageable. Different provers have taken different approaches to handling this problem, either dumping all derivation steps to an external medium, keeping a full record of all inferences in main memory, or utilizing invariants of the proof search algorithm that enable proof reconstruction from less extensive records.

Both SPASS and E are mainstream proof-producing provers available under open-source/free software licenses. E in particular implements the TPTP standard for proof output and includes a derivation not only for the saturation, but also for the clausification steps.

2 Calculi and Proof Systems

It was understood early on that showing the validity of a sentence in first-order logic can be reduced to demonstrating the unsatisfiability of a set of clauses. Indeed, for a long time "first order theorem proving" was nearly synonymous with "showing unsatisfiability of a formula in clause normal form". This, again, can be reduced to finding an unsatisfiable set of ground instances, or to deduce the empty clause (an explicit witness of unsatisfiability) from a set of clauses. Major early milestones were the original Davis-Putnam algorithm [10], which combined the generation of ground instances with a separate propositional satisfiability test, and Robinson's resolution [51], which uses unification to integrate instantiation and the search for an explicit contradiction in one simple inference process.

Resolution was the first major example of a saturating calculus. The search state is represented by a set of clauses. New clauses are systematically deduced using a set of *inference rules* and added to the search state. The aim is to eventually derive the empty clause. Resolution has proved to be an extremely productive line of research, and spawned a number of refinements, including ordered resolution [48] and hyper-resolution [52]. The general saturation principle and many of the inferences and techniques survive into current ATP systems.

Paramodulation [50] was introduced as a way to handle the important equality relation with an explicit inference rule. However, pure paramodulation was no significant improvement over resolution with an axiomatic description of equality. In 1970, Knuth and Bendix introduced *completion* [25] as a way to efficiently handle some pure unit equality problems, using a term ordering to transform a set of equations into a confluent rewrite rule system. This was later extended to *completion without failure* [20, 2], which provides a complete proof method for unit-equational theories. In contrast to pure resolution calculi, completion based methods make extensive use of *simplification*, in particular through rewriting. Simplification replaces a clause in the search state by a different, in some sense simpler, clause, which can be deduced from the original one (the *main premise*) and possibly additional clauses (the *side premises*).

Resolution and completion-based techniques have merged in the current generation of superposition calculi [3, 4, 40]. These calculi combine paramodulation and (possibly) resolution inferences restricted by literal selection and orderings on terms and literals with powerful redundancy elimination techniques, in particular rewriting and subsumption. Most practical implementations combine superposition with variants of resolution to handle non-equational literals, others (in particular E) encode non-equational literals and handle them in a uniform way via superposition and simplification.

Figure 1 shows examples of the most prolific generating inference rules (*superposition*) and the most important simplification rules (unconditional rewriting) as an example of the type of rules used in saturating calculi for first-order deduction. There are additional generating inference rules (in particular *equality factoring* and *equality resolution*) that are necessary for completeness of the calculus. However, in practical applications, typically more than 95% of generating inferences are superposition or resolution inferences. Simplification has been shown to be critical for the success of the proof search, and simplification effort dominates the overall effort of most saturating first-order provers.

$$\text{(SN)} \quad \frac{s \simeq t \vee S \quad u \not\simeq v \vee R}{\sigma(u[p \leftarrow t] \not\simeq v \vee S \vee R)} \qquad \begin{aligned} &\text{if } \sigma = mgu(u|_p, s),\ \sigma(s) \not< \sigma(t),\ \sigma(u) \not< \\ &\sigma(v),\ \sigma(s \simeq t) \text{ is eligible for paramodulation},\ \sigma(u \not\simeq v) \text{ is eligible for resolution, and} \\ &u|_p \notin V. \end{aligned}$$

$$\text{(RN)} \quad \frac{s \simeq t \quad u \not\simeq v \vee R}{s \simeq t \quad u[p \leftarrow \sigma(t)] \not\simeq v \vee R} \qquad \text{if } u|_p = \sigma(s) \text{ and } \sigma(s) > \sigma(t).$$

Note that superposition is a *generating inference* (symbolized by the single line separating premises and conclusion), while rewriting is a simplifying inference (the conclusions replace the premises in the search space). The rules are instantiated with a *simplification ordering* $<$ on terms, lifted to literals and clauses, and optionally a literal selection function. See e.g. [55] for details of the notation. These rules are complemented by dual rules for positive literals with slightly different conditions for the inference.

Fig. 1. Exemplary inference rules of the superposition calculus

Over time, there have been various other approaches to the CNF refutation problem. These include model elimination [32, 31] implemented, e.g., in SETHEO [30, 39] and leanCOP [44], model evolution [6] implemented in Darwin [5], and modern instantiation-based methods as implemented in iProver [27]. These do not naturally generate a derivation-based proof. However, such a proof can generally be extracted from information gathered during the proof search.

CNF translation has been performed using straightforward algorithms as, e.g., described in [33], more often than not by tools external to the main refutation prover. As a result, the clausification process was often not considered part of the proof search, and was not represented in any proof object. FLOTTER [70] first demonstrated that advanced clausification methods as described in [43] can significantly increase the class of first-order problems that can be solved by automated theorem provers. However, FLOTTER (and its accompanying prover SPASS) do not provide the clausification steps in a form useful for a proof object. E and Vampire implement clausifiers using similar techniques as FLOTTER, and are able to provide complete proof objects, including both clausification and saturation.

3 Proof Search

All mainstream saturating provers are based on some version of the *given-clause algorithm*. This algorithm represents the proof state by two distinct sets of clauses, the set P of *processed clauses* (initially empty) and the set U of *unprocessed clauses*. In its simplest version, it moves clauses, one at a time, from U to P, at each step adding all new clauses that can be derived from the *given clause* and other premises in P using a single inference to U. Thus it maintains the invariant that all direct inferences between clauses in P are represented in $P \cup U$. Provers differ in how they add simplification to this algorithm. The DIS-COUNT variant, first realized in the eponymous system [11] uses only clauses from P as side premises for simplification, and simplifies P, the given clause, and newly generated clauses. It is implemented in E, and, as an alternate method, in SPASS and Vampire. The other main variant is named after *Otter*. It uses all clauses in U and P as side premises for simplification. It is implemented e.g. in Otter, Prover9, and, as an alternate method, in SPASS and Vampire.

For both of these variants, one critical parameter is the order in which given clauses are selected from U. Completeness of the proof procedure requires a rather weak fairness criterion (usually implemented by making sure that no clause is allowed to remain in U forever). However, this leaves a large amount of freedom, and heuristics for clause selection have a large effect on the practical power of an ATP system.

A major challenge for reconstructing proof objects is simplification. Simplification modifies clauses in the search state or even removes them completely. Thus, derivations that reference clauses later affected by simplification are left with dangling references. On the other hand, simplification is crucial for the success of theorem provers. Section 5 discusses possible solutions.

4 Proof Formats

Historically, there has been a large number of languages for writing proof problems, and a different and only partially overlapping set of languages for writing proof objects.

Some languages, e.g., the LOP format [53], were designed for writing problems, and do not support writing solutions. Some languages for writing solutions are limited in scope, e.g., the PCL language [14] is limited to solutions to equational problems, and the OpenTheory language [21] is designed only to be a computer-processable form for systems that implement the HOL logic [17].

There are some general purpose languages that have features for writing derivations, e.g., Otter's `proof_object` format [37, 34] and the DFG syntax [18], but none of these (that we know of) also provide support for writing finite interpretations. Mark-up languages such as OmDoc [26], OpenMath [9], and MathML [9] are quite expressive (especially for mathematical content), but their XML based format is not suitable for human processing. Most of these languages have not seen much use outside the groups that originally developed them.

The current standard for writing first-order problems and solutions is the TPTP language [65]. The language was originally developed to realize the *Thousands of Problems for Theorem Provers* library [63]. Version 1 of the language supported only clause normal form (CNF), version 2 added support for full first-order logic (FOF), and version 3 unified the syntax for CNF and FOF, and added the ability to represent proof objects, derivations, and models. Version 3 has also been conservatively extended to cover other logics, in particular simply typed first-order logic and typed higher-order logic.

The language was designed to be suitable for writing both ATP problems and ATP solutions, to be flexible and extensible, and easily processed by both humans and computers. The syntax shares many features with Prolog, a language that is widely known in the ATP community. Indeed, with a few operator definitions, units of TPTP data can be read in Prolog using a single `read/1` call, and written with a single `writeq/1` call. The features were designed for writing derivations, but their flexibility makes it possible to write a range of DAG structures. Additionally, there are features of the language that make it possible to conveniently specify finite interpretations. The ability of the TPTP language to express solutions as well as problems, in conjunction with the simplicity of the syntax, sets it apart from other languages used in ATP. Overall, the TPTP language is more expressive and usable than other languages. Its use has been bolstered both by its use in the CADE ATP System Competition (CASC), and also by its support in many different provers and tools.

The TPTP language definition[3] uses a modified Backus-Naur Form (BNF) meta-language that separates semantic, syntactic, lexical, and character-macro rules. Syntactic rules use the standard : := separator, e.g.,

      `<source>`   `::= <general_term>`

When only a subset of the syntactically acceptable values for a non-terminal make semantic sense, a second rule for the non-terminal is provided using a : == separator, e.g.,

      `<source>`   `:== <dag_source> | <internal_source> | , etc.`

Any further semantic rules that may be reached only from the right hand side of a semantic rule are also written using the : == separator, e.g.,

      `<dag_source>`   `:== <name> | <inference_record>`

This separation of syntax from semantics eases the task of building a syntactic analyzer, as only the : := rules need be considered. At the same time, the semantic rules provide the detail necessary for semantic checking. The rules that produce tokens from the lexical level use a : :- separator, e.g.,

---

[3] `http://www.tptp.org/TPTP/SyntaxBNF.html`

```
<lower_word>   ::- <lower_alpha><alpha_numeric>*
```
with the bottom level character-macros defined by regular expressions in rules
using a ::: separator, e.g.,
```
<lower_alpha> ::: [a-z]
```
The top level building blocks of TPTP files are *annotated formulae, include di-
rectives*, and *comments*. An annotated formula has the form:

language(name, role, formula[, source[, useful_info]]).

The *languages* currently supported are `thf` - typed higher-order form, `tff` -
typed first-order form, `fof` - first order form, and `cnf` - clause normal form. The
role gives the user semantics of the *formula*, e.g., `axiom`, `lemma`, `conjecture`,
and hence defines its use in an ATP system - see the BNF for the list of recog-
nized roles and their meaning. The logical *formula* uses a consistent and easily
understood notation [64] that can be seen in the BNF. The *source* describes
where the formula came from, e.g., an input file or an inference. The *useful_info*
is a list of arbitrary useful information, as required for user applications. The
useful_info field is optional, and if it is not used then the *source* field becomes
optional. An example of a FOF formula, supplied from a file, is:

```
fof(formula_27,axiom,
    ! [X,Y] :
    ( subclass(X,Y) <=>
      ! [U] :
        ( member(U,X) => member(U,Y) )),
    file('SET005+0.ax',subclass_defn),
    [description('Definition of subclass'), relevance(0.9)]).
```

An example of an inferred CNF formula is:

```
cnf(175,lemma,
    ( rsymProp(ib,sk_c3)
    | sk_c4 = sk_c3 ),
    inference(factor_simp,[status(thm)],[
        inference(para_into,[status(thm)],[96,78,theory(equality)])]),
    [iquote('para_into,96.2.1,78.1.1,factor_simp')]).
```

The source field of an annotated formula is most commonly a `file` record or
an `inference` record. A `file` record stores the name of the file from which the
annotated formula was read, and optionally the name of the annotated formula
as it occurs in the file (this may be different from the name of the annotated
formula itself, e.g., if the ATP system renames the annotated formulae that
it reads in). An `inference` record stores three items of information about an
inferred formula: the name of the inference rule; a list of "useful information
items", and a list of the parents.

There currently is no fixed standard of supported inference rules, i.e. the inference rule is simply a name provided by the ATP system. However, the "useful
information" field allows the system to specify the logical relation between a
derived formula and its parents in the SZS ontology [64] – commonly, inferred
formulae are theorems of their parents, but in some cases the semantic relationship is weaker, as in Skolemization steps. The parents are either names of

existing clauses and formulas, nested inference records, or **theory** records. A theory record is used when the axioms of some theory are built into the inference rule.

A derivation is a directed acyclic graph (DAG) whose leaf nodes are formulae from the input, whose interior nodes are formulae inferred from parent formulae, and whose root nodes are the final derived formulae. For example, a proof of a FOF theorem from some axioms, by refutation of the CNF of the axioms and negated conjecture, is a derivation whose leaf nodes are the FOF axioms and conjecture, whose internal nodes are formed from the process of clausification and then from inferences performed on the clauses, and whose root node is the *false* formula.

The information required to record a derivation is, minimally, the leaf formulae, and each inferred formula with references to its parent formulae. More detailed information that may be recorded and useful includes: the name of the inference rule used in each inference step; sufficient details of each inference step to deterministically reproduce the inference; and the semantic relationships of inferred formulae with respect to their parents. The TPTP language is sufficient for recording all this, and more. A comprehensively recorded derivation provides the information required for various forms of processing, such as proof verification [60], proof visualization [59], and lemma extraction [15].

A derivation written in the TPTP language is a list of annotated formulae. Each annotated formula has a name, a role, and the logical formula. Each inferred formula has an **inference** record with the inference rule name, the semantic relationship of the formula to its parents as an SZS ontology value in a **status** record, and a list of references to its parent formulae. For example, consider the following toy FOF problem, to prove the **conjecture** from the **axioms**.

```
%--------------------------------------------------------------------
%----All (hu)men are created equal. John is a human. John got an F grade.
%----There is someone (a human) who got an A grade. An A grade is not
%----equal to an F grade. Grades are not human. Therefore there is a
%----human other than John.
fof(all_created_equal,axiom,(
    ! [H1,H2] : ( ( human(H1) & human(H2) ) => created_equal(H1,H2) ) )).
fof(john,axiom,(
    human(john) )).
fof(john_failed,axiom,(
    grade(john) = f )).
fof(someone_got_an_a,axiom,(
    ? [H] : ( human(H) & grade(H) = a ) )).
fof(distinct_grades,axiom,(
    a != f )).
fof(grades_not_human,axiom,(
    ! [G] : ~ human(grade(G)) )).
fof(someone_not_john,conjecture,(
    ? [H] : ( human(H) & H != john ) )).
%--------------------------------------------------------------------
```

Figure 4 shows a derivation recording a proof by refutation of the CNF, adapted (removing inferences that simply copy the parent formula) from the one produced by the ATP system E 1.8 [57].

5 Proof Production

As described above, current mainstream theorem provers combine a clausifier that converts a formula in full first-order logic into clause normal form, with a saturating refutation core that tries to derive the empty clause from the clause set. Proof objects are derivation graphs, showing at least how the empty clause was derived from the initial clause set, and should also show how the initial clauses were generated from the first-order axioms.

While this is fairly straightforward without simplification, it becomes much harder if simplification is present. For reasons of both time and space efficiency, most provers use destructive simplification (i.e., the old clause is modified in memory or discarded and replaced by the modified copy). In particular, some clauses that have been used in the proof may not be present in the final clause set. There are several approaches to dealing with this problem.

Older versions of E wrote all intermediate steps into a protocol file and extracted the needed inferences in a post-processing step. This results in about 100% to 200% overhead in proof time, and fails for proof searches beyond a few minutes because the amount of data becomes unmanagable even for modern computers. Recent versions of E ensure that versions of clauses that have participated in generating inferences or as side premises in simplifications (a proportionally very small number of clauses) are archived in memory. This resulted in barely measurable overhead [57]. SPASS retains the full history and all versions of each clause in memory, and pays an overhead of about 100%.[4] Prover9 has a concept of "kept clauses", i.e. clauses the system has decided it will use or consider for future inferences and simplifications. If a "kept" clause would be affected by simplification, it is not completely deleted, but deactivated and, if necessary, replaced in the proof state by a simplified copy.

6 Proof Applications

The first and original use of proof objects is the analysis of proofs by human users. This helps people to understand the proof and the application domain, to verify the correctness of the proof, but also to understand the behaviour of the ATP search process.

First-order ATP systems have directly been used for significant mathematical work, most famously for McCune's (and EQP's) proof of the Robbins problem [35]. In such cases, both manual and automatic *proof checking* increases the trust placed into the proof and hence the validity if the theorem. Proof checking, analysis, and visualization is supported by tools as described in the next section.

---

[4] Christoph Weidenbach, personal communication.

```
%----------------------------------------------------------------------
fof(c_0_0, conjecture,
    (?[X3]:(human(X3)&X3!=john)),
    file('CreatedEqual.p', someone_not_john)).
fof(c_0_1, axiom,
    (?[X3]:(human(X3)&grade(X3)=a)),
    file('CreatedEqual.p', someone_got_an_a)).
fof(c_0_2, axiom,
    (grade(john)=f),
    file('CreatedEqual.p', john_failed)).
fof(c_0_3, axiom,
    (a!=f),
    file('CreatedEqual.p', distinct_grades)).
fof(c_0_4, negated_conjecture,
    (~(?[X3]:(human(X3)&X3!=john))),
    inference(assume_negation,[status(cth)],[c_0_0])).
fof(c_0_5, negated_conjecture,
    (![X4]:(~human(X4)|X4=john)),
    inference(variable_rename,[status(thm)],
    [inference(fof_nnf,[status(thm)],[c_0_4])])).
fof(c_0_6, plain,
    ((human(esk1_0)&grade(esk1_0)=a)),
    inference(skolemize,[status(esa)],
    [inference(variable_rename,[status(thm)],[c_0_1])])).
cnf(c_0_7,negated_conjecture,
    (X1=john|~human(X1)),
    inference(split_conjunct,[status(thm)],[c_0_5])).
cnf(c_0_8,plain,
    (human(esk1_0)),
    inference(split_conjunct,[status(thm)],[c_0_6])).
cnf(c_0_9,plain,
    (grade(john)=f),
    inference(split_conjunct,[status(thm)],[c_0_2])).
cnf(c_0_10,negated_conjecture,
    (john=esk1_0),
    inference(spm,[status(thm)],[c_0_7, c_0_8])).
cnf(c_0_11,plain,
    (grade(esk1_0)=f),
    inference(rw,[status(thm)],[c_0_9, c_0_10])).
cnf(c_0_12,plain,
    (grade(esk1_0)=a),
    inference(split_conjunct,[status(thm)],[c_0_6])).
cnf(c_0_13,plain,(a!=f),
    inference(split_conjunct,[status(thm)],[c_0_3])).
cnf(c_0_14,plain,($false),
    inference(sr,[status(thm)],
    [inference(spm,[status(thm)],[c_0_11, c_0_12]), c_0_13],
    ['proof']).
%----------------------------------------------------------------------
```

Fig. 2. Example of a simple TPTP proof

First order ATP systems are increasingly integrated into higher-order inter-active proof assistants. A prominent example is the Sledgehammer tool [45, 8] in Isabelle [42]. Via the interactive system, ATP systems contribute to large scale projects like the formal proof of Kepler's conjecture in Flyspeck [19] or the verification of the L4 micro-kernel [24, 23].

In these applications, first-order proofs are used to guide the reconstruction of a proof in the native calculus of the embedding ITP system, which uses only the small set of trusted inferences of the very kernel of the system.

Another application is the validation and debugging not only of proofs, but also of specifications. Often large, manually assembled ontologies such as SUMO [41] or CYC [29, 47] contain unintended contradictions. Since most first-order systems are based on refutational calculi, they can be employed to find such contradictions, either directly on the full corpus, or a-posteriori, by check-ing if a given proof uses the negated conjecture to find the contradiction, or if it is based on a contradiction in the axioms. Because of the powerful goal-directed heuristics used by modern provers, the second approach often is more successful.

Finally, proofs reveal a large amount of information about the domain and reasoning strategies. As such, they have been mined for useful information to help further proof attempts. One approach is the learning of heuristic evaluation functions for the selection of the given clauses in the refutation procedure, e.g., by annotating *patterns* [13, 16, 54] and other abstractions [56] of clauses. While heuristic evaluation functions guide the selection of the given clauses, for domains with large background theories, processing the problem specification may alone overwhelm a theorem prover. To overcome this, machine learning techniques have been used to extract information from proofs to guide the selection of a subset of the axioms and assumptions that is likely to be useful in a subsequent proof search [1, 28]. A successful example of this has been the use of the MaLARea system [67] in emulating the development of the Flyspeck project [22].

Finding proofs can be resource-intensive. Many applications of proofs can profit by using existing proofs from a proof repository. The Thousands of So-lutions for Theorem Provers (TSTP) solution library [61, 62] contains proofs for many TPTP problems. In most cases, it contains proofs by several differ-ent ATP systems, thus facilitating comparisons of the different approaches to theorem proving.

7   Proof Consumption

Proof presentation is supported by two different kinds of tools. First, one can try to structure and present the proof as a sequential text, analogous to a clas-sical mathematical textbook proof. This has been particularly successful in the case of purely equational proofs, where the reasoning can be represented as an equational chain [12, 14]. To improve the presentation of a proof, and to make it more easily understandable, proofs by contradiction can be post-processed and converted into forward deductions of the conjecture from the axioms [7, 58].

The other approach is to visualize the proof as a DAG. IDV [66] is a tool for graphical rendering and analysis of TPTP format derivations. IDV provides an interactive interface that allows the user to quickly view features of the derivation, and access analysis facilities. The left hand side of Figure 3 shows the rendering of the derivation output by E for the TPTP problem PUZ001+1. PUZ001+1 provides axioms about three people – Charles, the butler, and Aunt Agatha – who live together in a mansion, one of whom killed Aunt Agatha, and a conjecture to prove that Aunt Agatha killed herself. The proof shown in Figure 3 converts the axioms and negated conjecture to clause normal form, and refutes the resultant clause set. The IDV window is divided into three panes: the top pane contains control buttons and sliders, the middle pane shows the rendered DAG, and the bottom pane gives the text of the annotated formula for the node pointed to by the mouse. The rendering of the derivation DAG uses shapes (e.g., inverted triangles for axioms), colors (e.g., to show which nodes lead to and from the node highlighted with the mouse), and tags (e.g., red dots inside nodes to indicate steps of non-logical consequence) to provide information about the derivation. The user can interact with the rendering in various ways using mouse-over and mouse clicks. The buttons and sliders in the control pane provide a range of manipulations on the rendering – zooming, hiding and displaying parts of the DAG, and access to GDV (see below) for verification. A particularly novel feature of IDV is its ability to provide a synopsis of a derivation by using the AGInTRater [46] to identify interesting lemmas, and hiding less interesting intermediate formulae. A synopsis is shown on the right hand side of Figure 3. The node highlighted with the mouse in the original derivation on the left hand side is derived from only axioms, and is thus a lemma. It states that the butler hates the person who killed Aunt Agatha. This is considered an interesting lemma, and is thus retained (highlighted again) in the synopsis.

Proof verification can be done on a syntactic level, or on a semantic level. On a syntactic level, a trusted checker reproduces each individual inference. Most current ATP systems do no export enough information to make that directly feasible, however, as stated in the previous section, many ITP proofs first reproduce the proof in their native format, and then validate this via their own trusted kernel.

The alternative is semantic verification, i.e. showing that each derived clause and formula is in the stated semantic relationship with its parents. GDV [60] is a tool that uses structural and then semantic techniques to verify TPTP format derivations. Structural verification checks that inferences have been done correctly in the context of the derivation, e.g., checking that the derivation is acyclic, checking that assumptions have been discharged (introduced in GDV for the cross-verification of the Mizar Mathematical Library [68]), checking that both sides of split derivations have been refuted (e.g., as often found in derivations generated by SPASS [71]), and checking that introduced symbols are distinct (e.g., as in Skolemization and introduction of new symbols in definitions [43]). Semantic verification checks the expected semantic relationship between the parents and inferred formula of each inference step. This is done by encoding

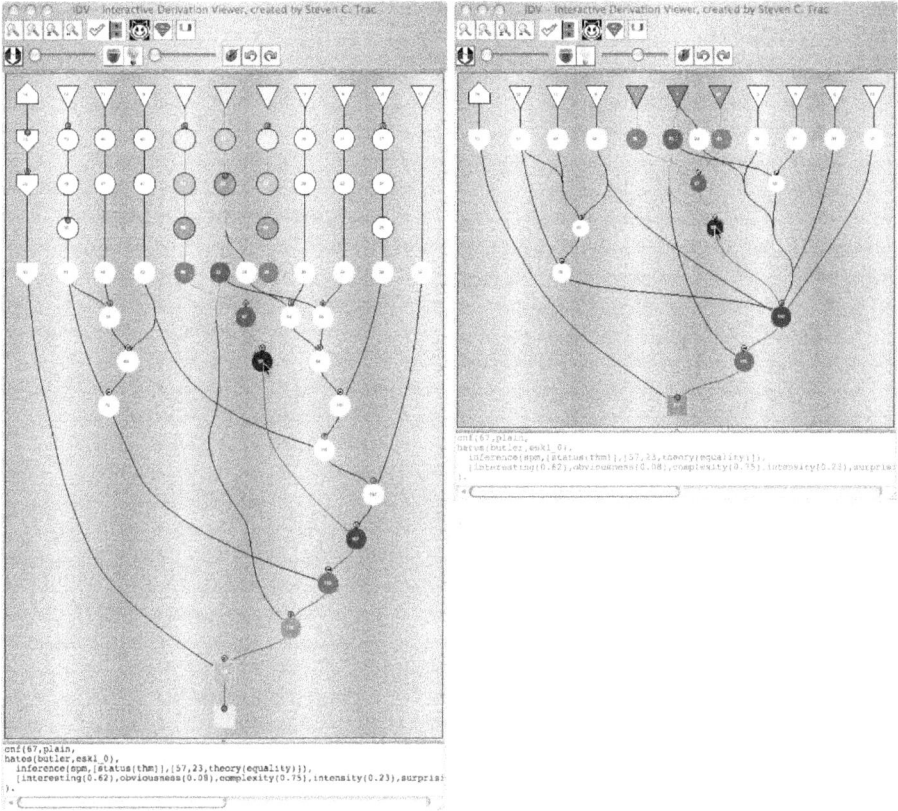

Fig. 3. E's proof by refutation of PUZ001+1

the expectation as a logical obligation in an ATP problem, and then discharging the obligation by solving the problem with trusted ATP systems. The expected semantic relationship between the parents and inferred formula of an inference step depends on the intent of the inference rule used. For example, deduction steps expect the inferred formula to be a theorem of its parent formulae. The expected relationship is recorded as an SZS value in each inferred formula of a derivation.

8 Conclusions

The generation of explicit proof objects has not originally been a primary focus for first-order ATP systems. However, by now it is an expected feature for systems that are widely used. Earlier ad-hoc formats are now strongly converging to the TPTP syntax, driven in part by the CASC competition, and in part by the increasing availability of tools that can process this information.

Proofs are used for several different applications:

- Human consumption
- Proof checking
- Embedding into interactive proofs
- Heuristics learning

The TPTP format is sufficiently detailed to support these applications. However, because of its high level of abstraction, TPTP proof objects do not always allow direct step-by-step reconstruction of the proof. The future will show if this feature is important enough to emerge despite the greater effort for both producers and consumers of proofs.

References

1. J. Alama, T. Heskes, D. Kuühlwein, E. Tsivtsivadze, and J. Urban. Premise selection for mathematics by corpus analysis and kernel methods. *Journal of Automated Reasoning*, 52(2):191–213, 2014.
2. L. Bachmair, N. Dershowitz, and D.A. Plaisted. Completion Without Failure. In H. Ait-Kaci and M. Nivat, editors, *Resolution of Equations in Algebraic Structures*, volume 2, pages 1–30. Academic Press, 1989.
3. L. Bachmair and H. Ganzinger. On Restrictions of Ordered Paramodulation with Simplification. In M.E. Stickel, editor, *Proc. of the 10th CADE, Kaiserslautern*, volume 449 of *LNAI*, pages 427—441. Springer, 1990.
4. L. Bachmair and H. Ganzinger. Rewrite-Based Equational Theorem Proving with Selection and Simplification. *Journal of Logic and Computation*, 3(4):217–247, 1994.
5. Peter Baumgartner, Alexander Fuchs, and Cesare Tinelli. Implementing the Model Evolution Calculus. *International Journal of Artificial Intelligence Tools*, 15(1):21–52, 2006.
6. Peter Baumgartner and Cesare Tinelli. The Model Evolution Calculus. In Franz Baader, editor, *Proc. of the 19th CADE, Miami*, volume 2741 of *LNCS*, pages 350–364. Springer, 2003.
7. Jasmin Christian Blanchette. Redirecting Proofs by Contradiction. In Jasmin C. Blanchette and Josef Urban, editors, *Proc. of the 3rd Workshop on Proof Exchange for Theorem Provers (PxTP-2013)*, volume 14 of *EPiC*, pages 11–26, 2013.
8. Sascha Böhme and Tobias Nipkow. Sledgehammer: Judgement Day. In Jürgen Giesel and Reiner Hähnle, editors, *Proc. of the 5th IJCAR, Edinburgh*, volume 6173 of *LNAI*, pages 107–121. Springer, 2012.
9. O. Caprotti and D. Carlisle. OpenMath and MathML: Semantic Mark Up for Mathematics. *ACM Crossroads*, 6(2), 1999.
10. M. Davis and H. Putnam. A Computing Procedure for Quantification Theory. *Journal of the ACM*, 7(1):215–215, 1960.
11. J. Denzinger, M. Kronenburg, and S. Schulz. DISCOUNT: A Distributed and Learning Equational Prover. *Journal of Automated Reasoning*, 18(2):189–198, 1997. Special Issue on the CADE 13 ATP System Competition.
12. J. Denzinger and S. Schulz. Analysis and Representation of Equational Proofs Generated by a Distributed Completion Based Proof System. Seki-Report SR-94-05, Universität Kaiserslautern, 1994.

13. J. Denzinger and S. Schulz. Learning Domain Knowledge to Improve Theorem Proving. In M.A. McRobbie and J.K. Slaney, editors, *Proc. of the 13th CADE, New Brunswick*, volume 1104 of *LNAI*, pages 62–76. Springer, 1996.
14. J. Denzinger and S. Schulz. Recording and Analysing Knowledge-Based Distributed Deduction Processes. *Journal of Symbolic Computation*, 21(4/5):523–541, 1996.
15. J. Denzinger and S. Schulz. Recording and Analysing Knowledge-Based Distributed Deduction Processes. *Journal of Symbolic Computation*, 21:523–541, 1996.
16. J. Denzinger and S. Schulz. Automatic Acquisition of Search Control Knowledge from Multiple Proof Attempts. *Journal of Information and Computation*, 162:59–79, 2000.
17. M. Gordon and T. Melham. *Introduction to HOL, a Theorem Proving Environment for Higher Order Logic*. Cambridge University Press, 1993.
18. R. Hähnle, M. Kerber, and C. Weidenbach. Common Syntax of the DFG-Schwerpunktprogramm Deduction. Technical Report TR 10/96, Fakultät für Informatik, Universät Karlsruhe, Karlsruhe, Germany, 1996.
19. T.C. Hales. Introduction to the Flyspeck project. In T. Coquand, H. Lombardi, and M.-F. Roy, editors, *Mathematics, Algorithms, Proofs*, number 05021 in Dagstuhl Seminar Proceedings, pages 1–11. Internationales Begegnungs- und Forschungszentrum für Informatik (IBFI), Schloss Dagstuhl, Germany, Dagstuhl, Germany, 2006.
20. J. Hsiang and M. Rusinowitch. On Word Problems in Equational Theories. In *Proc. of the 14th ICALP, Karlsruhe*, volume 267 of *LNCS*, pages 54–71. Springer, 1987.
21. J. Hurd and R. Arthan. OpenTheory. `http://www.cl.cam.ac.uk/~jeh1004/research/opentheory`.
22. C. Kaliszyk and J. Urban. Learning-assisted Automated Reasoning with Flyspeck. *Journal of Automated Reasoning*, 53(2):173–213, 2014.
23. Gerwin Klein, June Andronick, Kevin Elphinstone, Gernot Heiser, David Cock, Philip Derrin, Dhammika Elkaduwe, Kai Engelhardt, Rafal Kolanski, Michael Norrish, Thomas Sewell, Harvey Tuch, and Simon Winwood. seL4: Formal verification of an operating system kernel. *Communications of the ACM*, 53(6):107–115, 2010.
24. Gerwin Klein, Kevin Elphinstone, Gernot Heiser, June Andronick, David Cock, Philip Derrin, Dhammika Elkaduwe, Kai Engelhardt, Rafal Kolanski, Michael Norrish, Thomas Sewell, Harvey Tuch, and Simon Winwood. seL4: Formal verification of an OS kernel. In *Proc. 22nd ACM Symposium on Operating Systems Principles, Big Sky*, pages 207–220. ACM, 2009.
25. D.E. Knuth and P.B. Bendix. Simple Word Problems in Universal Algebras. In J. Leech, editor, *Computational Algebra*, pages 263–297. Pergamon Press, 1970.
26. M. Kohlhase. OMDOC: Towards an Internet Standard for the Administration, Distribution, and Teaching of Mathematical Knowledge. In J.A. Campbell and E. Roanes-Lozano, editors, *Proceedings of the Artificial Intelligence and Symbolic Computation Conference, 2000*, number 1930 in Lecture Notes in Computer Science, pages 32–52. Springer-Verlag, 2000.
27. Konstantin Korovin. iProver - An Instantiation-Based Theorem Prover for First-Order Logic (System Description). In A. Armando, P. Baumgartner, and G. Dowek, editors, *Proc. of the 4th IJCAR, Sydney*, volume 5195 of *LNAI*, pages 292–298. Springer, 2008.
28. D. Kuühlwein, J.C. Blanchette, C. Kaliszyk, and J. Urban. MaSh: Machine learning for Sledgehammer. In S. Blazy, C. Paulin-Mohring, and D. Pichardie, editors, *Proc. of the 4th International Conference on Interactive Theorem Proving (ITP13)*, volume 7998 of *LNCS*, pages 35–50. Springer, 2013.

29. D.B. Lenat. CYC: A Large-Scale Investment in Knowledge Infrastructure. *Communications of the ACM*, 38(11):35–38, 1995.

30. R. Letz, J. Schumann, S. Bayerl, and W. Bibel. SETHEO: A High-Performance Theorem Prover. *Journal of Automated Reasoning*, 1(8):183–212, 1992.

31. Reinhold Letz and Gernot Stenz. Model Elimination and Connection Tableau Procedures. In A. Robinson and A. Voronkov, editors, *Handbook of automated reasoning*, volume II, chapter 28, pages 2015–2112. Elsevier Science and MIT Press, 2001.

32. D.W. Loveland. Mechanical Theorem Proving by Model Elimination. *Journal of the ACM*, 15(2), 1968.

33. D.W. Loveland. *Automated Theorem Proving: A Logical Basis*. North Holland, Amsterdam, 1978.

34. W. McCune and O. Shumsky-Matlin. Ivy: A Preprocessor and Proof Checker for First-Order Logic. In M. Kaufmann, P. Manolios, and J. Strother Moore, editors, *Computer-Aided Reasoning: ACL2 Case Studies*, number 4 in Advances in Formal Methods, pages 265–282. Kluwer Academic Publishers, 2000.

35. William McCune. Solution of the Robbins problem. *Journal of Automated Reasoning*, 19(3):263–276, 1997.

36. William W. McCune. Prover9 and Mace4. http://www.cs.unm.edu/~mccune/prover9/, 2008. (acccessed 2009-10-04).

37. W.W. McCune. Otter 3.3 Reference Manual. Technical Report ANL/MSC-TM-263, Argonne National Laboratory, Argonne, USA, 2003.

38. W.W. McCune and L. Wos. Otter: The CADE-13 Competition Incarnations. *Journal of Automated Reasoning*, 18(2):211–220, 1997. Special Issue on the CADE 13 ATP System Competition.

39. M. Moser, O. Ibens, R. Letz, J. Steinbach, C. Goller, J. Schumann, and K. Mayr. SETHEO and E-SETHEO – The CADE-13 Systems. *Journal of Automated Reasoning*, 18(2):237–246, 1997. Special Issue on the CADE 13 ATP System Competition.

40. R. Nieuwenhuis and A. Rubio. Paramodulation-Based Theorem Proving. In A. Robinson and A. Voronkov, editors, *Handbook of Automated Reasoning*, volume I, chapter 7, pages 371–443. Elsevier Science and MIT Press, 2001.

41. Ian Niles and Adam Pease. Toward a Standard Upper Ontology. In Chris Welty and Barry Smith, editors, *Proc. 2nd International Conference on Formal Ontology in Information Systems (FOIS-2001)*, 2001.

42. Tobias Nipkow, Lawrence C. Paulson, and Markus Wenzel. *Isabelle/HOL: A Proof Assistant for Higher-Order Logic*, volume 2283 of *LNCS*. Springer, 2002.

43. A. Nonnengart and C. Weidenbach. Computing Small Clause Normal Forms. In A. Robinson and A. Voronkov, editors, *Handbook of Automated Reasoning*, volume I, chapter 5, pages 335–367. Elsevier Science and MIT Press, 2001.

44. Jens Otten and Wolfgang Bibel. leanCoP: Lean Connection-Based Theorem Proving,. *Journal of Symbolic Computation*, 36:139–161, 2003.

45. Lawrence C. Paulsson and Jasmin C. Blanchette. Three years of experience with Sledgehammer, a practical link between automatic and interactive theorem provers. In Geoff Sutcliff, Eugenia Ternovska, and Stephan Schulz, editors, *Procóf the 8th International Workshop on the Implementation of Logics (IWIL-2010), Yogyakarta, Indonesia*, volume 2 of *EPiC*, 2012.

46. Y. Puzis, Y. Gao, and G. Sutcliffe. Automated Generation of Interesting Theorems. In G. Sutcliffe and R. Goebel, editors, *Proceedings of the 19th International FLAIRS Conference*, pages 49–54. AAAI Press, 2006.

47. Deepak Ramachandran, Pace Reagan, and Keith Goolsbey. First-orderized ResearchCyc: Expressiveness and Efficiency in a Common Sense Knowledge Base. In Pavel Shvaiko, editor, *Proc. of the AAAI Workshop on Contexts and Ontologies: Theory, Practice and Applications (C&O-2005)*, 2005.

48. Raymod Reiter. Two Results on Ordering for Resolution with Merging and Linear Format. *Journal of the ACM*, 18(4):630–646, 1971.

49. A. Riazanov and A. Voronkov. Vampire 1.1 (System Description). In R. Goré, A. Leitsch, and T. Nipkow, editors, *Proc. of the 1st IJCAR, Siena*, volume 2083 of *LNAI*, pages 376–380. Springer, 2001.

50. G. Robinson and L. Wos. Paramodulation and Theorem Proving in First-Order Theories with Equality. In B. Meltzer and D. Michie, editors, *Machine Intelligence 4*. Edinburgh University Press, 1969.

51. J. A. Robinson. A Machine-Oriented Logic Based on the Resolution Principle. *Journal of the ACM*, 12(1):23–41, 1965.

52. J.A. Robinson. Automatic deduction with hyper-resolution. *International Journal of Computer Mathematics*, 1(3):227–234, 1965.

53. S. Schulz. LOP-Syntax for Theorem Proving Applications. http://www4.informatik.tu-muenchen.de/ schulz/WORK/lop.syntax.

54. S. Schulz. Learning Search Control Knowledge for Equational Theorem Proving. In F. Baader, G. Brewka, and T. Eiter, editors, *Proc. of the Joint German/Austrian Conference on Artificial Intelligence (KI-2001)*, volume 2174 of *LNAI*, pages 320–334. Springer, 2001.

55. S. Schulz. E – A Brainiac Theorem Prover. *Journal of AI Communications*, 15(2/3):111–126, 2002.

56. S. Schulz and F. Brandt. Using Term Space Maps to Capture Search Control Knowledge in Equational Theorem Proving. In A. N. Kumar and I. Russell, editors, *Proc. of the 12th FLAIRS, Orlando*, pages 244–248. AAAI Press, 1999.

57. Stephan Schulz. System Description: E 1.8. In Ken McMillan, Aart Middeldorp, and Andrei Voronkov, editors, *Proc. of the 19th LPAR, Stellenbosch*, volume 8312 of *LNCS*. Springer, 2013.

58. Steffen Juilf Smolka and Jasmin Christian Blanchette. Robust, Semi-Intelligible Isabelle Proofs from ATP Proofs. In Jasmin C. Blanchette and Josef Urban, editors, *Proc. of the 3rd Workshop on Proof Exchange for Theorem Provers (PxTP-2013)*, volume 14 of *EPiC*, pages 117–132, 2013.

59. G. Steel. Visualising First-Order Proof Search. In C. Aspinall, D. Lüth, editor, *Proceedings of User Interfaces for Theorem Provers 2005*, pages 179–189, 2005.

60. G. Sutcliffe. Semantic Derivation Verification. *International Journal on Artificial Intelligence Tools*, 15(6):1053–1070, 2006.

61. G. Sutcliffe. TPTP, TSTP, CASC, etc. In V. Diekert, M. Volkov, and A. Voronkov, editors, *Proceedings of the 2nd International Computer Science Symposium in Russia (CSR)*, number 4649 in Lecture Notes in Computer Science, pages 7–23. Springer, 2007.

62. G. Sutcliffe. The TPTP World - Infrastructure for Automated Reasoning. In E. Clarke and A. Voronkov, editors, *Proc. of the 16th International Conference on Logic for Programming Artificial Intelligence and Reasoning (LPAR)*, number 6355 in Lecture Notes in Artificial Intelligence, pages 1–12. Springer, 2010.

63. G. Sutcliffe and C.B. Suttner. The TPTP Problem Library: CNF Release v1.2.1. *Journal of Automated Reasoning*, 21(2):177–203, 1998.

64. G. Sutcliffe, J. Zimmer, and S. Schulz. TSTP Data-Exchange Formats for Automated Theorem Proving Tools. In Volker Sorge and Weixiong Zhang, editors,

Distributed Constraint Problem Solving And Reasoning In Multi-Agent Systems, Frontiers in Artificial Intelligence and Applications, pages 201–215. IOS Press, 2004.

65. Geoff Sutcliffe, Stephan Schulz, Koen Claessen, and Allen Van Gelder. Using the TPTP Language for Writing Derivations and Finite Interpretations . In Ulrich Fuhrbach and Natarajan Shankar, editors, *Proc. of the 3rd IJCAR, Seattle*, volume 4130 of *LNAI*, pages 67–81, 4130, 2006. Springer.

66. S. Trac, Y. Puzis, and G. Sutcliffe. An Interactive Derivation Viewer. In S. Autexier and C. Benzmüller, editors, *Proceedings of the 7th Workshop on User Interfaces for Theorem Provers, 3rd International Joint Conference on Automated Reasoning*, volume 174 of *Electronic Notes in Theoretical Computer Science*, pages 109–123, 2006.

67. J. Urban, G. Sutcliffe, P. Pudlak, and J. Vyskocil. MaLARea SG1: Machine Learner for Automated Reasoning with Semantic Guidance. In P. Baumgartner, A. Armando, and D. Gilles, editors, *Proceedings of the 4th International Joint Conference on Automated Reasoning*, number 5195 in Lecture Notes in Artificial Intelligence, pages 441–456. Springer-Verlag, 2008.

68. Josef Urban and Geofff Sutcliffe. ATP-based Cross Verification of Mizar Proofs: Method, Systems, and First Experiments. *Journal of Mathematics in Computer Science*, 2(2):231–251, 2009.

69. A. Voronkov. The Anatomy of Vampire: Implementing Bottom-Up Procedures with Code Trees. *Journal of Automated Reasoning*, 15(2):238–265, 1995.

70. C. Weidenbach, B. Gaede, and G. Rock. SPASS & FLOTTER Version 0.42. In M.A. McRobbie and J.K. Slaney, editors, *Proc. of the 13th CADE, New Brunswick*, volume 1104 of *LNAI*, pages 141–145. Springer, 1996.

71. Christoph Weidenbach, Dilyana Dimova, Arnaud Fietzke, Rohit Kumar, Martin Suda, and Patrick Wischnewski. SPASS Version 3.5. In Renate Schmidt, editor, *Proc. of the 22nd CADE, Montreal*, number 5663 in LNAI, pages 140–145. Springer, 2009.

72. Christoph Weidenbach, Renate Schmidt, Thomas Hillenbrand, Dalibor Topić, and Rostislav Rusev. SPASS Version 3.0. In Frank Pfenning, editor, *Proc. of the 21st CADE, Bremen*, volume 4603 of *LNAI*, pages 514–520. Springer, 2007.

Higher-Order Automated Theorem Provers

Christoph Benzmüller

Department of Mathematics and Computer Science
Freie Universität Berlin, Germany
c.benzmueller@fu-berlin.de

1 Introduction

The automation of simple type theory, also referred to as classical higher-order logic (HOL), has significantly progressed recently. This paper provides an allowedly slightly biased survey on these developments.

A distinguishing characteristic of HOL is its support for higher-order quantification, that is quantification over predicate and/or function variables. Higher-order quantification was developed first by Frege in his Begriffsschrift [61] and then by Russell in his ramified theory of types [90], which was later simplified by others, including Chwistek and Ramsey [89; 54], Carnap, and finally Church in his simple theory of types [53].

In addition to higher-order quantification, HOL gains expressivity by permitting formulas and logical connectives to occur within terms, which is prohibited in first-order logic. This intertwining of terms and formulas is achieved by using the special primitive type o to denote those simply typed terms that are the formulas of the logic. Additionally, λ-abstractions over formulas allow the explicit naming of sets and predicates, something that is achieved in set theory via the comprehension axioms. Mixing λ-terms and logic as is done in HOL permits capturing many aspects of set theory without direct reference to axioms of set theory. Moreover, the complex rules for quantifier instantiation at higher-types is completely explained via the rules of λ-conversion (the so-called rules of α-, β-, and η-conversion) which were proposed earlier by Church [51; 52].

Church first introduced elementary type theory (ETT), an extension of first-order logic with quantification at all simple types and with the term structure upgraded to be all simply typed λ-terms. He extended this logic into his simple type theory by adding further axioms, including the axioms for extensionality, description, choice, and infinity. Functional extensionality expresses that two functions are equal if and only if they are point-wise equal. Boolean extensionality says that two formulas are equal if and only if they are equivalent (or, alternatively, that there are not more than two truth values). The choice operator selects a member from a non-empty set and description selects a member from a set only if this set is a singleton set. Infinity asserts that a set with infinitely many objects exists.

Initially much of the work on the automation of higher-order logic strongly focused on the automation of elementary type theory. Recently, however, significant progress has also been made for the automation of extensional type theory, which is elementary type theory extended with functional and Boolean extensionality (and possibly choice or description). In the remainder we use the term HOL synonymous to extensional type theory; in [24] the abbreviation ExTT is used.

Interactive and automated theorem provers for HOL have been employed in a wide range of applications, for example, in mathematics and in hardware and software verification. Moreover, due to its expressivity, HOL is well suited as a meta-logic, and a range of (propositional and quantified) non-classical logics can be elegantly embedded in it. Exploiting this fact, automated theorem provers for HOL have recently been employed to automate reasoning within and about various non-classical logics.

There are several recommended sources providing more details on HOL and its applications [24; 6; 5; 7; 58; 73; 59]; the text presented here is partly based on [24].

1.1 Syntax of HOL

The set T of *simple types* in HOL is usually freely generated from a set of *basic types* $\{o, i\}$ using the function type constructor $\rightarrow$. o denotes the type of Booleans and i some non-empty domain of individuals. Further base types may be added.

Let $\alpha, \beta, o \in T$. The *terms* of HOL are defined by the grammar (c_α denotes typed constants and X_α typed variables distinct from c_α):

$$s, t ::= c_\alpha \mid X_\alpha \mid (\lambda X_\alpha s_\beta)_{\alpha \rightarrow \beta} \mid (s_{\alpha \rightarrow \beta} t_\alpha)_\beta \mid$$
$$(\neg_{o \rightarrow o} s_o)_o \mid (s_o \vee_{o \rightarrow o \rightarrow o} t_o)_o \mid (\Pi_{(\alpha \rightarrow o) \rightarrow o} s_{\alpha \rightarrow o})_o$$

Complex typed HOL terms are thus constructed via *abstraction* and *application*, and HOL terms of type o are called formulas.

The *primitive logical connectives* (chosen here) are $\neg_{o \rightarrow o}$, $\vee_{o \rightarrow o \rightarrow o}$ and $\Pi_{(\alpha \rightarrow o) \rightarrow o}$ (for each type α), and, additionally, *choice operators* $\epsilon_{(\alpha \rightarrow o) \rightarrow \alpha}$ (for each type α) or *primitive equality* $=_{\alpha \rightarrow \alpha \rightarrow \alpha}$ (for each type α), abbreviated as $=^\alpha$, may be added. From the selected set of primitive connectives, other logical connectives can be introduced as abbreviations: for example, $\varphi \wedge \psi$, $\varphi \rightarrow \psi$, and $\varphi \longleftrightarrow \psi$ abbreviate $\neg(\neg\varphi \vee \neg\psi)$, $\neg\varphi \vee \psi$, and $(\varphi \rightarrow \psi) \wedge (\psi \rightarrow \varphi)$, respectively.

Binder notation $\forall X_\alpha s_o$ is used as an abbreviation for $\Pi_{(\alpha \rightarrow o) \rightarrow o} \lambda X_\alpha s_o$. Type information as well as brackets may be omitted if obvious from the context.

Equality can be defined by exploiting Leibniz' principle, expressing that two objects are equal if they share the same properties. *Leibniz equality* $\doteq^\alpha$ at type α is thus defined as $s_\alpha \doteq^\alpha t_\alpha := \forall P_{\alpha \rightarrow o}(\neg Ps \vee Pt)$.

Each occurrence of a variable in a term is either bound by a λ or free. We use $free(s)$ to denote the set of free variables of s (i.e., variables with a free occurrence in s). We consider two terms to be *equal* if the terms are the same up to the names of bound variables (i.e., we consider α-conversion implicitly). A term s is closed if $free(s)$ is empty.

Substitution of a term s_α for a variable X_α in a term t_β is denoted by $[s/X]t$, where it is assumed that the bound variables of t avoid variable capture.

Well known operations and relations on HOL terms include $\beta\eta$-*normalization* and $\beta\eta$-*equality*, denoted by $s =_{\beta\eta} t$, β-*reduction* and η-*reduction*. A β-*redex* $(\lambda X s)t$ β-reduces to $[t/X]s$. An η-*redex* $\lambda X (sX)$ where variable X is not free in s, η-reduces to s. We write $s =_\beta t$ to mean s can be converted to t by a series of β-reductions and expansions. Similarly, $s =_{\beta\eta} t$ means s can be converted to t using both β and η.

For each simply typed λ-term s there is a unique β-*normal form* (denoted $s\!\downarrow_\beta$) and a unique $\beta\eta$-normal form (denoted $s\!\downarrow_{\beta\eta}$). From this fact we know $s \equiv_\beta t$ ($s \equiv_{\beta\eta} t$) if and only if $s\!\downarrow_\beta \equiv t\!\downarrow_\beta$ ($s\!\downarrow_{\beta\eta} \equiv t\!\downarrow_{\beta\eta}$).

Remember, that formulas are defined as terms of type o. A *non-atomic formula* is any formula whose β-normal form is of the form $[c\overline{\mathbf{A}^n}]$ where c is a logical constant. An *atomic formula* is any other formula.

1.2   Semantics of HOL

The following sketch of HOL semantics closely follows [7]; for a more detailed introduction see [18] and the references therein.

A *frame* is a collection $\{D_\alpha\}_{\alpha\in T}$ of nonempty sets called *domains* such that $D_o = \{T, F\}$ where T represents truth and F falsehood, $D_i \neq \emptyset$ is chosen arbitrary, and $D_{\alpha\to\beta}$ are collections of total functions mapping D_α into D_β.

An *interpretation* is a tuple $\langle\{D_\alpha\}_{\alpha\in T}, I\rangle$ where $\{D_\alpha\}_{\alpha\in T}$ is a frame and function I maps each typed constant symbol c_α to an appropriate element of D_α, which is called the *denotation* of c_α. The denotations of $\neg$, $\vee$ and $\Pi_{(\alpha\to o)\to o}$ (and $\epsilon_{(\alpha\to o)\to\alpha}$ and $=_{\alpha\to\alpha\to o}$) are always chosen as usual. A variable assignment σ maps variables X_α to elements in D_α.

An interpretation is a *Henkin model (general model)* if and only if there is a binary valuation function V such that $V(\sigma, s_\alpha) \in D_\alpha$ for each variable assignment σ and term s_α, and the following conditions are satisfied for all σ, variables X_α, constants c_α, and terms $l_{\alpha\to\beta}, r_\alpha, s_\beta$ (for $\alpha, \beta \in T$): $V(\sigma, X_\alpha) = \sigma(X_\alpha)$, $V(\sigma, c_\alpha) = I(c_\alpha)$, $V(\sigma, l_{\alpha\to\beta} r_\alpha) = (V(\sigma, l_{\alpha\to\beta})V(\sigma, r_\alpha))$, and $V(\sigma, \lambda X_\alpha s_\beta)$ represents the function from D_α into D_β whose value for each argument $z \in D_\alpha$ is $V(\sigma[z/X_\alpha], s_\beta)$, where $\sigma[z/X_\alpha]$ is that assignment such that $\sigma[z/X_\alpha](X_\alpha) = z$ and $\sigma[z/X_\alpha]Y_\beta = \sigma Y_\beta$ when $Y_\beta \neq X_\alpha$.

If an interpretation $H = \langle\{D_\alpha\}_{\alpha\in T}, I\rangle$ is a Henkin model, the function V is uniquely determined and $V(\sigma, s_\alpha) \in D_\alpha$ is called the *denotation* of s_α. H is called a *standard model* if and only if for all α and β, $D_{\alpha\to\beta}$ is the set of all functions from D_α into D_β. It is easy to verify that each standard model is also a Henkin model. A formula s of HOL is *valid* in a Henkin model H if and only if $V(\sigma, s) = T$ for all variable assignments σ. In this case we write $H \models^{HOL} s$. Formula s is (Henkin) valid, denoted as $\models^{HOL} s$, if and only if $H \models^{HOL} s$ for all Henkin models H.

While Church axiomatized the logical connectives in a rather conventional fashion (using, for example, negation, conjunction, and universal quantification as the primitive connectives), Henkin [66] and Andrews [2; 6] provided alternative formulations in which the sole logical connective was primitive equality (at all types). Not only was a formulation of logic using just this one logical connective perspicuous, it also improved on the notion of Henkin models. In fact, as Andrews shows in [2], the sets $\mathcal{D}_{\alpha \to o}$ may be so sparse when using a conventional set of logical connectives that Leibniz equality may denote a relation, which does not fulfill the functional extensionality principle. A solution is to presuppose the presence of the identity relations in all domains $\mathcal{D}_{\alpha \to \alpha \to o}$, which ensures the existence of unit sets $\{a\} \in \mathcal{D}_{\alpha \to o}$ for all elements $a \in \mathcal{D}_\alpha$. The existence of these unit sets in turn ensures that Leibniz equality indeed denotes the intended (fully extensional) identity relation. Syntactically, the existence of these sets can be enforced by working with primitive equality.

Modulo the above observation, Henkin models are *fully extensional*, that is, they validate the functional and Boolean extensionality principles. The construction of non-functional models for elementary type theory has been pioneered by Andrews [1]. In Andrews' so-called v-complexes, which are based on Schütte's semi-valuation method [91], both the functional and the Boolean extensionality principles fail. Assuming β-equality, functional extensionality splits into two weaker and independent principles η ($f \doteq \lambda X f X$, if X is not free in term f) and ξ (from $\forall X (f \doteq g)$ infer $(\lambda X f) \doteq (\lambda X g)$, where X may occur free in f and g). Conversely, $\beta\eta$-conversion, which is built-in in many modern implementations of HOL, together with ξ implies functional extensionality. Boolean extensionality, however, is independent of any of these principles. A whole landscape of respective notions of models structures for HOL between Andrews' v-complexes and Henkin semantics that further illustrate and clarify the above connections is developed in [9; 18; 45], and an alternative development and discussion has been contributed in [80].

2 Proof Systems and Proof-Theoretical Properties

2.1 Cut-free Sequent Calculi

Cut-free sequent calculi for elementary type theory and fragments of it have been studied by Takeuti [103], Schütte [91], Tait [101], Takahashi [102], Prawitz [88], and Girard [63]. Andrews [1] used the *abstract consistency principle* of Smullyan [92] in order to give a proof of the completeness of resolution in elementary type theory. Takeuti [105] presented a cut-free sequent calculus with extensionality that is complete for Henkin models. The abstract consistency proof technique, as used by Andrews, has been further extended and applied in [71; 9; 45; 18; 19; 20; 48] to obtain cut-elimination results for different systems between elementary type theory and HOL.

We here present the cut-free, sound and complete, one-sided sequent calculi for HOL (without choice) from [20]. In the context of this work, a sequent is a finite set Δ of β-normal closed formulas. A sequent calculus $\mathcal{G}$ provides an inductive definition for when $\vdash^{\mathcal{G}} \Delta$ holds. A sequent calculus rule

$$\frac{\Delta_1 \quad \cdots \quad \Delta_n}{\Delta} r$$

is *admissible* in $\mathcal{G}$ if $\vdash^{\mathcal{G}} \Delta$ holds whenever $\vdash^{\mathcal{G}} \Delta_i$ for all $1 \leq i \leq n$.

Definition 1 (Sequent calculi $\mathcal{G}_\beta$ and $\mathcal{G}_{\beta\mathfrak{fb}}$). *Let Δ and Δ' be finite sets of β-normal closed formulas of HOL and let Δ, s denote the set $\Delta \cup \{s\}$. The following sequent calculus rules are introduced:*

Basic Rules
$$\frac{\Delta, s}{\Delta, \neg\neg s} \; \mathcal{G}(\neg) \qquad \frac{\Delta, \neg s \quad \Delta, \neg t}{\Delta, \neg(s \vee t)} \; \mathcal{G}(\vee_-) \qquad \frac{\Delta, s, t}{\Delta, (s \vee t)} \; \mathcal{G}(\vee_+)$$

$$\frac{\Delta, \neg(sl){\downarrow}_\beta \quad l_\alpha \; closed \; term}{\Delta, \neg\Pi^\alpha s} \; \mathcal{G}(\Pi^l_-) \qquad \frac{\Delta, (sc){\downarrow}_\beta \quad c_\delta \; new \; symbol}{\Delta, \Pi^\alpha s} \; \mathcal{G}(\Pi^c_+)$$

Initialization
$$\frac{s \; atomic \; (and \; \beta\text{-}normal)}{\Delta, s, \neg s} \; \mathcal{G}(init)$$

$$\frac{\Delta, (s \doteq^o t) \quad s, t \; atomic}{\Delta, \neg s, t} \; \mathcal{G}(Init^{\doteq})$$

Extensionality
$$\frac{\Delta, (\forall X_\alpha sX \doteq^\beta tX){\downarrow}_\beta}{\Delta, (s \doteq^{\alpha \to \beta} t)} \; \mathcal{G}(\mathfrak{f}) \qquad \frac{\Delta, \neg s, t \quad \Delta, \neg t, s}{\Delta, (s \doteq^o t)} \; \mathcal{G}(\mathfrak{b})$$

Decomposition
$$\frac{\Delta, (s^1 \doteq^{\alpha_1} t^1) \;\cdots\; \Delta, (s^n \doteq^{\alpha_n} t^n) \quad \begin{array}{l} n \geq 1, \beta \in \{o, \iota\}, \\ h_{\overline{\alpha^n} \to \beta} \in \Sigma \end{array}}{\Delta, (h\overline{s^n} \doteq^\beta h\overline{t^n})} \; \mathcal{G}(d)$$

Sequent calculus $\mathcal{G}_\beta$ is defined by the rules $\mathcal{G}(init)$, $\mathcal{G}(\neg)$, $\mathcal{G}(\vee_-)$, $\mathcal{G}(\vee_+)$, $\mathcal{G}(\Pi^l_-)$ and $\mathcal{G}(\Pi^c_+)$. Sequent calculus $\mathcal{G}_{\beta\mathfrak{fb}}$ extends $\mathcal{G}_\beta$ by the additional rules $\mathcal{G}(\mathfrak{b})$, $\mathcal{G}(\mathfrak{f})$, $\mathcal{G}(d)$, and $\mathcal{G}(Init^{\doteq})$.

Theorem proving in these calculi works as follows: In order to prove that a (closed) conjecture formula c logically follows from a (possibly empty) set of (closed) axioms $\{a^1, \ldots, a^n\}$, we start from the initial sequent $\Delta := \{c, \neg a^1, \ldots, \neg a^n\}$ and reason backwards by applying the respective calculus rules. We are done, if all branches of the proof tree can be closed by an application of the $\mathcal{G}(init)$ rule. In this case $\vdash^{\mathcal{G}_\beta/\mathcal{G}_{\beta\mathfrak{fb}}} \Delta := \{c, \neg a^1, \ldots, \neg a^n\}$ holds, which means that the conjecture c logically follows from the axioms $a^1, \ldots, a^n$ within calculus $\mathcal{G}_\beta$, respectively $\mathcal{G}_{\beta\mathfrak{fb}}$.

Soundness and completeness results for $\mathcal{G}_\beta$ and $\mathcal{G}_{\beta\mathfrak{fb}}$ have been established in [20].

Theorem 1 (Soundness and Completeness).

1. *$\mathcal{G}_\beta$ is sound and complete for ETT:* $\qquad \models^{ETT} c$ *if and only if* $\vdash^{\mathcal{G}_\beta} \{c\}$
   ($\mathcal{G}_\beta$ is thus also sound for HOL).

66

2. $\mathcal{G}_{\beta\text{fb}}$ is sound and complete for HOL: $\models^{HOL} c$ if and only if $\vdash^{\mathcal{G}_{\beta\text{fb}}} \{c\}$

Similarly, $\{a^1, \ldots, a^n\} \models^{ETT/HOL} c$ if and only if $\vdash^{\mathcal{G}_\beta/\mathcal{G}_{\beta\text{fb}}} \{c, \neg a^1, \ldots, \neg a^n\}$.

Rule $\mathcal{G}(cut)$

$$\frac{\Delta, s \quad \Delta, \neg s}{\Delta} \, \mathcal{G}(cut)$$

is available neither in $\mathcal{G}_\beta$ nor in $\mathcal{G}_{\beta\text{fb}}$, and cut-elimination holds for both calculi [20].

Theorem 2 (Cut-elimination). *The rule $\mathcal{G}(cut)$ is admissible in $\mathcal{G}_\beta$ and $\mathcal{G}_{\beta\text{fb}}$.*

In spite of their cut-freeness, both calculi are obviously only mildly suited for automation. One reason is that they are blindly guessing instantiations l in rule $\mathcal{G}(\Pi^l_-)$. Another reason is that the treatment of equality in both calculi relies on Leibniz equality $\doteq$. Support for primitive equality is not provided. The problem with Leibniz equality (or other forms of defined equality) is that it threatens cut-freeness of the calculi by allowing for simulations (admissibility) of the cut rule. The problem of cut-simulation, which is problematic for effective proof automation, analogously applies to a wide range of other prominent HOL axioms. We therefore address this issue in more depth within the next subsection. In §4 we will then outline a proof procedure that is better suited for proof automation than sequent calculi $\mathcal{G}_\beta$ and $\mathcal{G}_{\beta\text{fb}}$ above.

2.2 Cut-Simulation

We illustrate why Leibniz equality implies cut-simulation. Assume we want to study in $\mathcal{G}_\beta$ or $\mathcal{G}_{\beta\text{fb}}$ whether a conjecture c logically follows from an equality axiom $l = r$ (where l and r are some arbitrary closed terms of type α). Since primitive equality is not available we formalize the axiom as $l \doteq^\alpha r$ and initialize the proof process with sequent $\Delta := \{c, \neg(l \doteq^\alpha r)\}$, that is, with $\Delta := \{c, \neg\Pi(\lambda P_{\alpha \to o}(\neg Pl \vee Pr))$.

Now consider the following derivation, where s is an arbitrary (cut) formula:

$$\frac{\dfrac{\dfrac{\Delta, s}{\Delta, \neg\neg s} \, \mathcal{G}(\neg) \quad \Delta, \neg s}{\Delta, \neg(\neg s \vee s)} \, \mathcal{G}(\vee_-)}{\Delta, \neg\Pi(\lambda P_{\alpha \to o}(\neg Pl \vee Pr))} \, \mathcal{G}(\Pi^{\lambda Xs}_-)$$

It is easy to see that this derivation introduces a cut on formula s; in the left branch s occurs positively and in the right branch negatively.

Cut-simulation is also enabled by the functional and Boolean extensionality axioms. The Boolean extensionality axiom (abbreviated as $\mathcal{B}_o$) is given as

$$\forall A_o \forall B_o (A \longleftrightarrow B) \to A \doteq^o B$$

67

The infinitely many functional extensionality axioms (abbreviated as $\mathcal{F}_{\alpha\beta}$) are parameterized over $\alpha, \beta \in T$. They are given as

$$\forall F_{\alpha\to\beta}\forall G_{\alpha\to\beta}(\forall X_\alpha FX \doteq^\beta GX) \to F \doteq^{\alpha\to\beta} G$$

Instead of the extensionality rules $\mathcal{G}(\mathfrak{f})$ and $\mathcal{G}(\mathfrak{b})$, as provided in calculus $\mathcal{G}_{\beta\mathfrak{f}\mathfrak{b}}$, we could alternatively postulate the validity of these axioms. For this we could replace the rules $\mathcal{G}(\mathfrak{f})$ and $\mathcal{G}(\mathfrak{b})$ in $\mathcal{G}_\beta$ by the following axiomatic extensionality rules $\mathcal{G}(\mathcal{F}_{\alpha\beta})$ and $\mathcal{G}(\mathcal{B})$:

$$\frac{\Delta, \neg\mathcal{F}_{\alpha\beta} \quad \alpha \to \beta \in T}{\Delta} \; \mathcal{G}(\mathcal{F}_{\alpha\beta}) \qquad \frac{\Delta, \neg\mathcal{B}_o}{\Delta} \; \mathcal{G}(\mathcal{B})$$

This calculus is still Henkin complete (even if rules $\mathcal{G}(d)$ and $\mathcal{G}(Init^{\doteq})$ are additionally removed) [20]. However, the modified calculus suffers severely from cut-simulation. For axiom $\mathcal{B}_o$ this is illustrated by the following derivation (a_o is new constant symbol):

derivable in 7 steps

$$\frac{\dfrac{\dfrac{\Delta, a \longleftrightarrow a}{\Delta, \neg\neg(a \longleftrightarrow a)} \, \mathcal{G}(\neg) \quad \dfrac{\Delta, s \quad \Delta, \neg s}{\vdots \text{ derivable in 3 steps, see above}}{\Delta, \neg(a \doteq^o a)}}{\dfrac{\Delta, \neg(\neg(a \longleftrightarrow a) \vee a \doteq^o a)}{\Delta, \neg\mathcal{B}_o} \, 2 \times \mathcal{G}(\Pi_-^a)} \, \mathcal{G}(\vee_-)}$$

The left branch is closed and on the right branch an arbitrary cut formula s is introduced. A similar derivation is enabled with axiom $\mathcal{F}_{\alpha\beta}$ (b_α is new constant symbol):

derivable in 3 steps

$$\frac{\dfrac{\dfrac{\dfrac{\Delta, fb \doteq^\beta fb}{\Delta, (\forall X_\alpha fX \doteq^\beta fX)} \, \mathcal{G}(\Pi_+^b)}{\Delta, \neg\neg\forall X_\alpha fX \doteq^\beta fX} \, \mathcal{G}(\neg) \quad \dfrac{\Delta, s \quad \Delta, \neg s}{\vdots \text{ derivable in 3 steps}}{\Delta, \neg(f \doteq^{\alpha\to\beta} f)}}{\dfrac{\Delta, \neg(\neg(\forall X_\alpha fX \doteq^\beta fX) \vee f \doteq^{\alpha\to\beta} f)}{\Delta, \neg\mathcal{F}_{\alpha\beta}} \, 2 \times \mathcal{G}(\Pi_-^f)} \, \mathcal{G}(\vee_-)}$$

In all cut-simulations above we have exploited the fact that predicate variables may be instantiated with terms that introduce arbitrary new formulas s. At these points the subformula property breaks. At the same time this offers the opportunity to mimic cut-introductions by appropriately selecting such instantiations for predicate variables. In addition to Leibniz equations and the Boolean and functional extensionality axioms, cut-simulations are analogously enabled by many prominent other axioms, including excluded middle, description, choice, comprehension, and induction. We may thus call these axioms cut-strong. More details on such cut-strong axioms are provided in [20].[1]

Cut-simulations have in fact been extensively used in literature. For example, Takeuti showed that a conjecture of Gödel could be proved without cut by using the induction principle instead [104]; [75] illustrates how the induction rule can be used to hide the cut rule; and [91] used excluded middle to similarly mask the cut rule.

For the development of automated proof procedures for HOL we thus learn an important lesson, namely that cut-elimination and cut-simulation should always be considered in combination: a pure cut-elimination result may indeed mean little if at the same time axioms are assumed that support effective cut-simulation. The challenge is to develop cut-free calculi for HOL that also try to avoid the pitfall of cut-simulations (as far as possible).

Church's use of the λ-calculus to build comprehension principles into the language can therefore be seen as a first step in the program to eliminate cut-strong axioms. Significant progress in the automation of HOL in existing prover implementations has been achieved after providing calculus level support for extensionality and also choice (avoiding cut-simulation effects). Respective extensionality rules have been provided for resolution [9; 10], expansion and sequent calculi [45; 46], and tableaux [48]. Similarly, choice rules have been proposed for the various settings: sequent calculus [78], tableaux [8] and resolution [38].

In §4 we outline the extensional RUE-resolution approach of the LEO-II theorem prover [39; 38]. In this approach some pragmatic improvements are offered regarding most pressing challenges for effective proof automation.

3 Proof-Theoretical Properties via Semantic Embeddings

Cut-elimination and cut-simulation in HOL have been addressed in the previous section. In particular, with sequent calculus $\mathcal{G}_{\beta fb}$ an example of a cut-free calculus for HOL has been provided.

The development of cut-free calculi for expressive logics is generally a non-trivial task. By modeling and studying these logics as fragments of HOL — a research direction proposed in [16] — existing results for HOL (for example, the cut-free sequent calculus $\mathcal{G}_{\beta fb}$) can easily be reused. The idea is illustrated next by choosing quantified conditional logics [95] as an example.

3.1 Quantified Conditional Logic

As an exemplary challenging logic we consider quantified conditional logic (QCL).

---

[1] Obviously, any universally quantified predicate variable (occuring negatively in the above approach) is a possible source for cut-simulation. The challenge thus is to avoid those predicate variables as far as possible. An axiomatic approach based on cut-strong axioms, as proposed by several authors including e.g. [68; 69], is therefore hardly a suitable option for the automation of HOL.

Let IV be a set of first-order (individual) variables, PV a set of propositional variables, and SYM a set of predicate symbols of any arity. Formulas of QCL are given by the following grammar (where $X^i \in$ IV, $P \in$ PV, $k \in$ SYM, and where $\Rightarrow$ represents conditionality):

$$\varphi, \psi ::= \ P \mid k(X^1, \ldots, X^n) \mid \neg\varphi \mid \varphi \vee \psi \mid \varphi \Rightarrow \psi \mid \forall^{co} X\varphi \mid \forall^{va} X\varphi \mid \forall P\varphi$$

From the selected set of primitive connectives, other logical connectives can be introduced as abbreviations: for example, $\varphi \wedge \psi$, $\varphi \to \psi$ (material implication), $\varphi \longleftrightarrow \psi$ and $\Box\varphi$ abbreviate $\neg(\neg\varphi \vee \neg\psi)$, $\neg\varphi \vee \psi$, $(\varphi \to \psi) \wedge (\psi \to \varphi)$ and $\neg\varphi \Rightarrow \varphi$, respectively. $\forall^{co}$ and $\forall^{va}$ are associated with constant domain and variable domain quantification. For $* \in \{co, va\}$, $\exists^* X\varphi$ abbreviates $\neg\forall^* X\neg\varphi$. Syntactically, QCL can be seen as a generalization of quantified multimodal logic where the index of modality $\Rightarrow$ is a formula of the same language. For instance, in $(\varphi \Rightarrow \psi) \Rightarrow \delta$ the subformula $\varphi \Rightarrow \psi$ is the index of the second occurrence of $\Rightarrow$.

Regarding semantics, different formalizations have been proposed (see [83]). Here we build on *selection function semantics* [95; 49], which is based on possible world structures and has been successfully used in [85] to develop proof methods for some propositional CLs.

An *interpretation* is a structure $M = \langle S, f, D, D', Q, I \rangle$ where, S is a set of items called possible worlds, $f : S \times 2^S \mapsto 2^S$ is the selection function, D is a non-empty set of *individuals* (the constant first-order domain), D' is a function that assigns a non-empty subset $D'(w)$ of D to each possible world w (the $D'(w)$ are the varying domains), Q is a non-empty collection of subsets of S (the propositional domain), and I is a classical interpretation function where for each n-ary predicate symbol k, $I(k, w) \subseteq D^n$.

A *variable assignment* $g = (g^i, g^p)$ is a pair of maps where, $g^i : $ IV $\mapsto D$ maps each individual variable in IV to an object in D, and $g^p : $ maps each propositional variable in PV to a set of worlds in Q.

Satisfiability of a formula φ for an interpretation $M = \langle S, f, D, D', Q, I \rangle$, a world $s \in S$, and a variable assignment $g = (g^i, g^p)$ is denoted as $M, g, s \models \varphi$ and defined as follows, where $[a/Z]g$ denote the assignment identical to g except that $([a/Z]g)(Z) = a$:

$M, g, s \models k(X^1, \ldots, X^n)$ if and only if $\langle g^i(X^1), \ldots, g^i(X^n) \rangle \in I(k, w)$

$M, g, s \models P$ if and only if $s \in g^p(P)$

$M, g, s \models \neg\varphi$ if and only if $M, g, s \not\models \varphi$ (that is, not $M, g, s \models \varphi$)

$M, g, s \models \varphi \vee \psi$ if and only if $M, g, s \models \varphi$ or $M, g, s \models \psi$

$M, g, s \models \forall^{co} X\varphi$ if and only if $M, ([d/X]g^i, g^p), s \models \varphi$ for all $d \in D$

$M, g, s \models \forall^{va} X\varphi$ if and only if $M, ([d/X]g^i, g^p), s \models \varphi$ for all $d \in D'(s)$

$M, g, s \models \forall P\varphi$ if and only if $M, (g^i, [p/P]g^p), s \models \varphi$ for all $p \in Q$

$M, g, s \models \varphi \Rightarrow \psi$ if and only if $M, g, t \models \psi$ for all $t \in S$ such that $t \in f(s, [\varphi])$

$\qquad\qquad$ where $[\varphi] = \{u \mid M, g, u \models \varphi\}$

An interpretation $M = \langle S, f, D, D', Q, I \rangle$ is a QCL *model* if for every variable assignment g and every formula φ, the set of worlds $\{s \in S \mid M, g, s \models \varphi\}$ is a member of Q. (This requirement, which is inspired by Fitting [60], Def. 3.5, ensures a natural correspondence to Henkin models in HOL.) As usual, a conditional formula φ is *valid in a QCL model* $M = \langle S, f, D, D', Q, I \rangle$, denoted with $M \models \varphi$, if and only if for all worlds $s \in S$ and variable assignments g holds $M, g, s \models \varphi$. A formula φ is *valid*, denoted $\models \varphi$, if and only if it is valid in every QCL model.

f is defined to take $[\varphi]$ (called the *proof set* of φ with respect to a given QCL model M) instead of φ. This approach has the consequence of forcing the so-called *normality* property: given a QCL model M, if φ and φ' are equivalent (i.e., they are satisfied in the same set of worlds), then they index the same formulas with respect to the $\Rightarrow$ modality. The axiomatic counterpart of the normality condition is given by the rule RCEA (which expresses a replacement property for equivalent formulas on the left-hand side of a conditional formula):

$$\frac{\varphi \leftrightarrow \varphi'}{(\varphi \Rightarrow \psi) \leftrightarrow (\varphi' \Rightarrow \psi)} \ (RCEA)$$

Moreover, it can be easily shown that the above semantics forces also the following rules to hold (RCEC expresses a right-hand side replacement property analogous to RCEA, and RCK expresses compatibility of the right-hand side of conditional formulas with conjunction):

$$\frac{\varphi \leftrightarrow \varphi'}{(\psi \Rightarrow \varphi) \leftrightarrow (\psi \Rightarrow \varphi')} \ (RCEC)$$

$$\frac{(\varphi_1 \wedge \ldots \wedge \varphi_n) \leftrightarrow \psi}{(\varphi_0 \Rightarrow \varphi_1 \wedge \ldots \wedge \varphi_0 \Rightarrow \varphi_n) \to (\varphi_0 \Rightarrow \psi)} \ (RCK)$$

We refer to QCK (cf. CK in [50]) as the minimal QCL closed under rules RCEA, RCEC and RCK. In what follows, only QCLs extending QCK are considered.

QCLs have many applications, including action planning, counterfactual reasoning, default reasoning, deontic reasoning, metaphysical modeling and reasoning about knowledge. While there is broad literature on propositional conditional logics only a few authors have addressed first-order extensions [56; 62]. Most interestingly, QCLs subsume normal modal logics ($\Box \varphi$ can be defined as $\neg \varphi \Rightarrow \varphi$, see [95]).

Modeling QCLs as fragments of HOL. Regarding the particular choice of HOL, we here assume a set of basic types $\{o, i, u\}$, where o denotes the type of Booleans as before. Without loss of generality, i is now identified with a (non-empty) set of worlds and u with a (non-empty) domain of individuals.

QCL formulas are now identified with certain HOL terms (predicates) of type $i \to o$. They can be applied to terms of type i, which are assumed to denote possible worlds. Type $i \to o$ is abbreviated as τ in the remainder.

The mapping $\lfloor \cdot \rfloor$ translates QCL formulas φ into HOL terms $\lfloor \varphi \rfloor$ of type τ. The mapping is recursively defined:

$$
\begin{aligned}
\lfloor P \rfloor &= P_\tau \\
\lfloor k(X^1, \ldots, X^n) \rfloor &= k_{u^n \to \tau}\, X_u^1 \ldots X_u^n \\
\lfloor \neg \varphi \rfloor &= \neg_\tau \lfloor \varphi \rfloor \\
\lfloor \varphi \vee \psi \rfloor &= \vee_{\tau \to \tau \to \tau} \lfloor \varphi \rfloor \lfloor \psi \rfloor \\
\lfloor \varphi \Rightarrow \psi \rfloor &= \Rightarrow_{\tau \to \tau \to \tau} \lfloor \varphi \rfloor \lfloor \psi \rfloor \\
\lfloor \forall^{co} X \varphi \rfloor &= \Pi^{co}_{(u \to \tau) \to \tau} \lambda X_u \lfloor \varphi \rfloor \\
\lfloor \forall^{va} X \varphi \rfloor &= \Pi^{va}_{(u \to \tau) \to \tau} \lambda X_u \lfloor \varphi \rfloor \\
\lfloor \forall P \varphi \rfloor &= \Pi_{(\tau \to \tau) \to \tau} \lambda P_\tau \lfloor \varphi \rfloor
\end{aligned}
$$

P_τ and $X_u^1, \ldots, X_u^n$ are variables and $k_{u^n \to \tau}$ is a constant symbol. $\neg_\tau$, $\vee_{\tau \to \tau \to \tau}$, $\Rightarrow_{\tau \to \tau \to \tau}$, $\Pi^{co,va}_{(u \to \tau) \to \tau}$ and $\Pi_{(\tau \to \tau) \to \tau}$ realize the QCL connectives in HOL. They abbreviate the following HOL terms:[2]

$$
\begin{aligned}
\neg_{\tau \to \tau} &= \lambda A_\tau \lambda X_i \neg (A\, X) \\
\vee_{\tau \to \tau \to \tau} &= \lambda A_\tau \lambda B_\tau \lambda X_i (A\, X \vee B\, X) \\
\Rightarrow_{\tau \to \tau \to \tau} &= \lambda A_\tau \lambda B_\tau \lambda X_i \forall V_i (f\, X\, A\, V \to B\, V) \\
\Pi^{co}_{(u \to \tau) \to \tau} &= \lambda Q_{u \to \tau} \lambda V_i \forall X_u (Q\, X\, V) \\
\Pi^{va}_{(u \to \tau) \to \tau} &= \lambda Q_{u \to \tau} \lambda V_i \forall X_u (eiw\, V\, X \to Q\, X\, V) \\
\Pi_{(\tau \to \tau) \to \tau} &= \lambda R_{\tau \to \tau} \lambda V_i \forall P_\tau (R\, P\, V)
\end{aligned}
$$

Constant symbol f in the mapping of $\Rightarrow$ is of type $i \to \tau \to \tau$. It realizes the selection function. Constant symbol eiw (for 'exists in world'), which is of type $(\tau \to u) \to \tau$, is associated with the varying domains. The interpretations of f and eiw are chosen appropriately below. Moreover, for the varying domains a non-emptiness axiom is postulated:

$$
\forall W_i \exists X_u (eiw\, W\, X) \tag{NE}
$$

Analyzing the validity of a translated formula $\lfloor \varphi \rfloor$ for a world represented by term t_i corresponds to evaluating the application $(\lfloor \varphi \rfloor\, t_i)$. In line with [21] (and analogous to [33; 34; 35]), we define $vld_{\tau \to o} = \lambda A_\tau \forall S_i (A\, S)$. With this definition, validity of a QCL formula φ in QCK corresponds to the validity of $(vld \lfloor \varphi \rfloor)$ in HOL, and vice versa.

Soundness and completeness of this embedding of QCL in HOL has been studied in [15].

Theorem 3 (Soundness and Completeness of QCL-embedding).
$\models^{QCL} \varphi$ *if and only if* $\{NE\} \models^{HOL} vld \lfloor \varphi \rfloor$

Combining Theorem 3 with Theorem 1 we obtain:

---

[2] Note the predicate argument A of f in the term for $\Rightarrow_{\tau \to \tau \to \tau}$ and the second-order quantifier $\forall P_\tau$ in the term for $\Pi_{(\tau \to \tau) \to \tau}$. FOL encodings of both constructs, if feasible, will be less natural.

Theorem 4 (Soundness and Completeness of $\mathcal{G}_{\beta\mathfrak{f}\mathfrak{b}}$ for QCL).
$\models^{QCL} \varphi$ if and only if $\vdash^{\mathcal{G}_{\beta\mathfrak{f}\mathfrak{b}}} \{vld \lfloor\varphi\rfloor, \neg NE\}$

Since $\mathcal{G}_{\beta\mathfrak{f}\mathfrak{b}}$ is cut-free (Theorem 2), we thus obtain a cut-elimination result for QCL for free. However, we need to point again to the subtle issue of cut-simulation. For example, when postulating additional axioms for the embedded logics in HOL (for example, QCL axiom ID: $\forall\varphi(\varphi \Rightarrow \varphi)$), cut-simulation effects may apply. In some cases the semantical conditions which correspond to such axioms can be postulated instead in order to circumvent the effect. This is for example possible for many prominent modal logic axioms. for example, the corresponding semantical condition for modal axiom T: $\forall\varphi(\Box\varphi \supset \varphi)$ is $\forall x(rxx)$ (where constant r denotes the associated accessibility relation). The latter axiom obviously does not support cut-simulations and should therefore be preferred. However, the semantical condition that corresponds to ID, $\forall A_\tau \forall W_i(f\,W\,A \subseteq A)$, unfortunately still introduces some problematic predicate variables.

3.2 Other Logic Embeddings in HOL

Recent work has shown that many other challenging logics can be characterized as HOL fragments via semantic embeddings. The logics studied so far comprise prominent non-classical logics, including modal logics, tense logics, intuitionistic logic, security logics, conditional logics, hybrid logics and logics for time and space [33; 34; 35; 12; 14; 22; 21; 32; 107; 108]. These fragments also comprise first-order and even higher-order extensions of non-classical logics, for which only little practical automation support has been available so far. Most importantly, however, combinations of embedded logics can be elegantly achieved in this approach. And, similar to above, cut-elimination results for these embedded logics can be obtained 'for free' by exploiting the results already achieved for HOL.

The embeddings approach bridges between the Tarski view of logics (for 'meta logic' HOL) and the Kripke view (for the embedded source logics) and exploits the fact that well known translations of logics, such as the relational translation [84], can be easily formalized in HOL. This way HOL-ATP systems can be uniformly applied to reason *within* and also *about* embedded logics and their combinations.

4 Pragmatic Properties

In this Section we outline the calculus and working principles of the higher-order automated theorem prover LEO-II [39; 38].

4.1 Extensional RUE-Resolution Calculus of Leo-II

LEO-II is an automated theorem prover for HOL. It supports primitive equality and choice.

In LEO-II, logical consequence of a conjecture c from a (possibly empty) set of axioms $\{a^1, \ldots, a^n\}$ is established by refuting the initial set of (non-normalized) unit clauses $\{[c]^{\mathrm{ff}}, [a^1]^{\mathrm{tt}}, \ldots, [a^n]^{\mathrm{tt}}\}$. Refuting means deriving the empty clause from this initial set by subsequent forward applications of the rules presented below. These rules operate on the clauses of a given clause set and they add their result clauses to this clause set. The superscripts tt and ff denote the polarity of clause literals. Clauses are generally depicted as $C \vee [s]^\alpha$ below, where $[s]^\alpha$ is a literal (for $\alpha \in \{\mathrm{tt}, \mathrm{ff}\}$) and where C is a clause rest.

LEO-II's calculus is based on an adaption of the RUE-resolution approach [57], originally developed for first-order logic with equality, to HOL [9; 10]. In RUE-resolution unification constraints are explicitly represented and manipulated as disagreement pairs, that is, negated equations. A unification constraint $[l = r]^{\mathrm{ff}}$ in clause $C \vee [l = r]^{\mathrm{ff}}$ can also be seen as a condition (obligation to make terms l and r equal) under which the clause rest C follows. In LEO-II, such unification constraints are amenable to resolution and factorization.

The calculus rules of LEO-II are presented next.

Definition 2 (Extensional Higher-order RUE-Resolution). *The following rules implicitly assume symmetry and associativity of the clause-level $\vee$-operator. Moreover, they assume that the formulas in clauses are always kept in $\beta\eta$-normal form.*

Basic Rules

$$\frac{C \vee [\neg s]^{\mathrm{tt}}}{C \vee [s]^{\mathrm{ff}}} \; \mathcal{R}(\neg_{\mathrm{tt}}) \qquad \frac{C \vee [\neg s]^{\mathrm{ff}}}{C \vee [s]^{\mathrm{tt}}} \; \mathcal{R}(\neg_{\mathrm{ff}})$$

$$\frac{C \vee [s \vee t]^{\mathrm{tt}}}{C \vee [s]^{\mathrm{tt}} \vee [t]^{\mathrm{tt}}} \; \mathcal{R}(\vee_{\mathrm{tt}}) \qquad \frac{C \vee [\Pi^\alpha s]^{\mathrm{tt}} \quad X_\alpha \; \text{fresh variable}}{C \vee [s\, X]^{\mathrm{tt}}} \; \mathcal{R}(\Pi^X_{\mathrm{ff}})$$

$$\frac{C \vee [s \vee t]^{\mathrm{ff}}}{C \vee [s]^{\mathrm{ff}} \quad C \vee [t]^{\mathrm{ff}}} \; \mathcal{R}(\vee_{\mathrm{ff}}) \qquad \frac{C \vee [\Pi^\alpha s]^{\mathrm{ff}} \quad \mathrm{sk}_\alpha \; \text{Skolem term}}{C \vee [s\, \mathrm{sk}_\alpha]^{\mathrm{ff}}} \; \mathcal{R}(\Pi^{\mathrm{sk}}_{\mathrm{tt}})$$

These rules deal with the normalization of clauses. The rules are straightforward – for instance, $\mathcal{R}(\vee_{\mathrm{tt}})$ lifts object-level disjunction to meta-level (i.e. clause-level) disjunction. Similarly, the rule $\mathcal{R}(\neg_{\mathrm{ff}})$ removes a dominant negation from a literal and flips the literal's polarity. Normalization cannot be treated as a pre-process as in first-order logic, since instantiations of predicate variables, for example with rules $\mathcal{R}(Subst)$ and $\mathcal{R}(PrimSubst)$ may introduce non-normal clauses and require subsequent clause normalization steps. Instead of blindly guessing instantiations as in sequent rule $\mathcal{G}(\Pi^l_-)$, LEO-II thus introduces free variables in the search space. The hope is that suitable instantiations for these variables can be determined by pre-unification within the subsequent proof search process. However, as we will explore further below, this idea can only be partly realized in HOL.

Resolution

$$\frac{C \vee [s]^{\mathrm{tt}} \quad D \vee [t]^{\mathrm{ff}}}{C \vee D \vee [s = t]^{\mathrm{ff}}} \; \mathcal{R}(Res) \qquad \frac{C \vee [s]^p \vee [t]^p}{C \vee [s]^p \vee [s = t]^{\mathrm{ff}}} \; \mathcal{R}(Fac)$$

In LEO-II, resolution and factorization are applied only to proper clauses, that is, clauses in which all literal formulas are atomic (or unification constraints).

Note that both rules introduce unification constraints $[s = t]^{\mathrm{ff}}$. On these unification constraints the pre-unification rules given below operate. Instead of the factorization rule as shown here, factorization in LEO-II is (for pragmatic reasons and at the cost of completeness) restricted to binary clauses only.

Extensionality

$$\frac{C \vee [s =^{\alpha \to \beta} t]^{\mathrm{tt}} \quad X_\alpha \; \textit{fresh variable}}{C \vee [sX =^{\beta} tX]^{\mathrm{tt}}} \; \mathcal{R}(\mathfrak{f}_{\mathrm{tt}}) \qquad \frac{C \vee [s =^{o} t]^{\mathrm{tt}}}{C \vee [s \longleftrightarrow t]^{\mathrm{tt}}} \; \mathcal{R}(\mathfrak{b}_{\mathrm{tt}})$$

$$\frac{C \vee [s =^{\alpha \to \beta} t]^{\mathrm{ff}} \quad \mathrm{sk}_\alpha \; \textit{Skolem term}}{C \vee [s \, \mathrm{sk} =^{\beta} t \, \mathrm{sk}]^{\mathrm{ff}}} \; \mathcal{R}(\mathfrak{f}_{\mathrm{ff}}) \qquad \frac{C \vee [s =^{o} t]^{\mathrm{ff}}}{C \vee [s \longleftrightarrow t]^{\mathrm{ff}}} \; \mathcal{R}(\mathfrak{b}_{\mathrm{ff}})$$

Conceptually the extensionality rules $\mathcal{R}(\mathfrak{f}_{\mathrm{tt}})$ and $\mathcal{R}(\mathfrak{b}_{\mathrm{tt}})$ belong to the normalization rules of LEO-II, while $\mathcal{R}(\mathfrak{f}_{\mathrm{ff}})$ and $\mathcal{R}(\mathfrak{b}_{\mathrm{ff}})$ are integrated with the pre-unification rules. Similar to sequent rules $\mathcal{G}(\mathfrak{f})$ and $\mathcal{G}(\mathfrak{b})$, these rules realize full extensionality reasoning while avoiding cut-simulation effects. The extensionality principles for Leibniz equality are implied.

Pre-Unification
$$\frac{C \vee [h\overline{s^n} =^{\beta} h\overline{t^n}]^{\mathrm{ff}} \quad n \geq 1, \beta \in \{o, \iota\}, h_{\overline{\alpha^n} \to \beta} \in \Sigma}{C \vee [s^1 =^{\alpha_1} t^1]^{\mathrm{ff}} \quad \cdots \quad C \vee [s^n =^{\alpha_n} t^n]^{\mathrm{ff}}} \; \mathcal{R}(d)$$

$$\frac{C \vee [s = s]^{\mathrm{ff}}}{C} \; \mathcal{R}(\textit{Triv}) \qquad \frac{C \vee [X = s]^{\mathrm{ff}} \quad X \notin \textit{free}(s)}{[s/X]C} \; \mathcal{R}(\textit{Subst})$$

$$\frac{C \vee [F\overline{s^n} =^{\beta} h\overline{t^m}]^{\mathrm{ff}} \quad n \geq 1, h_{\overline{\gamma^m} \to \beta} \in \Sigma, l \in \mathcal{AB}^{(h)}_{\overline{\alpha^n} \to \beta}}{C \vee [F = l]^{\mathrm{ff}} \vee [F\overline{s^n} =^{\beta} h\overline{t^m}]^{\mathrm{ff}}} \; \mathcal{R}(\textit{FlexRigid})$$

This set of rules implements pre-unification, cf. [67; 93]. LEO-II actually packages its unification steps into a single, abstract rule called **extuni**. This package also integrates the extensionality rules $\mathcal{R}(\mathfrak{f}_{\mathrm{ff}})$ and $\mathcal{R}(\mathfrak{b}_{\mathrm{ff}})$. The integration of these two rules, in particular, of $\mathcal{R}(\mathfrak{b}_{\mathrm{ff}})$, is the reason why pre-unification in LEO-II should rather be called extensional pre-unification; cf. [11]. Moreover, logical constants, such as disjunction, equality, etc. are interpreted and a special treatment is provided for them (an example would be symmetric decomposition in rule $\mathcal{R}(d)$ in case h is $\vee$ or $=^\alpha$).

In rule $\mathcal{R}(\textit{FlexRigid})$ (and again in rule $\mathcal{R}(\textit{PrimSubst})$ below), we use the symbol $\mathcal{AB}^{(h)}_{\overline{\alpha^n} \to \beta}$ to denote the set of approximating/partial bindings parametric to a type $\overline{\alpha^n} \to \beta$ and to a constant $h_{\overline{\gamma^m} \longrightarrow \beta}$. This is explained further next; see also [93]. Given a name k (where a name is either a constant or a variable) of type $\overline{\gamma^m} \longrightarrow \beta$, term l having form $\lambda \overline{X^n_{\alpha^n}}(k\overline{r^m})$ is a partial binding of type $\overline{\alpha^n} \to \beta$ and head k. Each $r^{i \leq m}$ has form $H^i \overline{X^n_{\alpha^n}}$ where $H^{i \leq m}$ are fresh variables typed $\overline{\alpha^n} \longrightarrow \gamma^{i \leq m}$. Projection bindings are partial bindings whose head k is one of $X^{i \leq l}$. Imitation bindings are partial bindings whose head k is identical to the given symbol h in the superscript of $\mathcal{AB}^{(h)}_{\overline{\alpha^n}\to\beta}$. $\mathcal{AB}^{(h)}_{\overline{\alpha^n}\to\beta}$ is the set of all projection and imitation bindings modulo type $\overline{\alpha^n}\to\beta$ and h.

LEO-II follows Huet [68; 69] in regarding flexflex clauses, that is, clauses consisting only of flexflex unification literals $[F\,\overline{s^n} =^\beta G\,\overline{t^m}]^{\text{ff}}$ (where both F and G are variables), to be empty.

Primitive Substitution $$\frac{C \vee [Q_{\overline{\alpha^n}\to o}\,\overline{s^n}]^p \quad l \in \mathcal{AB}^{(\neg,\vee,\Pi^\alpha,=^\alpha)}_{\overline{\alpha^n}\to o}}{[l/Q](C \vee [Q_{\overline{\alpha^n}\to o}\,\overline{s^n}]^p)}\ \mathcal{R}(PrimSubst)$$

In this rule, which is related to Huet's splitting rule [68; 69] and Andrews's primitive substitutions [4], $\mathcal{AB}^{(\neg,\vee,\Pi^\alpha,=^\alpha)}_{\overline{\alpha^n}\to o}$ stands for $\mathcal{AB}^{(\neg)}_{\overline{\alpha^n}\to o} \cup \mathcal{AB}^{(\vee)}_{\overline{\alpha^n}\to o} \cup \mathcal{AB}^{(\Pi^\alpha)}_{\overline{\alpha^n}\to o} \cup \mathcal{AB}^{(=^\alpha)}_{\overline{\alpha^n}\to o}$. This rule introduces a certain amount of blind guessing into the proof procedure. However, unlike in sequent calculus rule $\mathcal{G}(\Pi^l_-)$ only the top-level logical structure of the instantiation term l is guessed, while further decisions on l are delayed. The hope is that they can eventually be determined by pre-unification in subsequent resolution steps. Generally, however, subsequent applications of rule $\mathcal{R}(PrimSubst)$ are permitted and the deeper logical structure of l may thus be guessed later. It is an open challenge to suitably restrict this rule without threatening completeness.

A simple, prominent example to illustrate the need for splittings is $\exists P_o.P$. When using resolution the formula is first negated and then normalized to clause $[X_o]^{\text{ff}}$, where X is a predicate variable. There is no resolution partner for this clause available, hence the empty clause can not be derived. However, when guessing some top-level, logical structure for X, here the substitution $[\neg Y/X]$ is suitable, then $[\neg Y]^{\text{ff}}$ is derived, which normalizes into a new clause $[Y]^{\text{tt}}$. Now, resolution between the clauses $[X]^{\text{ff}}$ and $[Y]^{\text{tt}}$ with substitution $[Y/X]$ directly leads to the empty clause.

Choice $$\frac{[PX]^{\text{ff}} \vee [P(f_{(\alpha\to o)\to\alpha}P)]^{\text{tt}}}{\mathsf{CFs} \longleftarrow \mathsf{CFs} \cup \{f_{(\alpha\to o)\to\alpha}\}}\ \mathcal{R}(DetectChoice)$$

$$\frac{C := C' \vee [s[E_{(\alpha\to o)\to\alpha}t]]^p \quad \begin{matrix} E = \epsilon \text{ for } \epsilon \in \mathsf{CFs} \text{ or } E \in free(C), \\ free(t) \subseteq free(C), Y \text{ fresh} \end{matrix}}{[t\,Y]^{\text{ff}} \vee [t\,(\epsilon_{(\alpha\to o)\to\alpha}t)]^{\text{tt}}}\ \mathcal{R}(Choice)$$

Rule $\mathcal{R}(Choice)$ investigates whether a term $\epsilon_{(\alpha\to o)\to\alpha}t_{\alpha\to o}$ (where ϵ is a choice function, registered and memorized in a special set CFs, or a free variable) is contained as a subterm of a literal $[s]^p$ in a clause C. In this case it adds the instantiation of the choice axiom at type $(\alpha \to o) \to \alpha$ with term $t_{\alpha\to o}$ to the search space.[3] Side-conditions guard against unsound reasoning, such as the 'uncapturing' of free variables in t. Additionally, rule $\mathcal{R}(DetectChoice)$ detects and removes uninstantiated choice-axiom clauses from the search space (remember that they are cut-strong) and registers the corresponding choice function sym-

---

[3] Note that the instantiation of the choice axiom (scheme) $\forall F_{\tau\to o}((\exists Y_\tau FY) \to F(\epsilon_{(\tau\to o)\to\tau}F))$ for term t leads to the clause $[t\,Y]^{\text{ff}} \vee [t\,(\epsilon_{(\alpha\to o)\to\alpha}t)]^{\text{tt}}$.

bols f in CFs. By default, CFs contains at least one choice function symbol for each choice type. The rule does not describe a typical logical inference, since the conclusion of the rule indicates a side-effect which extends the set CFs of choice functions and which removes a clause. Both rules are obviously motivated by the idea to avoid cut-simulation effects. Moreover, it is easy to see that they are sound: $\mathcal{R}(DetectChoice)$ simply removes (cut-strong) clauses from the search space and registers choice functions, and for any registered choice function f, the rule $\mathcal{R}(Choice)$ only introduces new instances of the corresponding choice axiom.

Both choice rules can be disabled in LEO-II with the help of a special flag.

We now briefly summarize the organization of proof search in LEO-II.

In first-order logic, unification is decidable, and it is used as an eager filter during resolution. Unification in HOL is undecidable in general, so it is used more carefully. Therefore, LEO-II relies on a variant of Huet's pre-unification, which is a semi-decidable procedure. It works by accumulating flexflex unification pairs as unification constraints. When a clause consists only of flexflex constraints then it is considered to be empty, since, as Huet showed [67], such a system of equations always has solutions. An additional aspect of unification in LEO-II is the integration of the extensionality rule $\mathcal{R}(\mathfrak{b}_{tt})$. In addition to flexflex constraints, pre-unification in LEO-II may thus return negated equivalence literals.

Though it was originally intended as an alternative option for LEO-II's architecture, lazy unification has never been implemented. Instead eager unification is used in LEO-II, which works as follows: extensional pre-unification is applied to clauses with a predefined depth bound (for example, maximally five[4] nestings of the branching flex-rigid rule; modulo this depth-bound higher-order pre-unification becomes decidable, but at the cost of completeness; however, regarding the unification depth an iterative deepening approach is actually provided in LEO-II). The solved unification constraints are exhaustively applied in the resulting clauses, and any remaining flexflex unification pairs or negated equivalence literals generated by $\mathcal{R}(\mathfrak{b}_{tt})$ are kept as literals of the result clause.

LEO-II's calculus-level treatment of the axiom of choice is inspired by the work of Mints [78]. Choice is related to Skolemization. In HOL, Skolemization is not as straightforward as in first-order logic [77]. Naïve Skolemization is unsound with respect to Henkin models that invalidate choice, and incomplete with respect to Henkin models that validate choice [8; 17].

Like many other provers, LEO-II spends its time looping during its exploration of the search space — executing its *main loop*. By *search space* we mean the totality of clauses surveyed by LEO-II during its execution. Each iteration of this loop might generate new clauses, thus contributing to the representation of the search space that is kept by LEO-II. Each iteration does *not* change the satisfiability of the problem and its search space; this is an invariant of a prover's main loop.

---

[4] The pre-unification depth is a parameter in LEO-II that can be specified at the command line. By default LEO-II currently operates with values up to depth 8. So far there has been no exhaustive empirical investigation of the optimal setting of the pre-unification depth.

Unlike many provers Leo-II keeps an additional representation of the search space. This is used to store the input to external provers. The contents of this store are produced by translating the clauses in the main store. The source clauses consist of higher-order clauses, and the target clauses are encoded in the target logic. Since Leo-II currently only cooperates with first-order provers, the target clauses consist of first-order clauses.

The first-order clauses are accumulated during iterations of Leo-II's main loop, and are periodically sent to the external prover with which Leo-II is cooperating. If the external prover finds the first-order clauses to be inconsistent then, assuming that the translation was sound, it means that the original HOL clauses must also be inconsistent. This refutation is accepted by Leo-II, and presented to the user as a refutation of the initial conjecture.

Various translations from HOL to first-order logic are implemented in Leo-II [38]. These translations differ in the amount of information they encode in the resulting FO formulas. Encoding less information can lead to incompleteness. Leo-II also implements a method devised in [55], which describes an analysis on the cardinalities of types in order to safely erase some information. As part of this analysis, SAT problems are generated, and these are processed by MiniSat via an interface adapted from the Satallax prover [44; 47].

4.2 Leo-II's Proof Certificates

Running Leo-II on a problem can have several outcomes: the conjecture could be found to be a theorem, or found to be a non-theorem, or the prover could give up (because of a timeout, for instance). Leo-II conforms to the SZS standard ontology [97] for communicating the outcome of a proof attempt. This makes it easier for external tools to interpret this outcome.

In addition to this, Leo-II can also output a proof certificate. This details the justification for the outcome given by Leo-II, by providing the reasoning steps used by Leo-II to derive a refutation. This could then be used by an independent system to check Leo-II's reasoning, or to use that derivation in a bigger formalisation.

Leo-II can generate proof certificates in two levels of detail. When called with the option -po 1, Leo-II produces a proof containing the reasoning steps made by Leo-II alone — information on the reasoning made by the cooperating FO ATP are omitted. When called with option -po 2, Leo-II tries to merge the proof steps of the cooperating FO ATP with its own steps in order to return a joint THF-FOF proof object [96]. The '-po 2' mode is unfortunately still very brittle and therefore not yet recommended for extensive use.

Leo-II's proof certificates are encoded in the TPTP TSTP syntax [98; 99], in which each inference is encoded as an annotated formula. The inference's conclusion appears as the formula (e.g., in THF0 or FOF syntax), and the inference's hypotheses and other meta-data are referenced resp. encoded in the formula's annotations. Examples of proofs of both levels of detail are provided on the Leo-II website, at http://page.mi.fu-berlin.de/cbenzmueller/leo/download.html.

78

5 Tools and Provers

5.1 The TPTP THF Initiative

To foster the systematic development and improvement of higher-order auto-mated theorem proving systems the TPTP THF infrastructure [100] has been initiated (THF stands for *typed higher-order form*). This project, which was supported by several other members of the community, has introduced the THF syntax for higher-order logic, it has developed a library of benchmark and example problems, and it provides various support tools for the new THF0 language fragment. The THF0 language supports HOL with choice.

Version 6.0.0 of the TPTP library contains more than 3000 problems in the THF0 language. The library also includes the entire problem library of Andrews's TPS project, which, among others, contains formalizations of many theorems of his textbook [6]. The first-order TPTP infrastructure [99] provides a range of resources to support usage of the TPTP problem library. Many of these resources are now immediately applicable to the higher-order setting although some have required changes to support the new features of THF. The development of the THF0 language has been paralleled and significantly influenced by the development of the LEO-II prover. Several other provers have quickly adopted this language, leading to fruitful mutual comparisons and evaluations. Several implementation bugs in different systems have been detected this way.

5.2 Automated Theorem Provers for HOL

We briefly describe the currently available, fully automated theorem provers for HOL (with choice). These systems all support the new THF0 language and they can be employed online (avoiding local installations) via Sutcliffe's SystemOnTPTP facility [98; 99].[5] The descriptions below have been adapted from [24].

TPS The TPS prover can be used to prove theorems of ETT or HOL automatically, interactively, or semi-automatically. When searching for a proof automatically, TPS first searches for an expansion proof [76] or an extensional expansion proof [45] of the theorem. Part of this process involves searching for acceptable matings [3]. Using higher-order unification, a pair of occurrences of subformulas (which are usually literals) is mated appropriately on each vertical path through an expanded form of the theorem to be proved. Skolemization and pre-unification is employed, and calculus rules for extensionality reasoning are provided. The behavior of TPS is controlled by sets of flags, also called modes. About fifty modes have been found that collectively suffice for automatically proving virtually all the theorems that TPS has proved automatically thus far. A simple scheduling mechanism is employed in TPS to sequentially run these modes for a limited amount of time. The resulting fully automated system is called *TPS (TPTP)*.

---

[5] See also http://www.tptp.org/cgi-bin/SystemOnTPTP.

LEO-*II* [39; 38], the successor of LEO-I [23], has been described in more detail in §4. Communication between LEO-II and the cooperating first-order theorem prover uses the TPTP language and standards. LEO-II outputs proofs in TPTP TSTP syntax. An incremental communication between LEO-II and the first-order prover(s) would clearly make sense, and this has been on the list of planned improvements for quite some time. However, such a solution has not yet been implemented. The latest versions of LEO-II provide some modest (counter-)model finding capabilities.

Isabelle/HOL The Isabelle/HOL [82] system has originally been designed as an interactive prover. However, in order to ease user interaction several automatic proof tactics have been added over the years. By appropriately scheduling a subset of these proof tactics, some of which are quite powerful, Isabelle/HOL has in recent years been turned also into an automatic theorem prover, that can be run from a command shell like other provers. The latest releases of this automated version of Isabelle/HOL provide native support for different TPTP syntax formats, including THF0. The most powerful proof tactics that are scheduled by Isabelle/HOL include the *sledgehammer* tool [41], which invokes a sequence of external first-order and higher-order theorem provers, the model finder *Nitpick* [42], the equational reasoner *simp* [81], the untyped tableau prover *blast* [87], the simplifier and classical reasoners *auto*, *force*, and *fast* [86], and the best-first search procedure *best*. The TPTP incarnation of Isabelle/HOL does not yet output proof terms.

Satallax The higher-order automated theorem prover Satallax [44; 47] comes with model finding capabilities. The system is based on a complete ground tableau calculus for HOL (with choice) [8]. An initial tableau branch is formed from the assumptions of a conjecture and negation of its conclusion. From that point on, Satallax tries to determine unsatisfiability or satisfiability of this branch. Satallax progressively generates higher-order formulas and corresponding propositional clauses. Satallax uses the SAT solver MiniSat as an engine to test the current set of propositional clauses for unsatisfiability. If the clauses are unsatisfiable, the original branch is unsatisfiable. Satallax employs restricted instantiation and pre-unification, and it provides calculus rules for extensionality and choice. If there are no quantifiers at function types, the generation of higher-order formulas and corresponding clauses may terminate. In that case, if MiniSat reports the final set of clauses as satisfiable, then the original set of higher-order formulas is satisfiable (by a standard model in which all types are interpreted as finite sets). Satallax outputs proofs in different formats, including Coq proof scripts and Coq proof terms.

Nitpick and Refute These systems are (counter-)model finders for HOL. The ability of Isabelle to find (counter-)models using the *Refute* and *Nitpick* [42] commands has also been integrated into automatic systems. They provide the capability to find models for THF0 formulas, which confirm the satisfiability of axiom sets, or the counter-satisfiability of non-theorems. The generation of models is particularly useful for exposing errors in some THF0 problem encodings, and revealing bugs in the THF0 theorem provers. Nitpick employs Skolemization.

agsyHOL The agsyHOL prover [74] is based on a generic lazy narrowing proof search algorithm. Backtracking is employed and a comparably small search state is maintained. The prover outputs proof terms in sequent style which can be verified in the Agda system.

coqATP The coqATP prover [40] implements (the non-inductive) part of the calculus of constructions. The system outputs proof terms which are accepted as proofs by Coq (after the addition of a few definitions). The prover has axioms for functional extensionality, choice, and excluded middle. Propositional extensionality is not supported yet. In addition to axioms, a small library of basic lemmas is employed.

Satallax-MaLeS and LEO-*II-MaLeS* MaLeS is an automatic tuning framework for automatic theorem provers. It combines random-hill climbing based strategy finding with strategy scheduling via learned runtime predictions. The MaLeS system has been successfully combined with Satallax and with LEO-II [72].

6   Proof Applications

With respect to full proof automation the TPS system has long been the leading system, and the system has been employed to build up the TPS library of formalized and automated mathematical proofs. More recently, however, TPS is outperformed by several other THF0 theorem provers. Below we briefly point to some selected recent applications of the leading systems.

Both Isabelle/HOL and Nitpick have been successfully employed to check a formalization of a C++ memory model against various concurrent programs written in C++ (such as a simple locking algorithm) [43]. Moreover, Nitpick has been employed in the development of algebraic formal methods within Isabelle/HOL [64].

Isabelle/HOL, Satallax, and LEO-II performed well in recent experiments related to the Flyspeck project [65], in which a formalized proof of the Kepler conjecture has been developed (mainly) in HOL Light; cf. the experiments reported in [70], which inter alia investigated the potential of several ATPs for automating subgoals in the Flyspeck corpus. In these experiments the higher-order automated teorem provers performed better than many prominent first-order theorem provers, and they contributed many unique solutions.

Most recently, LEO-II, Satallax, and Nitpick were employed to achieve a formalization, mechanization, and automation of Gödel's ontological proof of the existence of God [32; 31; 27; 28; 30; 29; 26]. This work employs a semantic embedding of quantified modal logic in HOL [35], similar to the embedding presented in §2. LEO-II was the first prover to fully automate the four key steps of

Dana Scott's version of the proof [94]. These results were subsequently confirmed by Satallax, and consistency of the axioms was shown with Nitpick. Moreover, some previously unknown results were contributed by the provers.

Using the semantic embeddings approach, a wide range of propositional and quantified non-classical logics, including parts of their meta-theory and their combinations, can be automated with THF0 reasoners [14]. Automation is thereby competitive, as recent experiments for first-order modal logics show [37; 25]. In [13] HOL theorem provers have been successfully employed to reason about meta-level properties of modal logics. More precisely, the provers have been employed to verify the modal logic cube.

THF0 reasoners can also be fruitfully employed for reasoning in expressive ontologies [36]. Furthermore, the heterogeneous toolset HETS [79] employs THF0 to integrate the automated higher-order provers Satallax, LEO-II, Nitpick, Refute, and Isabelle/HOL.

7   Trends and Open Problems

The development of automated theorem provers for HOL has significantly progressed recently. At the same time the systems still lag far behind the theoretical and technical maturity that has been achieved in first-order automated theorem proving. Existing HOL provers have nevertheless already demonstrated their competitiveness in various applications.

Further challenges in this field include the development and systematic improvement of the theoretical foundations, in particular, of the underlying calculi, and better proof search organization.

The challenges also include the automation of reasoning with polymorphism, subtypes, type classes, and possibly even dependent types as supported in modern interactive proof assistants. The rationale here is that automated theorem provers for HOL should ideally be applicable directly to the logics as provided and employed in these proof assistants. It will be relevant to improve and adapt the TPTP THF languages accordingly.

Moreover, challenges include the suitable automation of any of the remaining, prominent cut-strong principles of HOL, such as induction and description.

There are also significant challenges regarding proof representation and user interaction. Experience shows that HOL proofs, for example, as produced by LEO-II or Satallax, can be quite hard to read and understand by humans. To foster the use of our systems in contexts where human understanding and the generation of human level answers is relevant, significant further improvements are thus required.

Acknowledgements This text is based on several earlier publications, as referenced throughout the paper, with various co-authors, including Chad Brown, Valerio Genovese, Michael Kohlhase, Dale Miller, Larry Paulson, Nik Sultana, Frank Theiss. I am very grateful to all of them. I am also grateful to the anonymous reviewers and Bruno Woltzenlogel Paleo for many useful comments and suggestions, and for proof reading this document. This work has been supported by the German National Research Foundation (DFG) under grants BE 2501/9-1 and BE 2501/11-1.

Bibliography

[1] Peter B. Andrews. Resolution in type theory. *Journal of Symbolic Logic*, 36(3):414–432, 1971.

[2] Peter B. Andrews. General models and extensionality. *Journal of Symbolic Logic*, 37(2):395–397, 1972.

[3] Peter B. Andrews. Theorem proving via general matings. *Journal of the ACM*, 28(2):193–214, 1981.

[4] Peter B. Andrews. On connections and higher order logic. *Journal of Automated Reasoning*, 5(3):257–291, 1989.

[5] Peter B. Andrews. Classical type theory. In Alan Robinson and Andrei Voronkov, editors, *Handbook of Automated Reasoning*, volume 2, chapter 15, pages 965–1007. Elsevier Science, Amsterdam, 2001.

[6] Peter B. Andrews. *An Introduction to Mathematical Logic and Type Theory: To Truth Through Proof*. Kluwer Academic Publishers, second edition, 2002.

[7] Peter B. Andrews. Church's type theory. In Edward N. Zalta, editor, *The Stanford Encyclopedia of Philosophy*. Stanford University, spring 2009 edition, 2009.

[8] Julian Backes and Chad Edward Brown. Analytic tableaux for higher-order logic with choice. *Journal of Automated Reasoning*, 47(4):451–479, 2011.

[9] Christoph Benzmüller. *Equality and Extensionality in Automated Higher-Order Theorem Proving*. PhD thesis, Saarland University, 1999.

[10] Christoph Benzmüller. Extensional higher-order paramodulation and RUE-resolution. In Harald Ganzinger, editor, *Automated Deduction - CADE-16, 16th International Conference on Automated Deduction, Trento, Italy, July 7-10, 1999, Proceedings*, number 1632 in LNCS, pages 399–413. Springer, 1999.

[11] Christoph Benzmüller. Comparing approaches to resolution based higher-order theorem proving. *Synthese*, 133(1-2):203–235, 2002.

[12] Christoph Benzmüller. Automating access control logic in simple type theory with LEO-II. In Dimitris Gritzalis and Javier López, editors, *Emerging Challenges for Security, Privacy and Trust, 24th IFIP TC 11 International Information Security Conference, SEC 2009, Pafos, Cyprus, May 18-20, 2009. Proceedings*, volume 297 of *IFIP*, pages 387–398. Springer, 2009.

[13] Christoph Benzmüller. Verifying the modal logic cube is an easy task (for higher-order automated reasoners). In Simon Siegler and Nathan Wasser, editors, *Verification, Induction, Termination Analysis - Festschrift for Christoph Walther on the Occasion of His 60th Birthday*, volume 6463 of *LNCS*, pages 117–128. Springer, 2010.

[14] Christoph Benzmüller. Combining and automating classical and non-classical logics in classical higher-order logic. *Annals of Mathematics and*

Artificial Intelligence (Special issue Computational logics in Multi-agent Systems (CLIMA XI)), 62(1-2):103–128, 2011.

[15] Christoph Benzmüller. Automating quantified conditional logics in HOL. In Francesca Rossi, editor, *23rd International Joint Conference on Artificial Intelligence (IJCAI-13)*, Beijing, China, 2013.

[16] Christoph Benzmüller. A top-down approach to combining logics. In *Proc. of the 5th International Conference on Agents and Artificial Intelligence (ICAART)*, Barcelona, Spain, 2013. SciTePress Digital Library.

[17] Christoph Benzmüller and Chad Brown. A structured set of higher-order problems. In Joe Hurd and Thomas F. Melham, editors, *Theorem Proving in Higher Order Logics, 18th International Conference, TPHOLs 2005, Oxford, UK, August 22-25, 2005, Proceedings*, number 3603 in LNCS, pages 66–81. Springer, 2005.

[18] Christoph Benzmüller, Chad Brown, and Michael Kohlhase. Higher-order semantics and extensionality. *Journal of Symbolic Logic*, 69(4):1027–1088, 2004.

[19] Christoph Benzmüller, Chad Brown, and Michael Kohlhase. Cut elimination with xi-functionality. In Christoph Benzmüller, Chad Brown, Jörg Siekmann, and Richard Statman, editors, *Reasoning in Simple Type Theory — Festschrift in Honor of Peter B. Andrews on His 70th Birthday*, Studies in Logic, Mathematical Logic and Foundations, pages 84–100. College Publications, 2008.

[20] Christoph Benzmüller, Chad Brown, and Michael Kohlhase. Cut-simulation and impredicativity. *Logical Methods in Computer Science*, 5(1:6):1–21, 2009.

[21] Christoph Benzmüller, Dov Gabbay, Valerio Genovese, and Daniele Rispoli. Embedding and automating conditional logics in classical higher-order logic. *Annals of Mathematics and Artificial Intelligence*, 66(1-4):257–271, 2012.

[22] Christoph Benzmüller and Valerio Genovese. Quantified conditional logics are fragments of HOL. In *The International Conference on Non-classical Modal and Predicate Logics (NCMPL)*, Guangzhou (Canton), China, 2011. The conference had no published proceedings; the paper is available as arXiv:1204.5920v1.

[23] Christoph Benzmüller and Michael Kohlhase. LEO – a higher-order theorem prover. In Claude Kirchner and Hélène Kirchner, editors, *Automated Deduction - CADE-15, 15th International Conference on Automated Deduction, Lindau, Germany, July 5-10, 1998, Proceedings*, number 1421 in LNCS, pages 139–143. Springer, 1998.

[24] Christoph Benzmüller and Dale Miller. Automation of higher-order logic. In Jörg Siekmann, Dov Gabbay, and John Woods, editors, *Handbook of the History of Logic, Volume 9 — Logic and Computation*. Elsevier, 2014. In print.

[25] Christoph Benzmüller, Jens Otten, and Thomas Raths. Implementing and evaluating provers for first-order modal logics. In *Proc. of the 20th*

European Conference on Artificial Intelligence (ECAI 2012), pages 163–168, Montpellier, France, 2012.

[26] Christoph Benzmüller, Leon Weber, and Bruno Woltzenlogel Paleo. Computer-assisted analysis of the anderson-hjek ontological controversy. In *1st World Congress on Logic and Religion*, 2015.

[27] Christoph Benzmüller and Bruno Woltzenlogel Paleo. Gödel's god in isabelle/hol. *Archive of Formal Proofs*, 2013, 2013.

[28] Christoph Benzmüller and Bruno Woltzenlogel Paleo. Gdel's god on the computer. In *10th International Workshop on the Implementation of Logics*, 2013.

[29] Christoph Benzmüller and Bruno Woltzenlogel Paleo. Formalization and automated verification of gdel's proof of god's existence. In *4th World Congress on the Square of Opposition*, 2014.

[30] Christoph Benzmüller and Bruno Woltzenlogel Paleo. On logic embeddings and gdel's god. In *22nd International Workshop on Algebraic Development Techniques*, 2014.

[31] Christoph Benzmüller and Bruno Woltzenlogel Paleo. Formalization, mechanization and automation of Gödel's proof of God's existence. 2013. Preprint available as arXiv:1308.4526.

[32] Christoph Benzmüller and Bruno Woltzenlogel Paleo. Automating Gödel's ontological proof of god's existence with higher-order automated theorem prover. In *Proc. of the 21st European Conference on Artificial Intelligence (ECAI 2014)*, pages 93–98, 2014. 2014.

[33] Christoph Benzmüller and Lawrence Paulson. Exploring properties of normal multimodal logics in simple type theory with LEO-II. In Christoph Benzmüller, Chad Brown, Jörg Siekmann, and Richard Statman, editors, *Reasoning in Simple Type Theory — Festschrift in Honor of Peter B. Andrews on His 70th Birthday*, Studies in Logic, Mathematical Logic and Foundations, pages 386–406. College Publications, 2008.

[34] Christoph Benzmüller and Lawrence Paulson. Multimodal and intuitionistic logics in simple type theory. *The Logic Journal of the IGPL*, 18(6):881–892, 2010.

[35] Christoph Benzmüller and Lawrence Paulson. Quantified multimodal logics in simple type theory. *Logica Universalis (Special Issue on Multimodal Logics)*, 7, 2013.

[36] Christoph Benzmüller and Adam Pease. Higher-order aspects and context in SUMO. *Journal of Web Semantics (Special Issue on Reasoning with context in the Semantic Web)*, 12-13:104–117, 2012.

[37] Christoph Benzmüller and Thomas Raths. HOL based first-order modal logic provers. In Kenneth L. McMillan, Aart Middeldorp, and Andrei Voronkov, editors, *Proceedings of the 19th International Conference on Logic for Programming, Artificial Intelligence and Reasoning (LPAR)*, volume 8312 of *LNCS*, pages 127–136, Stellenbosch, South Africa, 2013. Springer.

[38] Christoph Benzmüller and Nik Sultana. LEO-II version 1.5. In Jasmin Christian Blanchette and Josef Urban, editors, *PxTP 2013*, volume 14 of *EPiC Series*, pages 2–10. EasyChair, 2013.

[39] Christoph Benzmüller, Frank Theiss, Lawrence Paulson, and Arnaud Fietzke. LEO-II - a cooperative automatic theorem prover for higher-order logic (system description). In Alessandro Armando, Peter Baumgartner, and Gilles Dowek, editors, *Automated Reasoning, 4th International Joint Conference, IJCAR 2008, Sydney, Australia, August 12-15, 2008, Proceedings*, volume 5195 of *LNCS*, pages 162–170. Springer, 2008.

[40] Yves Bertot and Pierre Casteran. *Interactive Theorem Proving and Program Development - Coq'Art: The Calculus of Inductive Constructions*. Texts in Theoretical Computer Science. Springer, 2004.

[41] Jasmin Christian Blanchette, Sascha Böhme, and Lawrence C. Paulson. Extending Sledgehammer with SMT solvers. *Journal of Automated Reasoning*, 51(1):109–128, 2013.

[42] Jasmin Christian Blanchette and Tobias Nipkow. Nitpick: A counterexample generator for higher-order logic based on a relational model finder. In Matt Kaufmann and Lawrence C. Paulson, editors, *Proc. of ITP 2010*, volume 6172 of *LNCS*, pages 131–146. Springer, 2010.

[43] Jasmin Christian Blanchette, Tjark Weber, Mark Batty, Scott Owens, and Susmit Sarkar. Nitpicking C++ concurrency. In Peter Schneider-Kamp and Michael Hanus, editors, *Proceedings of the 13th International ACM SIGPLAN Conference on Principles and Practice of Declarative Programming, July 20-22, 2011, Odense, Denmark*, pages 113–124. ACM, 2011.

[44] Chad E. Brown. Satallax: an automatic higher-order prover. *Journal of Automated Reasoning*, pages 111–117, 2012.

[45] Chad E. Brown. *Set Comprehension in Church's Type Theory*. PhD thesis, Department of Mathematical Sciences, Carnegie Mellon University, 2004. See also Chad E. Brown, *Automated Reasoning in Higher-Order Logic*, College Publications, 2007.

[46] Chad E. Brown. Reasoning in extensional type theory with equality. In Robert Nieuwenhuis, editor, *Proc. of CADE-20*, volume 3632 of *LNCS*, pages 23–37. Springer, 2005.

[47] Chad E. Brown. Reducing higher-order theorem proving to a sequence of sat problems. *Journal of Automated Reasoning*, 51(1):57–77, 2013.

[48] Chad E. Brown and Gert Smolka. Analytic tableaux for simple type theory and its first-order fragment. *Logical Methods in Computer Science*, 6(2), 2010.

[49] Brian F. Chellas. Basic conditional logic. *Journal of Philosophical Logic*, 4(2):133–153, 1975.

[50] Brian F. Chellas. *Modal Logic: An Introduction*. Cambridge: Cambridge University Press, 1980.

[51] Alonzo Church. A set of postulates for the foundation of logic. *Annals of Mathematics*, 33(2):346–366, 1932.

[52] Alonzo Church. An unsolvable problem of elementary number theory. *American Journal of Mathematics*, 58:354–363, 1936.

[53] Alonzo Church. A formulation of the simple theory of types. *Journal of Symbolic Logic*, 5:56–68, 1940.

[54] Leon Chwistek. *The Limits of Science: Outline of Logic and of the Methodology of the Exact Sciences*. London: Routledge and Kegan Paul, 1948.

[55] Koen Claessen, Ann Lillieström, and Nicholas Smallbone. Sort it out with monotonicity. In *Proceedings of CADE-23*, volume 6803 of *LNAI*, pages 207–221. Springer, 2011.

[56] James P. Delgrande. On first-order conditional logics. *Artificial Intelligence*, 105(1-2):105–137, 1998.

[57] Vincent J. Digricoli. Resolution by unification and equality. In William H. Joyner, editor, *4th Workshop on Automated Deduction*, Austin, Texas, 1979.

[58] Herbert B. Enderton. Second-order and higher-order logic. In Edward N. Zalta, editor, *The Stanford Encyclopedia of Philosophy*. Stanford University, fall 2012 edition, 2012.

[59] William M. Farmer. The seven virtues of simple type theory. *Journal of Applied Logic*, 6(3):267–286, 2008.

[60] Melvin Fitting. Interpolation for first order S5. *Journal of Symbolic Logic*, 67(2):621–634, 2002.

[61] Gottlob Frege. *Begriffsschrift, eine der arithmetischen nachgebildete Formelsprache des reinen Denkens*. Halle, 1879. Translated in [106].

[62] Nir Friedman, Joseph Y. Halpern, and Daphne Koller. First-order conditional logic for default reasoning revisited. *ACM Transactions on Computational Logic*, 1(2):175–207, 2000.

[63] Jean-Yves Girard. Une extension de l'interpretation de Gödel à l'analyse, et son application à l'élimination des coupures dans l'analyse et la théorie des types. In J. E. Fenstad, editor, *2nd Scandinavian Logic Symposium*, pages 63–92. North-Holland, Amsterdam, 1971.

[64] Walter Guttmann, Georg Struth, and Tjark Weber. Automating algebraic methods in isabelle. In Shengchao Qin and Zongyan Qiu, editors, *Proc. of ICFEM 2011*, volume 6991 of *LNCS*, pages 617–632. Springer, 2011.

[65] Thomas Hales. Mathematics in the Age of the Turing Machine. *arXiv:1302.2898*, February 2013.

[66] Leon Henkin. A theory of propositional types. *Fundamatae Mathematicae*, 52:323–344, 1963.

[67] Gérard Huet. A Unification Algorithm for Typed Lambda-Calculus. *Theoretical Computer Science*, 1(1):27–57, 1975.

[68] Gérard P. Huet. *Constrained Resolution: A Complete Method for Higher Order Logic*. PhD thesis, Case Western Reserve University, 1972.

[69] Gérard P. Huet. A mechanization of type theory. In *Proceedings of the 3rd International Joint Conference on Artificial Intelligence*, pages 139–146, 1973.

[70] Cezary Kaliszyk and Josef Urban. Learning-assisted automated reasoning with flyspeck. *Journal of Automated Reasoning*, 53:173–213, 2014.

[71] Michael Kohlhase. *A Mechanization of Sorted Higher-Order Logic Based on the Resolution Principle*. PhD thesis, Saarland University, 1994.

[72] Daniel Külwein. *Machine Learning for Automated Reasoning*. PhD thesis, Radboud University Nijmegen, Netherlands, 2014.

[73] Daniel Leivant. Higher-order logic. In Dov M. Gabbay, C. J. Hogger, and J. A. Robinson, editors, *Handbook of Logic in Artificial Intelligence and Logic Programming*, volume 2, pages 229–321. Oxford University Press, 1994.

[74] Fredrik Lindblad. agsyHOL website. https://github.com/frelindb/agsyHOL, 2013.

[75] Raymond McDowell and Dale Miller. Reasoning with higher-order abstract syntax in a logical framework. *ACM Transactions on Computational Logic*, 3(1):80–136, 2002.

[76] Dale Miller. A compact representation of proofs. *Studia Logica*, 46(4):347–370, 1987.

[77] Dale A. Miller. *Proofs in Higher-Order Logic*. PhD thesis, Carnegie Mellon University, 1983. 81 pp.

[78] Grigori Mints. Cut-elimination for simple type theory with an axiom of choice. *Journal of Symbolic Logic*, 64(2):479–485, 1999.

[79] Till Mossakowski, Christian Maeder, and Klaus Lüttich. The heterogeneous tool set, Hets. In *Proceedings of TACAS 2007*, volume 4424 of *LNCS*, pages 519–522. Springer, 2007.

[80] Reinhard Muskens. Intensional models for the theory of types. *Journal of Symbolic Logic*, 72(1):98–118, 2007.

[81] Tobias Nipkow. Equational reasoning in Isabelle. *Sci. Comput. Program.*, 12(2):123–149, 1989.

[82] Tobias Nipkow, Markus Wenzel, and Lawrence C. Paulson. *Isabelle/HOL: A Proof Assistant for Higher-Order Logic*. Springer, 2002.

[83] D. Nute. *Topics in conditional logic*. Reidel, Dordrecht, 1980.

[84] Hans-Jürgen Ohlbach. Semantics Based Translation Methods for Modal Logics. *Journal of Logic and Computation*, 1(5):691–746, 1991.

[85] N. Olivetti, G.L. Pozzato, and C. Schwind. A sequent calculus and a theorem prover for standard conditional logics. *ACM Transactions on Computational Logic*, 8(4), 2007.

[86] Lawrence C. Paulson. *Isabelle - A Generic Theorem Prover (with a contribution by T. Nipkow)*, volume 828 of *LNCS*. Springer, 1994.

[87] Lawrence C. Paulson. A generic tableau prover and its integration with isabelle. *Journal of Universal Computer Science*, 5:51–60, 1999.

[88] Dag Prawitz. Hauptsatz for higher order logic. *Journal of Symbolic Logic*, 33:452–457, 1968.

[89] Frank P. Ramsey. The foundations of mathematics. In *Proceedings of the London Mathematical Society*, volume 25 of *2*, pages 338–384, 1926.

[90] Bertrand Russell. Mathematical logic as based on the theory of types. *American Journal of Mathematics*, 30:222–262, 1908.

[91] Kurt Schütte. Semantical and syntactical properties of simple type theory. *Journal of Symbolic Logic*, 25(4):305–326, 1960.

[92] Raymond M. Smullyan. A unifying principle for quantification theory. *Proc. Nat. Acad Sciences*, 49:828–832, 1963.

[93] Wayne Snyder and Jean H. Gallier. Higher order unification revisited: Complete sets of transformations. *Journal of Symbolic Computation*, 8(1-2):101–140, 1989.

89

[94] John H. Sobel. *Logic and Theism: Arguments for and Against Beliefs in God*, chapter Appendix B. Notes in Dana Scott's Hand, pages 145–146. Cambridge U. Press, 2004.

[95] Robert C. Stalnaker. A theory of conditionals. In *Studies in Logical Theory*, pages 98–112. Blackwell, 1968.

[96] Nik Sultana and Christoph Benzmüller. Understanding LEO-II's Proofs. In Konstantin Korovin, Stephan Schulz, and Eugenia Ternovska, editors, *IWIL 2012*, Easychair EPiC Series, 22:33-52, 2013.

[97] Geoff Sutcliffe. The SZS ontologies for automated reasoning software. In: The LPAR 2008 Workshops: KEAPPA and IWIL, CEUR Workshop Proceedings (http://ceur-ws.org/), vol. 418, 2008.

[98] Geoff Sutcliffe. The TPTP World - Infrastructure for Automated Reasoning. In Edmund M. Clarke and Andrei Voronkov, editors, *LPAR 2010*, volume 6355 of *LNCS*, pages 1–12. Springer, 2010.

[99] Geoff Sutcliffe. The TPTP problem library and associated infrastructure. *Journal of Automated Reasoning*, 43(4):337–362, 2009.

[100] Geoff Sutcliffe and Christoph Benzmüller. Automated reasoning in higher-order logic using the TPTP THF infrastructure. *Journal of Formalized Reasoning*, 3(1):1–27, 2010.

[101] William W. Tait. A nonconstructive proof of Gentzen's Hauptsatz for second order predicate logic. *Bulletin of the American Mathematical Society*, 72:980983, 1966.

[102] Moto-o Takahashi. A proof of cut-elimination theorem in simple type theory. *Journal of the Mathematical Society of Japan*, 19:399–410, 1967.

[103] Gaisi Takeuti. On a generalized logic calculus. *Japanese Journal of Mathematics*, 23:39–96, 1953. Errata: ibid, vol. 24 (1954), 149–156.

[104] Gaisi Takeuti. An example on the fundamental conjecture of GLC. *Journal of the Mathematical Society of Japan*, 12:238–242, 1960.

[105] Gaisi Takeuti. *Proof Theory*, volume 81 of *Studies in Logic and the Foundations of Mathematics*. Elsevier, 1975.

[106] Jean van Heijenoort. *From Frege to Gödel: A Source Book in Mathematics, 1879-1931*. Source books in the history of the sciences series. Harvard Univ. Press, Cambridge, MA, 3rd printing, 1997 edition, 1967.

[107] Max Wisniewski and Alexander Steen. Embedding of quantified higher-order nominal modal logic into classical higher-order logic. In *Proc. of the 1st International Workshop on Automated Reasoning in Quantified Non-Classical Logics (ARQNL)*, Vienna, 2014.

[108] Bruno Woltzenlogel Paleo. An Embedding of Neighbourhood-Based Modal Logics in HOL. Available at `https://github.com/Paradoxika/ModalLogic`.

Interactive Theorem Proving
from the perspective of Isabelle/Isar

Makarius Wenzel

http://sketis.net

Abstract. Interactive Theorem Proving (ITP) has a long tradition, going back to the 1970s when interaction was introduced as a concept in computing. The main provers in use today can be traced back over 20–30 years of development. As common traits there are usually strong logical systems at the bottom, with many layers of add-on tools around the logical core, and big applications of formalized mathematics or formal methods. There is a general attitude towards flexibility and open-endedness in the combination of logical tools: typical interactive provers use automated provers and disprovers routinely in their portfolio.

The subsequent exposition of ITP takes Isabelle/Isar as the focal point to explain concepts of the greater "LCF family", which includes Coq and various HOL systems. Isabelle itself shares much of the relatively simple logical foundations of HOL, but follows Coq in the ambition to deliver a sophisticated system to end-users, without requiring self-assembly of individual parts. Isabelle today is probably the most advanced proof assistant concerning its architecture and extra-logical infrastructure.

The Isar aspect of Isabelle refers first to the structured language for human-readable and machine-checkable proof documents, but also to the Isabelle architecture that emerged around the logical framework in the past 10 years. Thus Isabelle/Isar provides extra structural integrity beyond the core logical framework, with native support for parallel proof processing and asynchronous interaction in its Prover IDE (PIDE).

1 Introduction

1.1 Overview of ITP systems

Isabelle[1] is one of a handful of *Interactive Theorem Provers* or *Proof Assistants* that have reached a sufficient level of sophistication over some decades to support substantial applications. Today we see large formalization efforts like L4.verified [24], and the *Archive of Formal Proofs*[2] with entries for Isabelle/HOL.

Other notable systems in this category are Coq [48, §4], the HOL family [48, §1], PVS [48, §3], ACL2 [48, §8]. The volume [48] is generally useful as a reference to various interactive (and non-interactive) provers, although it reflects the state-of-the-art from 10 years ago.

---

[1] http://isabelle.in.tum.de
[2] http://afp.sf.net

91

Due to lack of expertise on PVS and ACL2, I need to restrict the perspective of ITP to the main European systems in the LCF tradition, namely the HOL family, Coq, and Isabelle. All of them share much of the heritage of LCF, such as strongly-typed foundations and programming in ML. In contrast, the US American prover tradition is mainly influenced by LISP, concerning both the syntax and type-system of the formal language.

Coq is probably the most popular and publicly visible ITP system at the moment. Its implementation language is OCaml, but user input works via a separate "vernacular" (Gallina) for specifications and a tactic language (Ltac) for proofs. Since Coq is based on constructive logic with a built-in notion of computation, it is customary to build tools for Coq within the logical language itself, which are then executed by the built-in normalization of proof terms. This trend to "internalize" all aspects of formal reasoning into one big language for logic and programming is typical for Coq.

1.2 Proof production in LCF-style provers

The HOL family, Coq, and Isabelle are all descendants of LCF, which was developed in Edinburgh and Cambridge in the 1980s by R. Milner. The LCF line was continued by M. Gordon (later HOL), L. C. Paulson (later Isabelle), and G. Huet (later Coq). Paulson and Huet were actually collaborating on the last version of Cambridge LCF, before the split into Isabelle and Coq happened. The core algorithm in Isabelle86 for higher-order unification is actually due to Huet.

Milner can be credited as the inventor of the key ideas for interactive theorem proving and the ML programming language, with the now famous Hindley-Milner type-inference [10]: it makes a statically-typed language almost as convenient as an untyped one. The so-called "LCF-approach" to interactive theorem proving means the following:

1. **Strong logical foundations.** Some well-understood logical basis is taken as starting point, and mathematical theories are explicitly constructed by reduction to first principles. This follows the tradition of "honest toil" in the sense of B. Russel: results are not just postulated as axioms, but derived from definitions as proper theorems.
2. **Free programmability and extensibility.** Derived proof tools can be implemented on top of the logical core, while retaining its integrity. This works thanks to the strong type-safety properties of the ML implementation platform of the prover.

1.3 Isabelle/HOL specifications

Originally, Isabelle was introduced as yet another *Logical Framework* [32, 33] when that idea was popular: this was motivated by frequently changing versions of Martin-Löf Type-Theory that Paulson wanted to support on the machine.

Later that flexibility of the Isabelle/Pure framework was used for Isabelle/ZF, which is a version of classical Zermelo-Fraenkel set-theory on top of first-order logic [34, 35].

In the past 15 years, Isabelle/HOL [31] has become the largest application of the Isabelle framework, with numerous add-on tools: advanced proof search, support for specific theory reasoning, integration with external ATPs and SMTs, counter-example generation etc. Moreover, any advanced specification mechanism like **datatype, inductive, fun** in Isabelle/HOL is implemented in the hard way, with proper definitions and proofs produced in ML. For example, some recursive function specification given by the user is turned by the system into suitable low-level definitions, and the expected characterization is derived as theorems under program control.

Conceptually, we have an alternation of theory definitions and ML modules that gradually build up a library of formalized mathematics. There is no theoretical limit to that, but in practice the library is eventually delivered to end-users as theory *Main* of Isabelle/HOL, and further libraries and applications require very little Isabelle/ML programming.

The subsequent examples give some taste of Isabelle/HOL for end-users. The theory snippet below defines an algebraic datatype of finite sequences, with a function for sequence concatenation, and proves some elementary facts.

datatype $'a\ seq = Empty \mid Seq\ 'a\ ('a\ seq)$

fun $concat :: 'a\ seq \Rightarrow 'a\ seq \Rightarrow 'a\ seq$
where
  $concat\ Empty\ ys = ys$
$\mid concat\ (Seq\ x\ xs)\ ys = Seq\ x\ (concat\ xs\ ys)$

theorem $concat$-$empty$: $concat\ xs\ Empty = xs$
  by $(induct\ xs)\ simp$-all

theorem $conc$-$assoc$: $concat\ (concat\ xs\ ys)\ zs = concat\ xs\ (concat\ ys\ zs)$
  by $(induct\ xs)\ simp$-all

On first sight that might look like a functional programming language in the style of ML or Haskell, with extensions for logic and proof. This is in fact the manner how Isabelle/HOL is introduced in the tutorial [30]. As a second approximation, it could be seen as theory axiomatization, as in algebraic specification or preludes of first-order automated theorem proving. In reality, the above is merely some surface syntax for a *definitional theory* in a double-sense:

1. The user writes specifications that define new formal entities based on existing formal entities, according to the classic mathematical approach of *definition–statement–proof.*

2. Complex definitional principles are internally built on simpler ones: the language elements **datatype** and **fun** above are defined in the library of Isabelle/HOL. Some Isabelle/ML implementation reduces user specifications

into primitive definitions, with full proofs for derived rules (e.g. freeness and induction for datatypes).

This strongly definitional approach is shared by all members of the LCF family, with the exception of LCF itself, which was still based on axiomatic specifications of types and constants, as a prelude to proper proofs. The introduction of Higher-Order Logic as new foundation for HOL [16, 15] made it feasible to apply the rigorous principles of "honest toil" (B. Russel) to mechanized reasoning on the machine. Note that Coq has its own variations on the same theme: its Calculus of Inductive Constructions starts as a relatively strong axiomatic basis for user theories that are usually definitional, except for situations where well-known axioms are added in applications that want to escape the constructive defaults of Coq.

1.4   Isabelle/Isar proofs

The example proofs above are terminal steps of the form **by** *initial-method terminal-method*, where the second argument is optional. That is the shortest possible structured Isar proof: a double-step to split the problem initially (e.g. by induction) and solve the remaining situation (e.g. by simplification). The Isar **by** command ensures the structural integrity of this reasoning scheme: there is no way to escape from the nested proof structure, which is explicit in the Isabelle/Isar proof state. Likewise, Coq has recently introduced "bullets" inspired by the Coq/SSReflect [14] extensions to impose more structure on tactic scripts, while retaining unstructured proof states.

Isabelle/Isar is sufficiently flexible to imitate unstructured proof scripts within its structured proof language, by so-called "improper language elements", notably **apply** and **done**:

> **theorem** *concat-empty′*: *concat xs Empty = xs*
>   **apply** (*induct xs*)
>   **apply** *simp*
>   **apply** *simp*
>   **done**
>
> **theorem** *conc-assoc′*: *concat (concat xs ys) zs = concat xs (concat ys zs)*
>   **apply** (*induct xs*)
>   **apply** *simp*
>   **apply** *simp*
>   **done**

Such liberality in the Isar proof language, outside its main scope for human-readable structured proofs, allows to import old Isabelle tactic scripts easily or to port applications from other tactical proof assistants. The price for that is some occasional confusion of users, who might mistake Isar proof methods like *simp* as "tactics", and the Isar proof text as some kind of "proof script". The primary input of Isabelle/Isar should be understood as a format for *proof documents*. In fact there is built-in support to render the result nicely in LaTeX, see also [40, chapter 4]. The present article is an example for that: it is a theory document in the context of *Main* from Isabelle/HOL, with a lot of unproven prose text and a few formally proven examples.

The subsequent example shows more structure, both in the specification context (via **class**) and the proofs (via calculational reasoning elements **also** and **finally**):

class *group* = *times* + *one* + *inverse* +
  assumes *group-assoc*: $(x * y) * z = x * (y * z)$
    and *group-left-one*: $1 * x = x$
    and *group-left-inverse*: *inverse* $x * x = 1$

theorem (**in** *group*) *group-right-inverse*: $x * inverse\ x = 1$
proof −
  have $x * inverse\ x = 1 * (x * inverse\ x)$
    by (*simp only*: *group-left-one*)
  also have $\ldots = 1 * x * inverse\ x$
    by (*simp only*: *group-assoc*)
  also have $\ldots = inverse\ (inverse\ x) * inverse\ x * x * inverse\ x$
    by (*simp only*: *group-left-inverse*)
  also have $\ldots = inverse\ (inverse\ x) * (inverse\ x * x) * inverse\ x$
    by (*simp only*: *group-assoc*)
  also have $\ldots = inverse\ (inverse\ x) * 1 * inverse\ x$
    by (*simp only*: *group-left-inverse*)
  also have $\ldots = inverse\ (inverse\ x) * (1 * inverse\ x)$
    by (*simp only*: *group-assoc*)
  also have $\ldots = inverse\ (inverse\ x) * inverse\ x$
    by (*simp only*: *group-left-one*)
  also have $\ldots = 1$
    by (*simp only*: *group-left-inverse*)
  finally show *?thesis* .
qed

theorem (**in** *group*) *group-right-one*: $x * 1 = x$
proof −
  have $x * 1 = x * (inverse\ x * x)$
    by (*simp only*: *group-left-inverse*)
  also have $\ldots = x * inverse\ x * x$
    by (*simp only*: *group-assoc*)
  also have $\ldots = 1 * x$
    by (*simp only*: *group-right-inverse*)
  also have $\ldots = x$
    by (*simp only*: *group-left-one*)
  finally show *?thesis* .
qed

The emphasis on structured proof texts — beyond mere "proof scripts" — goes back to the beginnings of Isar in 1999 [41] when *human-readable proof docu-*

ments were the main motivation. Rich structure helps to present small examples and big applications nicely: the user is taken seriously as proof author, although there is an extra effort to produce readable formal proofs in the first place. That investment is usually returned when big proof libraries need to be maintained or extended in the scope of application: adapting theories to new situations works better for well-structured proof texts than for unstructured proof scripts.

1.5 Parallel processing of structured proofs

Structured proofs also help the machine to process large formalizations efficiently. This is in contrast to the occasional misunderstanding of "high-level proofs" as "automated proofs". An important lesson learned from Isar is that human-readable structured proofs require only modest proof automation, namely the existing higher-order unification of Isabelle/Pure. Nonetheless it is possible to adjoin arbitrary proof tools as terminal proof steps.

Note that the mathematical proof language of Mizar [48, §2] follows a similar principle for its main proof structure: only a quite weak notion of "obvious inferences" performed by the machine. Mizar even lacks the possibilities to appeal to add-on tools proof, and still manages to support a substantial library of formalized mathematics[3] in a quasi-textbook style.

The key performance advantage of structured proofs is due to parallel proof processing on multi-core hardware, which is now ubiquitous. Initially, the strict modularity of Isar proofs was merely motivated by aesthetic principles. The advent of multicore hardware in the mass market introduced the demand to make high-performance applications work in parallel. This required some reforms of existing Isabelle/Isar document processing, to gain significant speedup of batch-processing of large formalizations [42, 29, 45]. For example, re-checking the *Archive of Formal Proofs* used to be an overnight job, but is today finished in ≈ 1h on 16 cores, with a real-world speedup factor of 10 or more.

Apart from parallel batch processing, interaction had to be reconsidered as well, to fit into the non-sequential paradigm. This is addressed in Isabelle/PIDE, by a timeless and stateless *document model* that integrates all aspects of proof editing and add-on tools within a uniform framework [43, 44, 46]. These advanced concepts of interactive theorem proving are specific to Isabelle: the HOL family still uses plain TTY sessions with copy-paste, while Coq adheres to the traditional model of Proof General [1], either directly in that Emacs mode or in its OCaml/Gtk clone CoqIDE.

1.6 HOL implementations for theoretical studies

HOL-Light[4] by J. Harrison is best-known for its very small code base, which greatly helps to give some impression how a classic LCF-style prover is implemented, with minimal system infrastructure. It is possible to browse through

---

[3] http://www.mizar.org/library
[4] https://code.google.com/p/hol-light

HOL-Light sources within a few hours, and learn many things from it. One needs to keep in mind, though, that HOL-Light lacks important architectural concepts of modern ITP systems, and the implementation of the LCF kernel is not bullet-proof due to weaknesses of its implementation language OCaml.

Isabelle/HOL is at the opposite end of the spectrum of technical sophistication and complexity. Its multi-layered kernel architecture is meant to provide certain guarantees to users, at the cost of considerable code complexity: thus it is harder to break, but a breakdown becomes more spectacular. The Isabelle/ML sources of the core context management have about the same size as the HOL-Light inference kernel, but that "nano kernel" of Isabelle introduces extra structural integrity that is absent in HOL-Light. The explicit context management of Isabelle is essential to allow unhindered parallel execution on multicore hardware, for example.

The CakeML[5] project is notable, since one of its many aspects is to reconstruct a fully verified HOL-Light kernel [25] in a safe variant of Standard ML, instead of unsafe OCaml. If a complex proof assistant like Isabelle would export all its reasoning to such a fully verified minimal system, we could gain levels of confidence beyond the classic LCF tradition so far.

2 Proof Systems

ITP systems usually don't commit to particular forms of mechanized reasoning, but the LCF family favours certain forms of λ-calculus and Natural Deduction. Since the systems are freely programmable in ML, arbitrary proof tools with different logical calculi may be implemented as well. Users and tool developers have done that routinely in the past decades, so varieties of resolution, sequent calculus, classical tableaux etc. may be all seen in the prover libraries. A typical example in Isabelle is the generic tableau prover *blast* [37]. The more elementary Classical Reasoner [36] reconsiders the natural deduction of Isabelle/Pure in the sense of classic sequent calculus, without requiring any changes to the underlying representations of proof states and rules, see also [40, §9.4].

2.1 Isabelle/Pure inferences

The subsequent inference systems are for Isabelle/Pure, which is the most elementary in the LCF family. It is based on *Minimal Higher-Order Logic*, i.e. a reduced version of HOL to support arbitrarily nested natural deduction proofs, yet without the axiomatic basis that is required for formalized mathematics. Thus Isabelle/Pure remains pure from logical presuppositions, and can be used e.g. for Isabelle/ZF. Isabelle/HOL adds its own bootstrap axioms, and commits to a particular classic interpretation of the logical framework.

Isabelle/Pure consists of three levels of λ-calculus: one for syntax and two for logical reasoning with the connectives $\bigwedge$ and $\Longrightarrow$. Figure 1 illustrates the

---

[5] https://cakeml.org

syntactic formation rules of the language of *types* and *terms*. Types have a simple first-order structure of type variables and type constructors, e.g. *bool*, *nat*, *'a list*. Terms are simply-typed λ-terms, with atoms (variables, constants), application, and abstraction.

type *fun* :: (*type*, *type*)*type*   function space $\alpha \Rightarrow \beta$
const *all* :: ($'a \Rightarrow prop$) $\Rightarrow prop$ universal quantification $\bigwedge x{::}\alpha.\ B\ x$
const *imp* :: $prop \Rightarrow prop \Rightarrow prop$ implication $A \Longrightarrow B$

$$\frac{}{a_\tau :: \tau} \qquad \frac{t :: \sigma}{(\lambda x_\tau.\ t) :: \tau \Rightarrow \sigma} \qquad \frac{t :: \tau \Rightarrow \sigma \quad u :: \tau}{t\ u :: \sigma}$$

Fig. 1. Isabelle/Pure: abstract syntax with simple types

Figure 2 illustrates the formal treatment of *context* in Isabelle/Pure. The background theory Θ helps to organize big libraries in a graph-structured manner, with merge and extend operations. The foundational order of logical specification is preserved: a theory graph may always be flattened into a linear order of primitives. In contrast to programming languages like Java or Haskell, there is no mutual recursion of modules!

Results proven in one theory may be transferred to a bigger theory, according to the sub-theory relation that is formally maintained by the system. LCF and the HOL family leave this context implicit in the ML runtime environment, which means there is only a single monotonically growing theory state, without the possibility to go back (undo) nor the possibility to let the system perform inferences within different theories at the same time. Coq does not have a theory context Θ, because its primary context Γ is sufficiently expressive to take over that role.

In Isabelle/Pure, the context Γ serves as local proof context, with slightly different characteristics than the background theory Θ. It allows to build a local situation for specifications and proofs, with operations to *export* (abstraction) and *interpret* (application) results. The system provides a notion of *morphism* for that, to transform arbitrary items along the structure of contexts, but morphisms are not part of the logic, see also [9, 17].

background theory Θ: polymorphic types, constants, axioms (definitions)
proof context Γ: fixed variables, assumptions

merge and extend: $\Theta_3 = \Theta_1 \cup \Theta_2 + \tau + c :: \tau + c \equiv t$
sub-theory relation: $\Theta \subseteq \Theta'$
transfer of results: if $\Theta \subseteq \Theta'$ and $\Theta, \Gamma \vdash \varphi$ then $\Theta', \Gamma \vdash \varphi$

Fig. 2. Isabelle/Pure: formal context

Figure 3 shows the rules for primitive inferences of Isabelle/Pure, within the formal theory and proof context. Apart from bookkeeping wrt. the context (via *axiom* and *assume*), the remaining rules are standard introductions and eliminations for the logical connectives $\bigwedge$ and $\Longrightarrow$. This is minimal Higher-Order Logic presented as Natural Deduction inference system.

$$\frac{A \in \Theta}{\vdash A} \ (axiom) \qquad \frac{}{A \vdash A} \ (assume)$$

$$\frac{\Gamma \vdash B[x] \quad x \notin \Gamma}{\Gamma \vdash \bigwedge x. \ B[x]} \ (\bigwedge\text{-}intro) \qquad \frac{\Gamma \vdash \bigwedge x. \ B[x]}{\Gamma \vdash B[a]} \ (\bigwedge\text{-}elim)$$

$$\frac{\Gamma \vdash B}{\Gamma - A \vdash A \Longrightarrow B} \ (\Longrightarrow\text{-}intro) \qquad \frac{\Gamma_1 \vdash A \Longrightarrow B \quad \Gamma_2 \vdash A}{\Gamma_1 \cup \Gamma_2 \vdash B} \ (\Longrightarrow\text{-}elim)$$

Fig. 3. Isabelle/Pure: primitive inferences

Isabelle/Pure also provides a notion of equality, which is axiomatized as $\alpha\beta\eta$-congruence of λ-calculus, and later used as foundation for definitional tools and equational reasoning.

Treatment of equality is particularly important for Type-Theory provers like Coq. Instead of naive "mathematical equality" seen in the HOL-based systems, constructive type-theory cares about more detailed distinction of computable vs. non-computable logical objects, observable vs. non-observable equality etc. In addition to regular $\alpha\beta\eta$-conversion, Coq has built-in notions of $\iota\delta\zeta$-conversion to expand definitions (recursive functions over inductive types or primitive definitions) inside the formal system. Isabelle and the HOL family perform the latter by more conventional equational reasoning, using tools for rewriting and normalization implemented outside the logical kernel.

Reconsidering the main inference systems of Isabelle/Pure (figure 1, figure 2, figure 3), it is important to understand that little of that is directly exposed to end-users, for a variety of reasons. Logical foundations merely provide the basis for system internals. Additional infrastructure outside the logic takes over responsibilities to provide high-level concepts for the convenience of users. For example, formation of well-typed terms (figure 1) works via Hindley-Milner type-inference [10], even though the logic has merely schematic type variables at the outer theory level and lacks proper polymorphism within the logic. Thus terms can be written with very little type information, and the system reconstructs the rest. Isabelle/HOL even provides an add-on module for the generic type-infrastructure of Isabelle, to work with *coercive sub-typing* [39]: suitable conversion functions are inserted to adjust to different types, e.g. to turn some non-fitting argument of type *nat* into *int*.

Even the main inference system of the Isabelle/Pure kernel (figure 3) is rarely encountered by users in that form. It serves as foundation to more abstract concepts, such as block-structured reasoning in nested contexts. The vacuous example below illustrates this in Isar notation:

notepad
begin
 — Fresh proof context Γ starts here.
 {
  — Lets fix some arbitrary hypothetical items ...
  fix $a\ b\ c\ ::\ {}'a$
  — ... and assume some propositions.
  assume $a = b$ **and** $b = c$
  — So in the current context we ...
  have $a = c$ **using** $\langle a = b \rangle$ **and** $\langle b = c \rangle$ **by** (*rule trans*)
 }

After leaving that context, the system performs some $\Longrightarrow$-*intro* and $\bigwedge-intro$ reasoning for us, so that the nested context elements are no longer fixed, but arbitrary. This means, we ...

 have $\bigwedge a\ b\ c.\ a = b \Longrightarrow b = c \Longrightarrow a = c$ **by** *fact*

end

That structured reasoning in Isar is not another logical inference system, but the operation of an interpreter for a proof language that reduces proof text to primitive steps of the logical kernel.

2.2   Suitability for advanced reasoning tools

One needs to understand that the implementation, maintenance, and application of major interactive theorem provers is an ongoing effort over several decades. That means the logical foundations are somewhat accidental, according to certain intentions of the original authors and "latest trends" from a long time ago. Such starting conditions are hard to change later, when a system has reached a critical mass of tools and applications that depend on exactly these foundations. Developers of ITP systems and add-on tools are used to cope with that routinely, and usually manage to turn seemingly old-fashioned structures into exciting new applications.

So in practice there is a combination of flexibility and stability: Whatever happens to be implemented as initial foundations is sufficiently open to be adapted gradually over time to changing demands and new requirements (or the prover in question would have died out already). Significant changes of the logical foundations are unlikely, just due to the weight and gravity of an implemented system with its existing applications.

For example, the notion of local context Γ in Isabelle/Pure started out as a modest concept of minimal Natural Deduction in 1989, but acquired further roles in structured Isar proofs and structured specifications in 1999, until it became the main infrastructure that underlies the *local theory* specifications of Isabelle today (since 2007). Local theories [18] provide generic means to implement module concepts in Isabelle, outside of the core logic. This unifies locales [23], locale interpretation [2], type classes and class instantiation [17], adhoc overloading etc. and allows to combine them with derived definitional principles like **definition, inductive, datatype, function, primrec, primcorec** etc.

2.3 Further trends

I see two contradictory trends happening at the same time.

1. **Convergence of different systems.** The demands from realistic applications and large-scale formalization efforts encourages quite different proof assistants to overcome their foundational biases and arrive at similar user-space tools and libraries. For example, Coq has built-in notions for inductive types and recursive functions over them, but they were not sufficient for general recursion so that was eventually re-implemented as add-on without changing the already complex kernel. Isabelle/HOL had nothing like that from the outset, so everything had to be implemented as add-on in userspace, leading to some complex tool suite in the library that achieves almost the same as Coq.

2. **Divergence within the same system.** Despite the huge efforts to push significant changes of the logical foundations through the real implementations, and the libraries in particular, there are occasional ambitions to invent new ways of formal logic on the computer. Coq is presently faced with the movement towards *Homotopy Type Theory*[6], which overthrows the starting conditions of the original Calculus of Inductive Constructions as foundation for Coq. HOL is generally less interesting from a theoretical viewpoint, and thus more robust against experiments with alternative axiomatizations, but it sometimes happens on private side-branches of HOL [21].

3 Proof Search

3.1 Particular proof tools in Isabelle

LCF-style proof assistants have traditionally ignored the question of proof search, and merely provided the logical core with free programmability (in ML) to let the user implement arbitrary proof tools. After some decades, numerous tools have accumulated. The standard distribution of Isabelle/HOL today provides plenty of possibilities to apply proof search in practice, without asking the user to do ML programming again.

The classic portfolio of Isabelle proof tools may be categorized as follows:

---

[6] http://homotopytypetheory.org

- Single rule application, with application-specific declarations and concrete syntax, notably:
  - method *rule* for generic Natural Deduction (with higher-order unification);
  - method *cases* for elimination, syntactic representation of datatypes, inversion of inductive sets and predicates;
  - methods *induct* and *coinduct* for induction and coinduction over types, sets, and predicates.
- Equational reasoning by the Simplifier, with possibilities for add-on tools: simplification procedures, loopers, solvers. The main entry points are *simp* and *simp-all*, see also [40, §9.3].
- Classical reasoning with simple proof search procedures, based on classification of rules in the library: *intro*, *elim*, *dest*. Due to this instrumentation, relatively simple proof tools work quite effectively, even with large theory libraries. There are two standard implementations available.
  1. The Classical Reasoners with variations on search strategy via *fast*, *safe*, *clarify*, *best*, see also [36] and [40, §9.4].
  2. The Classic Tableau Prover *blast*, see [37].
- Various combinations of the Simplifier and Classical Reasoner tools, notably *auto*, *force*, *fastforce*.
- Metis[7] as medium-strength ATP inside Isabelle/HOL, with full proof reconstruction like the other standard tools. This is particularly important to Sledgehammer [5], which uses arbitrary external ATPs (and SMTs) for proof search, but needs to produce a proper LCF-style proof in the end. Note that *metis* is *not* accepting any instrumentation of the library: it works on a limited collection of facts that are given on the spot, usually produced by Sledgehammer for inlining into the proof text.
- Special tools for special situations, with some (semi)decision procedures for well-known theories like Presburger Arithmetic (methods *arith* or *presburger*).
- Integration of Z3 from MS Research[8] as general-purpose SMT, via the proof method *smt*, which performs explicit proof reconstruction for a large subset of Z3 proofs [7], using standard Isabelle tools like *simp*, *blast*, *auto* internally.

To get an overview of the zoo of Isabelle proof methods, one needs to keep the general tool frameworks behind them in mind, and avoid confusion of tools of similar names in other provers. For example, `simpl` in Coq refers to the builtin notion of computation of inductive definitions and recursive functions over them, which is a relatively powerful principle, but not as flexible as the Isabelle Simplifier method *simp*. Moreover, `auto` in Coq refers to elementary rule application with depth-first search, while *auto* in Isabelle is a relatively strong combination of the Simplifier and Classical Reasoner with additional instrumentation provided by the library.

---

[7] http://www.gilith.com/software/metis
[8] http://z3.codeplex.com/

There is further potential for confusion concerning the syntactic (and conceptual) categories of language elements. Isabelle/Isar clearly distinguishes *theorem attributes* (for small forward rules or context declarations), *proof methods* (for structured backwards refinement), and *proof commands* (for primary proof structure). In Coq, almost everything is just a "tactic", including `apply` — but in Isabelle **apply** is a proof command to apply an arbitrary proof method in an unstructured situation.

Tools for *disproving* a pending statement have become increasingly important in recent years. Current Isabelle routinely provides Quickcheck [8] and Nitpick [6]. These tools are also integrated into the Prover IDE, such that the system can provide information about faulty statements (with proposals for alternatives) produced asynchronously, while the user is working on the proof document in the vicinity.

3.2 Further trends

The general trend, which has already a long tradition in Isabelle, is to combine many tools in the library: automated provers and disprovers, general proof search and special decision procedures, connection to external proof tools with explicit proof reconstruction.

This leads to a considerable distance of Isabelle/HOL experienced by end-users, and the foundations of Isabelle/Pure sketched above (§2).

4 Proof Formats

The classic *input* format for theories and proofs in LCF-style proof assistants is just the ML code that constructs the intended results. By storing the sources and replaying them eventually, the user can return to some particular situation in the formal context. Thus the input format produced by one user also serves as *output* format given to other users.

An alternative, but less portable format, is the "dumped world image" that the ML runtime system might provide. Dumped images were once popular in the LISP world (later also in Smalltalk), and might see renewed popularity due to some trends in general IDE design to support "live programming" with direct manipulation of the running system while it is developed incrementally.[9]

4.1 The problem of standard formats

The OpenTheory[10] project aims at a well-defined independent exchange format for HOL-based proof assistants [22]. This may be understood as HOL "object-

---

[9] E.g. see the blog of Bret Victor `http://worrydream.com`, which also discusses old Smalltalk concepts to some extent.
[10] `http://www.gilith.com/research/opentheory`

code", so it lacks high-level structures of the original sources. It is also difficult to exchange proof tools, but there are some moves in that direction [26].

Fixed formats for interactive proof assistants are rare, due to the openess of the LCF approach. Although both Coq and Isabelle have some surface syntax, this "vernacular" of theory specifications and proofs is merely some default notation, with unlimited possibilities for extensions. In Isabelle, this open-endedness is particularly extreme: only the commands **theory** and **ML-file** are primitive, to start a new theory and load ML modules, and the other Isabelle/Pure and Isabelle/HOL language elements are bootstrapped from that: **theorem**, **definition**, **inductive**, **datatype**, **function** etc.

In principle, Isabelle theory content is just a sequence of semantic operations to update the toplevel state consecutively (in a purely functional manner). The surface syntax is part of the command implementation, using computationally complete mechanisms (e.g. parser combinators) instead of a fixed grammar.

This flexibility makes it difficult to build tools that operate on theory sources of LCF-style proof assistants. The situation is particularly difficult for the HOL family, where the input consists of unrestricted ML code. Nonetheless, in the past 5 years, the Isabelle/PIDE framework has managed to provide access to aspects of the prover and its internal data structures, with continuous feedback in the text editor [44]. The PIDE approach of Isabelle/jEdit gives the user some illusion of direct manipulation of formal source text, despite the conceptual distance of the concrete syntax to the abstract entities behind it.

4.2   YXML/XML/ML data exchange

It is always possible to invent specific formats for specific tools, and implement them in ML. Current Isabelle actually uses a hybrid (two-legged) approach: Isabelle/ML for hard-core logical tools, and Isabelle/Scala for systems-programming on the JVM. For example, a simple exchange-format with parsers for concrete syntax could be implemented either in Isabelle/ML or Isabelle/Scala, using existing libraries that are available on both sides.

Bypassing official standards of XML, the exchange of tree-structured data in Isabelle works conveniently via YXML/XML/ML data representation[11], which is already available for SML, OCaml, and Scala, and may be easily ported to other languages on the spot. So if there is any standard exchange format in Isabelle, it is YXML as transfer syntax for untyped XML trees [47, §6.11], which is used to encode algebraic datatypes of ML. Actual content needs to be defined by the tools themselves.

4.3   Primary proofs versus primitive proofs

We have already seen various Isabelle/Isar proof texts in §1 of this document, which is a pretty-printed version of a formal theory in Isabelle. It is a PDF-LaTeX document with relatively high type-setting quality, not a verbatim listing.

---

[11] https://bitbucket.org/makarius/yxml

The subsequent example of the Knaster-Tarski fixed-point theorem shall illustrate the different notions of proofs further. According to the textbook [11, pages 93–94], the Knaster-Tarski fixpoint theorem is as follows:

> **The Knaster-Tarski Fixpoint Theorem.** Let L be a complete lattice and $f\colon L \to L$ an order-preserving map. Then $\bigsqcap\{x \in L \mid f(x) \le x\}$ is a fixpoint of f.
>
> **Proof.** Let $H = \{x \in L \mid f(x) \le x\}$ and $a = \bigsqcap H$. For all $x \in H$ we have $a \le x$, so $f(a) \le f(x) \le x$. Thus $f(a)$ is a lower bound of H, whence $f(a) \le a$. We now use this inequality to prove the reverse one (!) and thereby complete the proof that a is a fixpoint. Since f is order-preserving, $f(f(a)) \le f(a)$. This says $f(a) \in H$, so $a \le f(a)$.

This proof is formalized in Isabelle/Isar as follows, streamlining the reasoning a bit to achieve structured top-down decomposition of the problem at the outer level, while only the inner steps of reasoning are done in a forward manner.

```
theorem Knaster-Tarski:
  fixes f :: 'a::complete-lattice ⇒ 'a
  assumes mono f
  shows ∃ a. f a = a
proof
  let ?H = {u. f u ≤ u}
  let ?a = ⊓ ?H
  show f ?a = ?a
  proof (rule order-antisym)
    show f ?a ≤ ?a
    proof (rule Inf-greatest)
      fix x
      assume x ∈ ?H
      then have ?a ≤ x by (rule Inf-lower)
      with ⟨mono f⟩ have f ?a ≤ f x ..
      also from ⟨x ∈ ?H⟩ have … ≤ x ..
      finally show f ?a ≤ x .
    qed
    show ?a ≤ f ?a
    proof (rule Inf-lower)
      from ⟨mono f⟩ and ⟨f ?a ≤ ?a⟩ have f (f ?a) ≤ f ?a ..
      then show f ?a ∈ ?H ..
    qed
  qed
qed
```

This format is called *primary proof text* in Isabelle/Isar jargon. It is notable that the reasoning merely composes some elementary rules of lattice theory into a structured proof outline, according to Natural Deduction rule composition in Isabelle/Pure. The only "proof automation" used here is unification, to search for suitable rules and instantiations from the context (the background theory or current proof).

The *primitive proof term* is normally not seen nor stored in memory. Here it is reproduced with special system options that are normally disabled:

λ(*H*: -) *Ha*: -.
 exI · (λx. *?f x = x*) · - · (*thm* · *H*) ·
  (*order-antisym* · - · - · (*thm* · *H*) ·
   (*complete-lattice-class.Inf-greatest* · - · - · *H* ·
    (λx *Hb*: -.
      order-trans-rules-23 · - · - · - · (*thm* · *H*) ·
       (*order-class.monoD* · *?f* · - · - · (*thm* · *H*) · (*thm* · *H*) · *Ha* ·
        (*complete-lattice-class.Inf-lower* · - · - · *H* · *Hb*)) ·
       (*iffD1* · - · - · (*thm.Set.mem-Collect-eq* · - · (λu. *?f u $\leq$ u*) · (*thm* · *H*)) ·
Hb))) ·
   (*complete-lattice-class.Inf-lower* · - · - · *H* ·
    (*iffD2* · - · - · (*thm.Set.mem-Collect-eq* · - · (λa. *?f a $\leq$ a*) · (*thm* · *H*)) ·
     (*order-class.monoD* · *?f* · - · - · (*thm* · *H*) · (*thm* · *H*) · *Ha* ·
      (*complete-lattice-class.Inf-greatest* · - · - · *H* ·
       (λx *Hb*: -.
         order-trans-rules-23 · - · - · - · (*thm* · *H*) ·
          (*order-class.monoD* · *?f* · - · - · (*thm* · *H*) · (*thm* · *H*) · *Ha* ·
           (*complete-lattice-class.Inf-lower* · - · - · *H* · *Hb*)) ·
          (*iffD1* · - · - · (*thm.Set.mem-Collect-eq* · - · (λu. *?f u $\leq$ u*) · (*thm* · *H*))
· *Hb*)))))))

Some further examples are available in `Isabelle2014/src/HOL/Proofs/ ex/Hilbert_Classical.thy`, but note that this requires the special session of `HOL-Proofs` where proof terms are enabled by default. Despite a few interesting applications here and there, proof terms are irrelevant to the mainstream of Isabelle projects.

4.4 Guiding principles of Isar

The guiding principles of the primary proof format — the Isar proof text — are as follows:

- Input produced by one user coincides with the output given to other users.
- Human-readability is emphasized, but it should coincide with efficient machine-checking.
- Structure is more important than strong automation, but arbitrary proof tools may be included on demand.
- Proofs are documents (with proper type-setting), not "proof scripts" (nor "listings").

4.5 Further trends

I see a general trend towards more structure. The Isar proof language was a relatively significant reform of the main Isabelle input format 15 years ago.

Likewise, Coq/SSReflect and classic Coq/Ltac have gradually provided a bit more explicit information about the intended meaning of proof sources.

Another trend is to go beyond static proof formats, and understand the proof development process as genuine interaction of the user with the prover, which needs to be supported explicitly by some document-model in the Prover IDE. Internal aspects of the prover are externalized on the spot, to let the editor participate. Isabelle/jEdit [44] is an important step towards that, but advanced structural editing of proof documents is still lacking.

5 Proof Production

5.1 Foundations of derived proof tools

The LCF approach demands proper foundations of all reasoning performed in the system: proof tools need to produce some explicit proof record, and have the core logic accept it. There are two main approaches to that problem:

1. **Explicit replay.** The proof tool reduces its reasoning at run-time to primitive inferences of the LCF kernel. The results of the proof search are turned into a proof trace of some form, and passed through the abstract datatype *thm* to infer proper theorems (*without* storing a proof trace by default). The separation of proof discovery versus justification is important: finding a solution is usually much harder than checking it by explicit inferences.
2. **Reflection.** The proof tool is implemented within the logical language of the prover itself and formally proven correct for all input values (e.g. by induction). Tool applications then turn some symbolic representation of the problem into a theorem, but without performing explicit inferencing again. The proof is finished beforehand, to establish the correctness of the implemented algorithm once and for all.

Explicit replay is the standard approach of LCF, the HOL family, and Isabelle: tools themselves are not proven, but produce explicit proof steps at run-time. Thus the tool might fail or diverge unexpectedly, but cannot return wrong results.

Reflection is routinely used in Coq, since its stronger type-theory facilitates that and the access to ML as tool implementation language is more difficult than in the other systems: the general philosophy is to internalize all aspects of computation within the main logic.

After many years of both approaches side-by-side, there is no clear "winner". Explicit proof replay might sound less efficient in theory, but performs quite well in practice: as may be experienced in big applications e.g. of AFP[12].

For reflection there is first the hurdle to produce a formal correctness proof of substantial tool implementations. Afterwards its symbolic evaluation needs to be performed by built-in normalization of the logical language.

---

[12] `http://afp.sf.net`

5.2 Costs of explicit proof production

By default, a true LCF-style proof assistant does not need to store proof objects, since all inferences are "correct by construction". An explicit proof record may be requested nonetheless, e.g. for external checking with some independent tool. The cost for that is typically an order of magnitude: a factor 2–10 in time and space requirements. Naive proof storage would require much more space, but extra time can be invested to maintain reasonably compact proof terms [4].

For properly implemented proof tools that use the LCF-style inference kernel as intended, there is no difference in the algorithm. It is merely a matter to change some system flags, and to spend more machine resources to turn the platonistic idea of a proof into a materialized proof object.

To select various degrees of thoroughness in formal proof checking, the main options of Isabelle are as follows:

- **Skip proofs:** the user merely wants to see some checking of theory specifications, without going into proofs.
- **Quick-and-dirty:** tools are allowed to omit explicit inferences or proper definitions, and replace them by ad-hoc oracle invocations or axioms.
- **Default LCF mode:** all inferences go through the LCF kernel (*without* storing proof traces) and definitions are done properly.
- **Proof terms:** LCF kernel inferences are stored as compact λ-terms for later inspection, extraction etc.

5.3 Further trends

With routine support for parallel LCF-style proof checking in Isabelle, the default LCF mode has regained its central role, and the weaker and stronger options are becoming less important. On the one hand, parallel proof checking is fast enough to render most quick-and-dirty shortcuts obsolete. On the other hand, the practical irrelevance of proof terms (their inaccessibility to the formal derivation) is important for scalability of really big proof developments. Retrospectively, the classic LCF-approach from 1979 fits nicely to multicore programming with shared memory from 2009, see also [45].

So far there are only Isabelle and ACL2 with realistic support for parallel processing [38]. Hopefully, other proof assistants will manage to catch-up eventually, otherwise that will become a dividing line for slow sequential systems vs. fast parallel ones: theoretically the difference is only a constant factor, but in practice that is something like 10 or more, i.e. a full order of magnitude.

6 Proof Consumption

Interactive theorem provers are a natural sink for proofs produced by other sources, internal or external tools. Prominent examples in Isabelle/HOL are the integration of Z3 via the *smt* proof method [7] and Sledgehammer [5], which appeals to E Prover[13] and SPASS[14] in particular. The external tools are used here merely for search: internal LCF-style proofs are reconstructed from the result.

Whenever such a project to connect an external prover is started, there are usually technical problems and genuine doubts about the proof trace of the other tool, which usually require several iterations to incorporate resulting proofs reliably.

Tools integrated into an interactive prover like Isabelle are usually for genuine production use in the target system, to support the user in big formalizations. So the main purpose is to make an external tool look like an internal one. Independent checking of the external tools could in principle be done as well, but the practical relevance is much lower.

6.1 Further trends

Big ITP systems are becoming more and more a "workbench" for many different proof tools, both internal and external ones. Thus the classic proof assistants become the integrating environment for proofs produced by other systems, but there is also the possibility to export their reasoning again as a trace of the inference kernel.

Integration of tools has two aspects in practice: (1) the actual work on the tool and its logic, (2) technical side-conditions to glue together quite dissimilar programs into a usable system. The Isabelle environment is particularly polished to accommodate tools stemming from very mixed background, with uniform support for the main operating system families: Linux, Windows, Mac OS X.

7 Proof Applications

7.1 Some application domains and big formalization projects

ITP systems are open-ended and the scope of applications depends on the imagination of their users. Despite some big and publicly visible applications, the main potential is probably still unexplored, due to traditional obscurity and inaccessibility of the ITP systems (which was inherited from LCF). Applications today are typically about formalized mathematics, modelling computer-science concepts, or specification and verification of software / hardware systems. Here is a list of some large-scale proof development projects (project leaders and proof assistant in parentheses).

- The **Flyspeck Project** https://code.google.com/p/flyspeck (Tom Hales, HOL-Light): formal proof of Kepler's Conjecture [19, 20].

---

[13] http://www.eprover.org
[14] http://www.spass-prover.org

- The **L4.verified project** `http://ertos.nicta.com.au/research/l4.verified` (Gerwin Klein, Isabelle/HOL): formally correct operating system kernel [24].
- The **Feit-Thompson Odd Order Theorem** `http://www.msr-inria.fr/news/feit-thomson-proved-in-coq` (Georges Gonthier, Coq/SSReflect) [13, 12].
- The **CompCert verified compiler** `http://compcert.inria.fr/doc` (Xavier Leroy, Coq): optimizing C-compiler for various assembly languages, written and proven in the functional language of Coq [27, 28].

There are also some efforts to build libraries of formalized mathematics. The Feit-Thompson proof depends on the *Mathematical Components* library that was developed by the same research group. For Isabelle/HOL, the *Archive of Formal Proof*[15] has accumulated quite substantial material over 10 years, from many different sources and different degrees of quality and sophistication.[16]

7.2 Proof quality

To discuss the question of quality of different proofs for the same theorem, we need to reconsider the different notions of formal proof in Isabelle/Isar:

- The **primitive proof** (or **proof object**) is some way to ensure that results produced by the system are reliable records of logical reasoning.
  The standard LCF approach does not require to record proof objects explicitly, because the abstract datatype of theorems ensures that its elements enjoy certain properties by construction (thanks to strongly-typed virtues of Standard ML).
- The **primary proof** (or **proof document**) is the main source text that is processed by the system to accept a formal proof, and to tell the human reader that the reasoning makes sense. Proof documents are suitable for "archival publication" of formal proofs, and have a chance to be useful for other proof assistants or other versions of the same proof assistant even after some time.

Isabelle/Isar is mostly concerned about proof documents, i.e. the real sources of some formal reasoning, in a form that is both human-readable and machine checkable (to turn it into an abstract proof object that is not stored).

The relation of "proof document P is better than Q" is not formally defined, but may be understood as a natural continuation of quality standards for program source text, with the following criteria:

- Readability for humans, not just the machine.

---

[15] `http://afp.sf.net`

[16] At the time of writing (November 2014), AFP consists of more than 200 articles with more than 50 MB of theory sources.

- Maintainability and stability: small changes of theory definitions lead to reasonably small changes of proofs.
- Generality: existing proofs for one theory may be easily re-used in a more general theory by reworking the proof text.
- Performance of proof checking: Isabelle/Isar is faster in checking nicely structured proof documents than unstructured tactic scripts.

Of course, people have also applied a standard repertoire from Proof Theory to the internal proof objects of various ITP systems, including Isabelle [4, 3]. This is interesting in its own right, but merely a niche of the full spectrum of ITP applications.

7.3 Further trends

For the near future I see the need for more systematic tool support to maintain the growing repositories of formalized mathematics. There is also a need to make proof assistants more accessible to more users, to allow easy browsing and editing of such libraries.

Current Isabelle addresses that by its default Prover IDE: Isabelle/jEdit is intended for beginners and experts alike. It allows to edit theory libraries directly, with continuous proof checking on multiple cores. "The ACL2 Sedan"[17] is a similar approach to Eclipse-based IDE support for ACL2, with the notable difference that it is meant as a comfortable "family car", while the traditional Emacs interface remains available as "race car" for experts. In contrast, the classic Proof General Emacs support has already been removed from Isabelle at the time of writing (November 2014).

Watching the ITP community over almost 20 years, my impression is that the logical foundations at the bottom matter very little today. For example, Coq started out very constructionistic, but the majority of its users don't understand much of that. In applications, Coq users often include a prelude of classic axioms for their application theory. e.g. for *real* real numbers in classic analysis. Even if the application is as constructive and computational as some floating-point arithmetic algorithms, the mathematics of the problem domain works more smoothly with all the tools of classical reasoning that have accumulated over the centuries.

8 Conclusions

Taking even just a single ITP system like Isabelle, it is difficult to pin down its boundaries precisely, to say what it is and what it does. Too many aspects of formal logic and proof development have accumulated over the decades. Today one could characterize Isabelle and its Prover IDE as an integrative platform

---

[17] http://acl2s.ccs.neu.edu/acl2s/doc

for domain-specific formal languages and tools around them, but it depends on many side-conditions how the system is perceived in a particular application scenario. For example, Isabelle/jEdit could be also used as IDE for Standard ML, without any theory or proof development.

Each proof assistant has its own tradition and culture. New users should spend time to develop a sense for more than one accidental candidate, before making a commitment: substantial time will be required to become proficient with any of these systems. Old users should try to learn how other proof assistants actually work, and what are their specific strengths and weaknesses. There are very few people who are proficient with more than one ITP system, and the exchange of ideas within the various brands of proof assistants is often slow and cumbersome.

This article hopes to contribute to such a better understanding of ITP systems as a whole, by outlining the situation from the viewpoint of Isabelle/Isar.

References

1. D. Aspinall. Proof General: A generic tool for proof development. In S. Graf and M. Schwartzbach, editors, *European Joint Conferences on Theory and Practice of Software (ETAPS)*, volume 1785 of *LNCS*. Springer, 2000.
2. C. Ballarin. Interpretation of locales in Isabelle: Theories and proof contexts. In J. M. Borwein and W. M. Farmer, editors, *Mathematical Knowledge Management (MKM 2006)*, LNAI 4108, 2006.
3. U. Berger, S. Berghofer, P. Letouzey, and H. Schwichtenberg. Program extraction from normalization proofs. *Studia Logica*, 82(1):25–49, 2006.
4. S. Berghofer and T. Nipkow. Proof terms for simply typed higher order logic. In J. Harrison and M. Aagaard, editors, *Theorem Proving in Higher Order Logics: TPHOLs 2000*, volume 1869 of *LNCS*, pages 38–52. Springer, 2000.
5. J. C. Blanchette. *Hammering Away: A User's Guide to Sledgehammer for Isabelle/HOL*. http://isabelle.in.tum.de/website-Isabelle2014/dist/Isabelle2014/doc/sledgehammer.pdf.
6. J. C. Blanchette and T. Nipkow. Nitpick: A counterexample generator for higher-order logic based on a relational model finder. In *Interactive Theorem Proving, First International Conference, ITP 2010, Edinburgh, UK, July 11-14, 2010. Proceedings*, volume 6172 of *Lecture Notes in Computer Science*, pages 131–146. Springer, 2010.
7. S. Böhme and T. Weber. Fast LCF-Style proof reconstruction for Z3. In M. Kaufmann and L. C. Paulson, editors, *Interactive Theorem Proving, First International Conference, ITP 2010, Edinburgh, UK, July 11-14, 2010. Proceedings*, volume 6172 of *Lecture Notes in Computer Science*, pages 179–194. Springer, 2010.
8. L. Bulwahn. The new Quickcheck for Isabelle – random, exhaustive and symbolic testing under one roof. In C. Hawblitzel and D. Miller, editors, *Certified Programs and Proofs - Second International Conference, CPP 2012, Kyoto, Japan, December 13-15, 2012. Proceedings*, volume 7679 of *Lecture Notes in Computer Science*, pages 92–108. Springer, 2012.

9. A. Chaieb and M. Wenzel. Context aware calculation and deduction — ring equalities via Gröbner Bases in Isabelle. In M. Kauers, M. Kerber, R. Miner, and W. Windsteiger, editors, *Towards Mechanized Mathematical Assistants (CALCULEMUS 2007)*, volume 4573 of *LNAI*. Springer, 2007.

10. L. Damas and H. Milner. Principal type schemes for functional programs. In *ACM Symp. Principles of Programming Languages*, 1982.

11. B. A. Davey and H. A. Priestley. *Introduction to Lattices and Order*. Cambridge University Press, 1990.

12. G. Gonthier. Engineering mathematics: the odd order theorem proof. In R. Giacobazzi and R. Cousot, editors, *The 40th Annual ACM SIGPLAN-SIGACT Symposium on Principles of Programming Languages, POPL '13, Rome, Italy - January 23 - 25, 2013*. ACM, 2013.

13. G. Gonthier, A. Asperti, J. Avigad, Y. Bertot, C. Cohen, F. Garillot, S. L. Roux, A. Mahboubi, R. O'Connor, S. O. Biha, I. Pasca, L. Rideau, A. Solovyev, E. Tassi, and L. Théry. A machine-checked proof of the odd order theorem. In S. Blazy, C. Paulin-Mohring, and D. Pichardie, editors, *Interactive Theorem Proving - 4th International Conference, ITP 2013, Rennes, France, July 22-26, 2013. Proceedings*, volume 7998 of *Lecture Notes in Computer Science*, pages 163–179. Springer, 2013.

14. G. Gonthier and A. Mahboubi. An introduction to small scale reflection in Coq. *J. Formalized Reasoning*, 3(2), 2010.

15. M. J. C. Gordon and T. F. Melham, editors. *Introduction to HOL: A theorem proving environment for higher order logic*. Cambridge University Press, 1993.

16. M. J. C. Gordon, R. Milner, and C. P. Wadsworth. *Edinburgh LCF: A Mechanized Logic of Computation*, volume 78 of *LNCS*. Springer, 1979.

17. F. Haftmann and M. Wenzel. Constructive type classes in Isabelle. In T. Altenkirch and C. McBride, editors, *Types for Proofs and Programs (TYPES 2006)*, volume 4502 of *LNCS*. Springer, 2007.

18. F. Haftmann and M. Wenzel. Local theory specifications in Isabelle/Isar. In S. Berardi, F. Damiani, and U. de Liguoro, editors, *Types for Proofs and Programs, TYPES 2008*, volume 5497 of *LNCS*. Springer, 2009.

19. T. C. Hales. Formalizing the proof of the kepler conjecture. In K. Slind, A. Bunker, and G. Gopalakrishnan, editors, *Theorem Proving in Higher Order Logics, 17th International Conference, TPHOLs 2004, Park City, Utah, USA, September 14-17, 2004, Proceedings*, volume 3223 of *Lecture Notes in Computer Science*, page 117. Springer, 2004.

20. T. C. Hales, J. Harrison, S. McLaughlin, T. Nipkow, S. Obua, and R. Zumkeller. A revision of the proof of the kepler conjecture. *Discrete & Computational Geometry*, 44(1):1–34, 2010.

21. P. V. Homeier. The HOL-Omega logic. In S. Berghofer, T. Nipkow, C. Urban, and M. Wenzel, editors, *Theorem Proving in Higher Order Logics, 22nd International Conference, TPHOLs 2009, Munich, Germany, August 17-20, 2009. Proceedings*, volume 5674 of *Lecture Notes in Computer Science*, pages 244–259. Springer, 2009.

22. J. Hurd. The OpenTheory standard theory library. In M. G. Bobaru, K. Havelund, G. J. Holzmann, and R. Joshi, editors, *NASA Formal Methods – Third International Symposium, NFM 2011, Pasadena, CA, USA, April 18-20, 2011. Proceedings*, volume 6617 of *Lecture Notes in Computer Science*, pages 177–191. Springer, 2011.

23. F. Kammüller, M. Wenzel, and L. C. Paulson. Locales: A sectioning concept for Isabelle. In Y. Bertot, G. Dowek, A. Hirschowitz, C. Paulin, and L. Thery, editors,

Theorem Proving in Higher Order Logics (TPHOLs 1999), volume 1690 of *LNCS*. Springer, 1999.

24. G. Klein, J. Andronick, K. Elphinstone, T. C. Murray, T. Sewell, R. Kolanski, and G. Heiser. Comprehensive formal verification of an os microkernel. *ACM Trans. Comput. Syst.*, 32(1):2, 2014.

25. R. Kumar, R. Arthan, M. O. Myreen, and S. Owens. HOL with definitions: Semantics, soundness, and a verified implementation. In G. Klein and R. Gamboa, editors, *Interactive Theorem Proving (ITP 2014)*, volume 8558 of *LNCS*. Springer, 2014.

26. R. Kumar and J. Hurd. Standalone tactics using OpenTheory. In L. Beringer and A. P. Felty, editors, *Interactive Theorem Proving - Third International Conference, ITP 2012, Princeton, NJ, USA, August 13-15, 2012. Proceedings*, volume 7406 of *Lecture Notes in Computer Science*, pages 405–411. Springer, 2012.

27. X. Leroy. Formal certification of a compiler back-end or: programming a compiler with a proof assistant. In J. G. Morrisett and S. L. P. Jones, editors, *Proceedings of the 33rd ACM SIGPLAN-SIGACT Symposium on Principles of Programming Languages, POPL 2006, Charleston, South Carolina, USA, January 11-13, 2006*, pages 42–54. ACM, 2006.

28. X. Leroy. Formal verification of a realistic compiler. *Commun. ACM*, 52(7):107–115, 2009.

29. D. C. J. Matthews and M. Wenzel. Efficient parallel programming in Poly/ML and Isabelle/ML. In *ACM SIGPLAN Workshop on Declarative Aspects of Multicore Programming (DAMP 2010), co-located with POPL*. ACM Press, January 2010.

30. T. Nipkow. *Programming and Proving in Isabelle/HOL*. http://isabelle.in.tum.de/website-Isabelle2014/dist/Isabelle2014/doc/prog-prove.pdf.

31. T. Nipkow, L. C. Paulson, and M. Wenzel. *Isabelle/HOL — A Proof Assistant for Higher-Order Logic*, volume 2283 of *LNCS*. Springer, 2002.

32. L. C. Paulson. The foundation of a generic theorem prover. *Journal of Automated Reasoning*, 5(3):363–397, 1989.

33. L. C. Paulson. Isabelle: The next 700 theorem provers. In P. Odifreddi, editor, *Logic and Computer Science*, pages 361–386. Academic Press, 1990.

34. L. C. Paulson. Set theory for verification: I. from foundations to functions. *J. Autom. Reasoning*, 11(3):353–389, 1993.

35. L. C. Paulson. Set theory for verification: II: induction and recursion. *J. Autom. Reasoning*, 15(2):167–215, 1995.

36. L. C. Paulson. Generic automatic proof tools. *CoRR*, cs.LO/9711106, 1997.

37. L. C. Paulson. A generic tableau prover and its integration with isabelle. *J. UCS*, 5(3):73–87, 1999.

38. D. L. Rager, W. A. Hunt, and M. Kaufmann. A parallelized theorem prover for a logic with parallel execution. In S. Blazy, C. Paulin-Mohring, and D. Pichardie, editors, *Interactive Theorem Proving (ITP 2013)*, volume 7998 of *LNCS*. Springer, 2013.

39. D. Traytel, S. Berghofer, and T. Nipkow. Extending Hindley-Milner Type Inference with Coercive Structural Subtyping. In H. Yang, editor, *APLAS 2011*, volume 7078 of *LNCS*, pages 89–104, 2011.

40. M. Wenzel. *The Isabelle/Isar Reference Manual*. http://isabelle.in.tum.de/website-Isabelle2014/dist/Isabelle2014/doc/isar-ref.pdf.

41. M. Wenzel. Isar — a generic interpretative approach to readable formal proof documents. In Y. Bertot, G. Dowek, A. Hirschowitz, C. Paulin, and L. Thery, editors, *Theorem Proving in Higher Order Logics (TPHOLs 1999)*, volume 1690 of *LNCS*. Springer, 1999.

42. M. Wenzel. Parallel proof checking in Isabelle/Isar. In G. Dos Reis and L. Théry, editors, *ACM SIGSAM Workshop on Programming Languages for Mechanized Mathematics Systems (PLMMS 2009)*. ACM Digital Library, 2009.

43. M. Wenzel. Isabelle as document-oriented proof assistant. In J. H. Davenport et al., editors, *Conference on Intelligent Computer Mathematics (CICM 2011)*, volume 6824 of *LNAI*. Springer, 2011.

44. M. Wenzel. Isabelle/jEdit — a Prover IDE within the PIDE framework. In J. Jeuring et al., editors, *Conference on Intelligent Computer Mathematics (CICM 2012)*, volume 7362 of *LNAI*. Springer, 2012.

45. M. Wenzel. Shared-memory multiprocessing for interactive theorem proving. In S. Blazy, C. Paulin-Mohring, and D. Pichardie, editors, *Interactive Theorem Proving (ITP 2013)*, volume 7998 of *Lecture Notes in Computer Science*. Springer, 2013.

46. M. Wenzel. Asynchronous user interaction and tool integration in Isabelle/PIDE. In G. Klein and R. Gamboa, editors, *Interactive Theorem Proving (ITP 2014)*, volume 8558 of *LNCS*. Springer, 2014.

47. M. Wenzel and S. Berghofer. *The Isabelle System Manual*. `http://isabelle.in.tum.de/website-Isabelle2014/dist/Isabelle2014/doc/system.pdf`.

48. F. Wiedijk, editor. *The Seventeen Provers of the World*, volume 3600 of *LNAI*. Springer, 2006.

Introduction to the
Calculus of Inductive Constructions

Christine Paulin-Mohring[1]

LRI, Univ Paris-Sud, CNRS and INRIA Saclay - Île-de-France, Toccata, Orsay
F-91405
Christine.Paulin@lri.fr

1 Introduction

The Calculus of Inductive Constructions (CIC) is the formalism behind the interactive proof assistant Coq [24, 5]. It is a powerful language which aims at representing both functional programs in the style of the ML language and proofs in higher-order logic. Many data-structures can be represented in this language: usual data-types like lists and binary trees (possibly polymorphic) but also infinitely branching trees. At the logical level, inductive definitions give a natural representation of notions like reachability and operational semantics defined using inference rules.

Inductive definitions in the context of a proof language were formalized in the early 90's in two different settings. The first one is Martin-Löf's Type Theory [18]. This theory was originally presented with a set of rules defining basic notions like products, sums, natural numbers, equality. All of them (except for functions) are an instance of a general scheme of inductive definitions which has been studied by P. Dybjer [14]. The second one is the pure Calculus of Constructions. This is a typed polymorphic functional language, which is powerful enough to encode inductive definitions [6, 23], but this encoding has some drawbacks: efficiency of computation of functions over these data-types and some natural properties that cannot be proven. The extension of the formalism with primitive inductive definitions [13, 20] was consequently a natural choice. In proof assistants based on higher-order logic (HOL), an impredicative encoding of inductive definitions is used: this is made possible by the existence of a primitive infinite type (including integers) and the fact that HOL is only concerned by extensional properties of objects (not computations) [22].

Inductive definitions, as a primitive or derived notion are one of the main ingredients of the languages for interactive theorem proving both for representing objects and logical notions.

In this paper, we give a quick overview of the Calculus of Inductive Constructions, the formalism behind the Coq proof assistant. In section 2, we present the language and the typing rules. We start with the pure functional part and then continue with the inductive declarations. We shall then briefly discuss the properties of this language in section 3, both from the theoretical and pragmatic points of view. We shall then conclude with examples of applications, the description of some of the trends in sections 4,5 and 6.

2 Proof System

2.1 The Calculus of Constructions

The Calculus of Constructions which is the purely functional language underlying the Calculus of Inductive Constructions has been introduced by Coquand and Huet [11, 12]. It can be defined as a pure type system (PTS). A PTS is a typed lambda-calculus with a unique syntactic language describing both terms and types. Terms include variables, (typed) abstractions (written $\mathbf{fun}\, x : A \Rightarrow t$) and applications (written $t\, u$) as in ordinary lambda-calculus. The type of an abstraction is a (dependent) product $\Pi x : A, B$ making possible for the type B of t to depend on the variable x. The notation $A \rightarrow B$ for the type of functions from type A to type B is just an abbreviation in the special case where B does not depend on x. Types are themselves typed objects, the type of a type will be a special constant called a *sort*. There is at least one sort called Type. Different PTS depend on the set of sorts we start with (each sort corresponds to a certain universe of objects) and also which products can be done (in which universes).

For instance, with A a type, the identity function $\mathbf{fun}\, x : A \Rightarrow x$ is a term of type $A \rightarrow A$. With A being a type variable, we may build the polymorphic identity $\mathbf{fun}\, A : \mathsf{Type} \Rightarrow \mathbf{fun}\, x : A \Rightarrow x$ of type $\Pi A : \mathsf{Type}, A \rightarrow A$.

In the case of the Calculus of Constructions, we have an infinite set of sorts $\mathcal{S} \overset{\mathrm{def}}{=} \{\mathsf{Prop}\} \cup \bigcup_{i \in \mathbb{N}} \{\mathsf{Type}_i\}$. The sort Prop captures the type of expressions which denote logical propositions. We follow the Curry-Howard correspondence where a proposition A is represented by a type (namely the type of proofs of A) and a proof of the logical proposition A will correspond to an object t of type A. If A and B are two types corresponding to logical propositions, then the proposition $A \Rightarrow B$ will be represented by the type $A \rightarrow B$ of functions transforming proofs of A into proofs of B, the proposition $A \wedge B$ will be represented by the type $A \times B$ of pairs built with a proof of A and a proof of B. Given a type T, the type $\Pi x : T, B$ will represent the type of dependent functions which given a term $t : T$ computes a term of type $B[t/x]$ corresponding to proofs of the logical proposition $\forall x : T, B$. Because types represent logical propositions, the language will contain empty types corresponding to unprovable propositions.

Notations. We shall freely use the notation $\forall x : A, B$ instead of $\Pi x : A, B$ when B represents a proposition. We write $t[u/x]$ for the term t in which the variable x has been replaced by the term u. The term $t\, u_1 \ldots u_n$ represents $(\ldots (t\, u_1) \ldots u_n)$ and $\mathbf{fun}\, (x_1 : A_1) \ldots (x_n : A_n) \Rightarrow t$ (resp. $\Pi(x_1 : A_1) \ldots (x_n : A_n), B$) is the same as $\mathbf{fun}\, x_1 : A_1 \Rightarrow \ldots \mathbf{fun}\, x_n : A_n \Rightarrow t$ (resp. $\Pi x_1 : A_1, \ldots \Pi x_n : A_n, B$). The term $A \rightarrow B \rightarrow C$ should be understood as $A \rightarrow (B \rightarrow C)$ and $\Pi x : A, B \rightarrow C$ is the same as $\Pi x : A, (B \rightarrow C)$. We sometimes omit the type of the variable in abstractions and products when they are clear from the context.

In higher-order logic, propositions and objects are written using the same functional language with abstractions and applications. We may want to introduce a binary relation on a type A as a variable R. This variable will have type $A \rightarrow A \rightarrow \mathsf{Prop}$ (this type replaces an arity declaration in first-order logic). We can build a predicate **fun** $x : A \Rightarrow R\,x\,x$ representing the set of objects which are in relation with themselves, this predicate will have type $A \to$ Prop (this expression shall not be confused with the type $\forall x : A, R\,x\,x$ of type Prop, expressing the reflexivity of R).

We need $A \to A \to$ Prop to be a well-formed type, which means it is typed with a sort. This sort will be Type_1. If we want to iterate constructions on Type_1, we shall need this sort itself to be well-typed, that will require introducing a new sort Type_2 to be the type of the object Type_1. The need for an infinite hierarchy of universes comes from the fact that the more naive system where we have only one sort Type of type Type is inconsistent.

The Calculus of Inductive Constructions manipulates judgments which are of the form

$$x_1 : A_1, \ldots, x_n : A_n \vdash t : A$$

In this judgment, $x_1 : A_1, \ldots, x_n : A_n$ is called the context and the part $t : A$ on the right-hand side of the $\vdash$ sign is called the conclusion. In the context, x_i is a variable declared of type A_i representing the name of an object. Following the propositions as types paradigm, when A_i denotes a logical proposition, x_i will be a name given to the hypothesis that A_i holds. The judgment can be read as: under the assumption that we have objects x_i of type A_i, the term t is well-formed of type A.

For instance, in order to reason on Peano integers, it is possible to introduce a signature with a type variable $N :$ Type for representing the type of Peano integers, an object $z : N$ (for zero) and an object $S : N \to N$ for the successor function. We call Γ_S the context $N :$ Type$, z : N, S : N \to N$. The following judgments are derivable :

$$\Gamma_S \vdash z : N \qquad \Gamma_S \vdash S\,z : N \qquad \Gamma_S \vdash S\,(S\,z) : N$$

It is also possible to introduce a binary relations le which represents the natural order on N. We shall add a new symbol $le : N \to N \to$ Prop and hypotheses like $\forall x : N, le\,z\,n$ and $\forall x\,y : N, le\,x\,y \to le\,(S\,x)\,(S\,y)$. We introduce a new context Γ_N defined as:

$$\Gamma_S, le : N \to N \to \mathsf{Prop}, lez : (\forall x : N, le\,z\,x), leS : (\forall x\,y : N, le\,x\,y \to le\,(S\,x)\,(S\,y))$$

We shall be able to derive:

$$\Gamma_N \vdash lezz : le\,z\,z \qquad \Gamma_N \vdash leSz\,z\,(lezz) : le\,(S\,z)\,(S\,z)$$

It is easy, for every natural number n to find a term l_n such that

$$\Gamma_N \vdash l_n : le\,(S^n z)\,(S^n z)$$

We can do it (at the meta level) via a simple induction on n. It would be nice to prove internally the property $\forall x : N, le\,x\,x$ but our context is too weak for that because it contains no induction property, so N might contain more objects than the ones written $S^n\,z$. In first-order logic, the induction principle for a property

118

P is added for every possible property P, leading to an infinite set of axioms. In a higher-order logic like the Calculus of Constructions it is sufficient to add a single proposition as an axiom that will contain a quantification over all possible properties P. We introduce a new context Γ_P defined as

$$\Gamma_N, ind : (\forall P : N \to \mathsf{Prop}, P\, z \to (\forall x : N, P\, x \to P\, (S\, x)) \to \forall x : N, P\, x)$$

and we can derive the judgment

$$\Gamma_P \vdash ind\,(\mathbf{fun}\, x : N \Rightarrow le\, x\, x)(le z\, z)(\mathbf{fun}\, x\, (I : le\, x\, x) \Rightarrow le S\, x\, x\, I) : \forall x : N, le\, x\, x$$

Type-checking proof-terms can be tricky and cumbersome. The user usually does not want and does not need to do it, because proofs are built just interacting with the properties that have to be proven using high-level programs called tactics (see section 3.2). However, the proof-term is always built underneath by the system and given for type checking (which is also proof-checking) to a trusted kernel of the system that will guaranty the absence of flaw in the proof. This proof term can be checked independently by different programs and the information contained in the proof term can also be used for instance for analyzing dependency and intelligent printing.

Formally a CIC judgment will be of the form $\Gamma \vdash t : A$ with Γ a context and t and A two terms (A being also a type). The weaker judgment $\Gamma \vdash$ states that the context Γ is well-formed.

The inference rules corresponding to the pure functional part of the Calculus of Constructions are given in figure 1, with $\mathcal{S} \stackrel{\text{def}}{=} \{\mathsf{Prop}\} \cup \bigcup_{i \in \mathbb{N}} \{\mathsf{Type}_i\}$.

$$\frac{\Gamma \vdash}{\Gamma \vdash \mathsf{Prop} : \mathsf{Type}_1} \qquad \frac{\Gamma \vdash}{\Gamma \vdash \mathsf{Type}_i : \mathsf{Type}_{i+1}}$$

$$\frac{\Gamma \vdash \quad x : A \in \Gamma}{\Gamma \vdash x : A} \qquad \frac{\Gamma \vdash A : s \quad x \notin \Gamma \quad s \in \mathcal{S}}{\Gamma, x : A \vdash}$$

$$\frac{\Gamma, x : A \vdash t : B}{\Gamma \vdash \mathbf{fun}\, x : A \Rightarrow t : \Pi x : A, B} \qquad \frac{\Gamma \vdash f : \Pi x : A, B \quad \Gamma \vdash a : A}{\Gamma \vdash f\, a : B[a/x]}$$

$$\frac{\Gamma, x : A \vdash B : \mathsf{Prop}}{\Gamma \vdash \Pi x : A, B : \mathsf{Prop}} \qquad \frac{\Gamma, x : A \vdash B : \mathsf{Type}_i \quad \Gamma \vdash A : \mathsf{Type}_i}{\Gamma \vdash \Pi x : A, B : \mathsf{Type}_i}$$

$$\frac{\Gamma \vdash t : A \quad \Gamma \vdash B : s \quad A \preceq B}{\Gamma \vdash t : B}$$

Fig. 1. Inference rules for the purely functional part of CIC

We comment some rules in figure 1. There are two different rules for typing a product $\Pi x : A, B$. In both cases the term B should be well-typed in a context where we have $x : A$ and its type should be a sort s. When s is Prop, then the product stays in Prop even if A itself lies in a bigger universe. So $\Pi X :$

Prop, $X \to X$ has type Prop. We say that Prop is an impredicative sort, because one can build new objects in Prop using a universal quantification on the class of all objects in Prop (including the one currently defined). Impredicative systems are powerful but also very fragile in the sense that impredicativity does not interact very well with other features leading rapidly to inconsistent systems. For instance we have Prop of type Type_1 but Prop only can be impredicative, the impredicativity of Type_1 gives an inconsistent system [10], so if we want to build the type $\Pi X : \mathsf{Type}_1, X \to X$ we need a bigger universe Type_2.

Another interesting rule is the last one, called the conversion rule. It says that a term of type A can be also seen as a term of type B, given B is well-formed and the relation $A \preceq B$ holds. This relation serves two purposes. First it implements the universe cumulativity, $\mathsf{Prop} \preceq \mathsf{Type}_1$ and $\mathsf{Type}_i \preceq \mathsf{Type}_{i+1}$: any term typed in one universe can be considered as an element of a bigger universe. Second, it implements the fact that types are considered modulo computation (namely β-equivalence in the current system). The β-reduction rule implements function computation namely the fact that an abstraction of a term t over a variable x applied to a term a behaves like the term t in which x has been replaced by a: $(\mathbf{fun}\, x : A \Rightarrow t)\, a \simeq t[a/x]$. An important aspect of this rule is that this is the same term which is typed by A and by B, so these computation steps are done automatically and do not leave traces in the proof-term. We shall come back to this feature in section 3.2.

Using this purely functional part, it is possible to encode many interesting notions. For instance $\forall C : \mathsf{Prop}, C$ is a logical proposition (a term of type Prop) which encodes absurdity ($\bot$) : there is no closed term of type $\forall C : \mathsf{Prop}, C$ (so no proof of $\bot$ without hypothesis) and also from a proof t of $\forall C : \mathsf{Prop}, C$ one can build a proof $t\, C$ of an arbitrary proposition C so the natural deduction rule for eliminating $\bot$ is derivable in the logic.

It is also possible to encode an existential quantification (using a universal quantification in a negative position)

$$\exists x : A, B \stackrel{\text{def}}{=} \forall C : \mathsf{Prop}, (\forall x : A, B \to C) \to C$$

Both the introduction and elimination rules for existential quantification in natural deduction (first column) are derivable (CIC typed terms in second column)

$$\frac{\Gamma \vdash B[t/x]}{\Gamma \vdash \exists x : A, B}$$

$$\frac{\Gamma \vdash p : B[t/x]}{\Gamma \vdash \mathbf{fun}\, C\,(H : \forall x : A, B \to C) \Rightarrow H\, t\, p : (\exists x : A, B)}$$

$$\frac{\Gamma \vdash \exists x : A, B \quad \Gamma, B \vdash C \quad x \notin \Gamma, C}{\Gamma \vdash C}$$

$$\frac{\Gamma \vdash t : \exists x : A, B \quad \Gamma, x : A, p : B \vdash u : C \quad x \notin \Gamma, C}{\Gamma \vdash t\, C\,(\mathbf{fun}\,(x : A)(p : B) \Rightarrow u) : C}$$

One has to be careful that CIC implements a constructive logic and not a classical one. So the stronger form of elimination of absurdity namely

$$\frac{\Gamma, \neg C \vdash \bot}{\Gamma \vdash C}$$

is not provable in general and also one can prove $\exists x, B \vdash \neg\forall x, \neg B$ but not the opposite direction. The meaning of an existential quantification in CIC is stronger than the one in classical logic in the sense that from a proof of $\exists x : A, B$ one will always be able to extract a term t such that $B[t/x]$ is provable while in classical logic one only gets (via Herbrand's theorem) the existence of a finite number of terms $t_1, \ldots, t_k$ such that $B[t_1/x] \vee \ldots \vee B[t_k/x]$ is provable for existential formulas.

Higher-order quantification can also be used to represent logical relations. For instance Leibniz equality $x = y$ with x and y of type A can be encoded using the proposition $\forall P : A \to \mathsf{Prop}, P\,x \to P\,y$. The two rules for introduction and elimination are derivable in CIC

$$\frac{}{\Gamma \vdash t = t} \qquad \frac{\Gamma \vdash t = u \quad \Gamma, x : A \vdash B : \mathsf{Prop} \quad \Gamma \vdash B[t/x]}{\Gamma \vdash B[u/x]}$$

2.2 Inductive Definitions

Inductive definitions are introduced on top of the pure Calculus of Constructions. Their main purpose is to provide an efficient representation of data-types.

A new inductive definition can be added to the environment: it requires to specify its name, its arity (the type of the inductive definition) and the set of its *constructors*.

Examples For instance, natural numbers defined with two constructors z and S, can be introduced as before with the following declaration:

```
Inductive N : Type := z : N | S : N → N.
```

This declaration adds new typed objects $N : \mathsf{Type}, z : N, S : N \to N$ in the context. But unlike in the first-order case, the logic captures the fact that N is the *initial algebra* relatively to these two operations, consequently we shall be able to build functions of type $N \to A$ using the initiality property and to derive properties like $\forall x : N, S\,x \neq z$ and $\forall x\,y : N, S\,x = S\,y \to x = y$. Also the induction principle is provable.

Inductive definitions are also used for defining relations. The order on natural numbers can be defined using the properties we gave as axioms before:

```
Inductive le : N → N → Prop :=
  lez:∀x, le z x
| leS:∀x y, le x y → le (S x) (S y).
```

This declaration introduces $le : N \to N \to \mathsf{Prop}$, $lez : (\forall x : N, le\, z\, x)$, $leS : (\forall x\, y : N, le\, x\, y \to le\,(S\, x)\,(S\, y))$ in the context, we shall also be able to prove the fact that le is the smallest relation R such that $\forall x : N, R\, z\, x$ and $\forall x\, y : N, R\, x\, y \to R\,(S\, x)\,(S\, y)$. It gives us a powerful way to prove lemmas of the form $\forall x\, y : N, le\, x\, y \to R\, x\, y$.

Inductive definitions can be defined with parameters. For instance the reflexive-transitive closure of a binary relation R on a type A can be defined as an inductive definition RT with A and R as parameters:

```
Inductive RT A (R : A → A → Prop) : A → A → Prop :=
  RTrefl:∀ x, RT A R x x
| RTR:∀ x y, R x y → RT A R x y
| RTtran:∀ x y z, RT A R x z → RT A R z y → RT A R x y.
```

We have three constructors for this definition: *RTrefl* states that the reflexive-transitive closure is reflexive, *RTR* says it contains R and *RTtran* that it is transitive. For instance, we can prove that the reflexive-transitive closure of the successor relation on N is equivalent to the previously defined le relation:

```
∀ x y, le x y ↔ RT N (fun x y ⇒ y=S x)
```

Inductive definitions are also used to encode logical operations like absurdity (a definition with no constructor) existential quantification or equality.

```
Inductive False : Prop := .
Inductive ex A (P:A→Prop) : Prop := exists : ∀ x, P x → ex A P.
Inductive eq A (x:A) : A → Prop := eqrefl : eq x x.
```

In first-order logic, when axioms are introduced to form a theory, there is always a risk that it has no model, and consequently everything can be proven. This cannot happen with inductive definitions: there are syntactic restrictions on the type of constructors that ensures the existence of a model.

General rules. The general pattern for declaring an inductive definition is

```
Inductive I pars : Ar := ...
  | c : Π(x₁ : A₁)..(xₙ : Aₙ), I pars u₁..uₚ
  | ...
```

Coq allows the declaration of mutually inductive definitions but, for the sake of simplicity, we shall not give the details here. We introduce some terminology

- *pars* are called the *parameters* of the inductive definition and will be the same for all definitions;
- *Ar* is called the *arity*;
- u_i is an *index*;
- $\Pi(x_1 : A_1)..(x_n : A_n), I\, pars\, u_1..u_p$ is a *type of constructor*
- A_i is a *type of argument* of constructor

There are conditions to accept that the definition is well-formed:

- An arity has the form $\Pi(y_1 : B_1)..(y_p : B_p), s$, with s a sort which is called the *sort* of the inductive definition.
- Type of constructors C are well-typed:

$$(I : \Pi pars, Ar), pars \vdash C : s$$

  - if s is predicative (not Prop) then the condition above on C requires the type of arguments of constructors to be in the same universe: for all i, $A_i : s$ or $A_i : \mathsf{Prop}$
  - if s is Prop, we distinguish between *predicative* definitions where $A_i : \mathsf{Prop}$ for all i and *impredicative* definitions otherwise, meaning there is at least one i such that A_i has type Type_i and not Prop.

There is also a positivity condition: occurrences of I should only occur strictly positively in types of arguments of constructors A_i which means that we are in one of these cases:

- non-recursive case: I does not occur in A_i
- simple case: $A_i = I\, t_1 \ldots t_p$ (not necessarily the same parameters), $I \notin t_k$
- functional case: $A_i = \Pi z : B_1, B_2$ with $I \notin B_1$ and I strictly positive in B_2
- nested case: $A_i = J\, t_1 \ldots t_q$ with J another inductive definition with parameters $X_1 \ldots X_r$. When $t_1 \ldots t_r$ are substituted for $X_1 \ldots X_r$ in the types of constructors of J, the strict positivity condition should still be satisfied. We also need $I \notin t_k$ for $r < k \leq q$.

The language of the PTS is extended with access to the inductive definition and its constructors plus two new constructions for pattern-matching and fixpoint.

The inductive definition itself is a new constant, its type is given by its arity and is generalized with respect to the parameters.

$$I : \Pi pars, Ar$$

The Calculus of Constructions follows the logical rules of natural deduction where each concept is associated with introduction and elimination rules. A computation rules explains how a combination of introduction and elimination rules for the same notion (a cut) can be eliminated.

In the case of inductive definitions, introduction rules are given by the constructors.

Given that c is the i-th constructor of an inductive definition I with parameters $pars$ and type of constructor C, we have:

$$c \equiv \mathtt{Constr}(i, I) : \Pi pars, C$$

Elimination rules use two different notions: a pattern-matching rule extended for dependent types (each branch can have a different type, depending on the constructor) and a (restricted) fixpoint construction for recursive definitions.

The primitive rule for pattern-matching comes in a very primitive way: it covers one level of constructors and should be complete (one branch for each constructor):

$$
\frac{
\begin{array}{l}
t : I \; pars\, t_1 \ldots t_p \\
y_1 \ldots y_p, x : I \; pars\, y_1 \ldots y_p \vdash P : s' \\
\{x_1 : A_1 \ldots x_n : A_n \vdash f : P[u_1/y_1, \ldots, u_p/y_p, (c\,x_1 \ldots x_n)/x]\}_c
\end{array}
}{
\begin{array}{l}
\mathbf{match}\, t \,\mathbf{as}\, x \,\mathbf{in}\, I \,\_\, y_1 \ldots y_p \,\mathbf{return}\, P \\
\mathbf{with} \ldots |\, c\, x_1 \ldots x_n \Rightarrow f | \ldots \\
\mathbf{end} : P[t_1/y_1, \ldots, t_p/y_p, t/x]
\end{array}
}
$$

Reduction rule. The reduction rule (called ι) applies when t starts with a constructor and is as expected (reduces to the corresponding branch after instantiating the pattern variables with the arguments of the constructor).

Examples. This unique rule covers many different situations. In Coq high-level language that we shall use in the examples, we can generally omit the information **as** x, **in** $I \,\_\, y_1 \ldots y_p$ and **return** P.

The **match** construction can be used to define a function by pattern-matching like the predecessor function.

```
Definition pred (x:N) : N := match x with z ⇒ z | S n ⇒ n end.
```

in this case the return predicate P is just the type N. The same match rule can be used to reason by case analysis on integer and prove the following principle:

$$
\forall P : N \rightarrow \mathsf{Prop}, P\,z \rightarrow (\forall x, P\,(S\,x)) \rightarrow (\forall x, P\,x)
$$

Given P, $Hz : P\,z$ and $HS : \forall x, P\,(S\,x)$ the following **case** has type $\forall x, P\,x$

```
Definition case (x:N) : P x
   := match x with z ⇒ Hz | S n ⇒ HS n end.
```

In this example the return predicate is just $P\,n$ and the two branches have two different types: $P\,z$ for the first one and $P\,(S\,n)$ for the second.

The two previous constructions are available in most formalisms. One specificity of CIC is to allow also the definition of new types by pattern-matching. An example can be found when we want to study the semantic of a program and model an environment mapping variables to values. We can represent variables by natural numbers and an environment by a function of type $N \rightarrow T$. However with this simple function type, all values should be in the same type. Assume we know a more precise type for some of the variables (the variable 0 as type N and the variable 1 is $N \rightarrow N$, other variables are in a default type T), we may want to capture this in the type of environments. We can do it with a construction similar to *pred* but returning a Type as the result instead of a number:

```
Definition env (x:N) : Type
   := match x with z ⇒ N | S z ⇒ N → N | S (S x) ⇒ T end.
```

In this example we use a more elaborate pattern-matching construction that Coq compiles into the primitive forms (using two nested simple pattern-matching). The reduction rule gives us the following equivalence between expressions:

$$env\, z \simeq N \qquad env\,(S\,z) \simeq N \to N \qquad env\,(S\,(S\,x)) \simeq T$$

We need again to use the **match** construction to build a specific environment, but now each branch as a different type, the return predicate is $env\,x$.

```
Definition e1 (x:N) : env x :=
  match x with
  | z ⇒ S (S z) | S z ⇒ (fun x : N ⇒ z) | S (S x) ⇒ t
  end.
```

The result $S\,(S\,z)$ in the first branch (constructor z) has type N which is equivalent to the expected type $env\,z$. The result $(\textbf{fun}\, x : N \Rightarrow z)$ in the second branch (constructor $S\,z$) has type $N \to N$ which is equivalent to the expected type $env\,(S\,z)$, the last branch returns a default value t in type T which is equivalent to the expected type $env\,(S\,(S\,x))$.

The **match** operation is also available for inductive relations. For instance, for the equality, it gives us the elimination principle we mentioned before

$$\frac{\Gamma \vdash e : t = u \qquad \Gamma, x : A \vdash B : \mathsf{Prop} \qquad \Gamma \vdash f : B[t/x]}{\Gamma \vdash \textbf{match}\, e \,\textbf{in}\, \_ = x \,\textbf{return}\, B \,\textbf{with}\, eqrefl \Rightarrow f \,\textbf{end} : B[u/x]}$$

Actually the **match** rule gives us a stronger rule, called a dependent elimination, which allows to replace (in certain contexts) a proof e of $t = u$ by $eqrefl\,t$.

If we take the definition of le, the **match** operation will be useful to prove a property like $le\,(S\,x)\,z \to \bot$. Intuitively this is because no constructor can give a proof of an instance of $le\,(S\,x)\,z$. Building the proof term requires to define an invariant relation $inv\,x\,y$ that will be true for all x, y such that $le\,x\,y$ but will be false for $(S\,x)\,z$. It is easily done using **match** to define a proposition which is $\bot$ for $(S\,x)\,z$ and $\top$ for all the other values. We use Coq high-level syntax.

```
Definition inv (x y:N) : Prop :=
  match x,y with S _, z → ⊥ | _,_ ⇒ ⊤ end.
```

We then can build the proof (the term I is a trivial proof of $\top$)

```
Definition leinv n (H:le (S n) z) : ⊥ :=
  match H in le x y return inv x y with
  | lez x ⇒ I | leS x y p ⇒ I end.
```

Type-checking conditions. The main restriction lies in the relation between the sort s of the inductive definition and the sort s' of the pattern-matching.

When s is Type, which means that we have a predicative inductive definition, then we can have any possible sorts s' for case analysis.

When s is Prop however, the question is a bit more tricky for several reasons:

- Prop is an impredicative sort, so uncareful elimination can easily introduce paradoxes;
- it is sometimes useful to add an axiom of proof irrelevance for propositions (which says that two different proofs of the same property can be considered as equal) so while it is good to be able to prove that for instance $\mathtt{true} \neq \mathtt{false}$, a similar mechanism that will lead to two terms (representing proofs of) in $A \vee B$ that are provably different is less desirable;
- Prop is used for program extraction: any term in $A : \mathsf{Prop}$ is removed during extraction so should not be needed for computing the informative part. When a pattern-matching is done on a term in an inductive definition in Prop, but with the result being used for computing, then we need to be able to execute the match without executing the head, which is only feasible in specific cases.

For an inductive definition of sort Prop, the only elimination allowed is on the sort Prop itself. There are exceptions where any elimination is allowed: in the specific case where I is a *predicative* definition with only zero or one constructor with all its arguments $A_i : \mathsf{Prop}$. The exception covers cases like absurdity, equality, conjunction of two propositions, accessibility. . .

Fixpoints. Fixpoint constructions in Coq are mainly introduced via global declarations.

```
Fixpoint f (x_1:A_1)... (x_m:A_m){struct x_n}:B:=t.
```

they correspond to an internal fixpoint construction

```
fix f (x_1:A_1)... (x_n:A_n):Π(x_{n+1} : A_{n+1})... (x_m : A_m),B
   := fun x_{n+1}... x_m ⇒ t.
```

In general, an expression **fix** $f(x_1 : A_1)\dots(x_n : A_n) : B := t$ is well typed of type B when

- t is well-typed of type B in an environment containing $(f : \Pi(x_1 : A_1)\dots(x_n : A_n), B)$ and $(x_1 : A_1)\dots(x_n : A_n)$;
- t satisfies an extra syntactic condition that recursive calls to $(f\, u_1 \dots u_n)$ in t are made on terms u_n *structurally* smaller than x_n.

The reduction rule is the usual fixpoint reduction except that in order to avoid infinite loops, it is only activated when the n-th argument of the fixpoint starts with a constructor.

3 Properties

3.1 Proof-Theoretical Properties

The main property of the system is decidability of type-checking: we need to be able to say that a proof is correct. It requires the relation $A \preceq B$ to be decidable, which is done by showing the system is strongly normalizing with respect to the

126

computation rules. Another important property is to prove that any closed term with type an inductive definition will reduce to a value starting with one of the constructors of the inductive definition. Consistency can be derived as a consequence because they cannot be a proof without hypothesis of $\perp$ which has no constructor.

Proof theoretical properties of systems like the Calculus of Constructions are complex to perform in full detail, first because these systems are logically powerful (due to the impredicativity, the hierarchy of universes, the type dependency) and second because there are many syntactic properties to be established like the Church-Rosser property or subject-reduction which are made even more complicated because of the general pattern of inductive definitions. Several proofs covering subsystems of Coq exists, Bruno Barras in his thesis [3] formalized and proved meta-theoretical properties (including typing decidability assuming normalization) of a Calculus of Constructions with Inductive Definitions.

3.2 Pragmatic Properties

Dependent types. The type system of CIC allows types to depend on objects in many different ways. We have seen propositions parameterized by objects either defined inductively (*le*) or in a computational way using pattern-matching and fixpoints (*inv*).

It is also possible to mix computational and logical parts. For instance one can build the type of even numbers by defining a predicate even and then introducing the type

Inductive Ne : Type := Nec : $\Pi n : N$, even n $\to$ Ne.

An object in the type Ne will be a pair containing a natural number plus a proof that this number is even. This definition looks like the one for the existential quantifier given in section 2 except that it is a type and not a proposition and consequently it is possible to define a projection function of type $Ne \to N$.

Another possibility to add logical information inside a type is to associate an index to the type declaration representing this information. A classical example is the one of vectors of length n.

Inductive vec (A: Type) : N $\to$ Type
    := v0 : vec A z | vS : A $\to$ vec A n $\to$ vec A (S n).

We could also associate a predicate to a list describing the set of its elements, we can even use this predicate to ensure that there are no duplicate elements in the list.

Inductive L (A: Type) : (*set* A) $\to$ Type
    := L0 : vec $\emptyset$
    | L1 : $\Pi(P : set A)$ $(x : A)$, $(x \notin P) \to$ L A P $\to$ L A $(P \cup \{x\})$.

However manipulating terms in these dependent types might become tricky because the system will generally see the types $vec\,A\,n$ and $vec\,A\,m$ as different, even when n and m can be proven equal if they are not identical with respect to CIC internal equivalence.

Declarative specifications. Inductive definitions of relations can be used for a declarative style of specifications. They are a natural way to encode relations like reachability, or semantics of programming languages or transition systems. Proofs can be done using a resolution-like mechanism.

Let us take an example. Assume we want to study a Post correspondence problem given by the three pairs of words on the alphabet $\{a, b\}$ (example taken from Wikipedia):

$$P \stackrel{\text{def}}{=} \{(a, baa); (ab, aa); (bba, bb)\}$$

We first introduce a data-type in Coq to encode words on the alphabet. We have three constructors: one for the empty word and two unary constructors to add the letter a (resp. b) in front of a word.

```
Inductive word: Type := emp | a : word → word | b : word → word.
```

An inductive relation is used to define a binary relation between words called post: two words u and v are related is there exists a sequence of indexes $i_1, \ldots, i_k$ such that $u = u_{i_1} \ldots u_{i_k}$ and $v = v_{i_1} \ldots v_{i_k}$ and $(u_{i_j}, v_{i_j}) \in P$. The constructors of the inductive definition correspond to the basic case where the sequence (and consequently the words u and v) is empty plus one constructor for each pair of words in P.

```
Inductive post : word → word → Prop :=
    start : post emp emp
  | R1 : ∀ x y, post x y → post (a x) (b (a (a y)))
  | R2 : ∀ x y, post x y → post (a (b x)) (a (a y))
  | R3 : ∀ x y, post x y → post (b (b (a x))) (b (b y)).
```

Now a solution is given by a word x which is not empty and such that $(post\,x\,x)$ is provable. A non-empty word can be written ax or bx. One defines the proposition which states that there exists a solution (either starting with a or with b).

```
Inductive sol : Prop :=
  sola : ∀ x, post (a x) (a x) → sol
 |solb : ∀ x, post (b x) (b x) → sol.
```

Now finding a solution is just finding a proof of *sol*. It can be done using automatic tactics. First we declare the constructors of the two definitions *post* and *sol* in the Hint database.

```
Hint Constructors post.
Hint Constructors sol.
```

We can now solve the goal using an automatic tactic which tries to apply the constructors using unification and backtracking:

```
Lemma ok : sol.
eauto 6.
Defined.
```

The **Defined** command at the end builds the proof term which is type-checked by the kernel of Coq verifying that it is indeed a valid term of CIC. The term can be printed

```
ok =
solb (b (a (a (b (b (b (a (a emp)))))))))
  (R3 (a (b (b (b (a (a emp)))))) (a (a (b (b (b (a (a emp)))))))
    (R2 (b (b (a (a emp)))) (b (b (b (a (a emp)))))
      (R3 (a emp) (b (a (a emp))) (R1 emp emp start))))
    : sol
```

We see that the proof contains both the witness word, in this case $u \overset{\text{def}}{=} bbaabbbaa$ (the first b comes from the fact that the constructor $solb$ is used), and a proof of $post\,u\,u$ where we find the sequence 3231 corresponding to the decomposition.

3.3 Computation and Reflection

Inductive definitions help represent data-structures without extra encoding and provide primitives to define recursive functions on these data. The Coq language contains a mini ML sub-language (with no side effect and only terminating functions). It is convenient for formalizing complex programs which can then be proven, the most impressive example being the CompCert project of an optimizing compiler for C programs [17]. Computation is part of the Coq kernel, many efforts have been made to make it more efficient using compiler technologies.

We can use the underlying language to implement inside Coq various decision procedures. For instance we can have a concrete data structures R to represent symbolic ring expressions with variables, constants, and ring operations. The ring equality eqR can be defined using an inductive definition. Then a simplification function $simpl$ of type $R \to R$ can be defined inside Coq and proven correct with respect to equality $\forall x : R, eqR\,x\,(simpl\,x)$.

We can then prove a scheme of reflection which allows to use the result of the execution of the procedure in order to build proofs of complex facts. If we have in Coq another type C with some operations which have a Ring structure, then it is easy to define recursively a function val of type $R \to C$ which interprets a symbolic expression as an element of C, given an environment which maps variables to elements of C. We have to prove that $\forall (x\,y : R)(\rho : X \to C), eqR\,x\,y \to val\,\rho\,x = val\,\rho\,y$, which is just the justification that our structure C is indeed a ring.

If we put the decision procedure on R and the C interpretation together we can build a proof of

$$\forall (x\,y : R)(\rho : X \to C), val\,\rho\,(simpl\,x) = val\,\rho\,(simpl\,y) \to val\,\rho\,x = val\,\rho\,y$$

This lemma states the (partial)-correctness of the procedure. It is proven once but can be instantiated on many different problems.

Let us illustrate that on our ring example. We want to prove a goal of the form $t = u$ between two Coq expressions in the concrete type C. It might require many rewritings involving associativity, commutativity, distributivity, resulting in very large proof terms. Instead we take a detour via our symbolic expressions in R and then use computation. We first write a tactic which guesses by pattern-matching two symbolic expressions p and q (closed terms in R) and an environment ρ such that $val\,\rho\,p \simeq t$ and $val\,\rho\,q \simeq u$. There is always at least one trivial solution with p a variable and $\rho = \{x \mapsto t\}$ for the environment. But of course we want to capture in R as much structure as possible. Now the expressions $simpl\,p$ and $simpl\,q$ can be computed. If they happen to be the same, then $val\,\rho\,(simpl\,x)$ and $val\,\rho\,(simpl\,y)$ are the same Coq term v and so reflexivity justifies that $val\,\rho\,(simpl\,x) = val\,\rho\,(simpl\,y)$. Using the correctness of the procedure we get a proof of $t = u$ that may require a lot of computations but which corresponds to a very simple proof term $correct\,p\,q\,\rho\,(eqrefl\,v)$.

This principle can be used for complex procedures but also for simple reasoning steps. The popular Ssreflect [16] (for small scale reflection) environment (including a tactic language and libraries) which has been successfully used for formalizing in Coq the four colour theorem and the Feit-Thompson theorem uses intensively this computational capability of the Coq system mainly on the type of booleans.

4 Proof Applications

Coq is developed for more than 30 years now and there has been a lot of impressive examples formalized using it.

Many interesting proofs combine advanced algorithms and non-trivial mathematics like the proof of the four-colour theorem by Gonthier & Werner at INRIA and Microsoft-Research [15], a primality checker using Pocklington and Elliptic Curve Certificates developed by Théry et al. at INRIA [25] and the proof of a Wave Equation Resolution Scheme by Boldo et al. [7]. Coq can also be used to certify the output of external theorem provers like in the work on termination tools by Contejean and others [9], or the certification of traces issued from SAT & SMT solvers done by Grégoire and others [2]. Coq is also a good framework for formalizing programming environments: the Gemalto and Trusted Logic companies obtained the highest level of certification (common criteria EAL 7) for their formalization of the security properties of the JavaCard platform [8]; as mentioned earlier Leroy and others developed in Coq a certified optimizing compiler for C (Leroy et al.) [17]. Barthe and others used Coq to develop Certicrypt, an environment of formal proofs for computational cryptography [4]. G. Morrisett and others also developed on top of Coq the YNOT library for proving imperative programs using separation logic [19]. These represent typical examples of what can be achieved using Coq.

5 Trends and Open Problems

The current inductive definitions of Coq present certain drawbacks. The syntactic condition for accepting fixpoints is very sensitive and not well-suited when developing a proof using tactics. Different approaches using type annotations have been proposed instead (for instance [1]) but none of them is yet available for Coq. Also the primitive pattern-matching is not the natural expected rule when dealing with elimination of a particular instance of an inductive definition, where you expect some cases to disappear or to be partially instantiated.

In systems like Coq there always has been a trade-off between keeping the language and the kernel small enough to ensure correctness and use encodings for more high-level constructions or include these constructions directly in the language.

In general the defined equality in the Calculus of Inductive Constructions does not have all the expected properties. The current work on Homotopy Type Theory [26] is an attempt to solve this problem. It includes a notion of generalized inductive definitions where the equality definition is included in the declaration, making the definition of quotient types more direct.

6 Conclusions

The Calculus of Inductive Constructions provides a powerful language for the interactive development of proofs and programs. It includes a mini functional programming languages that is sufficient for programming complex data-structures and programs. The specification language itself can use a declarative style with (almost) no limit to the expressiveness. It is integrated in a proof environment (the Coq system) which provides many other functionalities like modules, libraries, notations, type inference, tactics which are essential for a practical use of the formalism. We refer the interested reader to more extensive descriptions of the system like [5, 21].

References

1. Andreas Abel. *A Polymorphic Lambda-Calculus with Sized Higher-Order Types.* PhD thesis, Ludwig-Maximilians-Universität München, 2006.
2. M. Armand, G. Faure, B. Grégoire, Ch. Keller, L. Théry, and B. Werner. A Modular Integration of SAT/SMT Solvers to Coq through Proof Witnesses. In Jean-Pierre Jouannaud and Zhong Shao, editors, *Certified Programs and Proofs, CPP 2011*, Lecture Notes in Computer Science. Springer, 2011.
3. B. Barras. *Auto-validation d'un système de preuves avec familles inductives.* Thèse de doctorat, Université Paris 7, 1999.
4. G. Barthe, B. Grégoire, and S. Zanella Béguelin. Formal certification of code-based cryptographic proofs. In *36th ACM SIGPLAN-SIGACT Symposium on Principles of Programming Languages, POPL 2009*, pages 90–101. ACM, 2009. See also: CertiCrypt http://www.msr-inria.inria.fr/projects/sec/certicrypt.
5. Y. Bertot and P. Castéran. *Interactive Theorem Proving and Program Development.* Springer-Verlag, 2004.
6. C. Böhm and A. Berarducci. Automatic synthesis of typed λ-programs on term algebras. *Theoretical Computer Science*, 39, 1985.

7. S. Boldo, F. Clément, J.-C. Filliâtre, M. Mayero, G. Melquiond, and P. Weis. Formal Proof of a Wave Equation Resolution Scheme: the Method Error. In Matt Kaufmann and Lawrence C. Paulson, editors, *Proceedings of the first Interactive Theorem Proving Conference*, volume 6172 of *LNCS*, pages 147–162, Edinburgh, Scotland, July 2010. Springer. Extended version `http://hal.inria.fr/hal-00649240/PDF/RR-7826.pdf`.

8. B. Chetali and Q.-H. Nguyen. About the world-first smart card certificate with EAL7 formal assurances. Slides 9th ICCC, Jeju, Korea, September 2008. `www.commoncriteriaportal.org/iccc/9iccc/pdf/B2404.pdf`.

9. E. Contejean, A. Paskevich, X. Urbain, P. Courtieu, O. Pons, and J. Forest. A3pat, an approach for certified automated termination proofs. In John P. Gallagher and Janis Voigtländer, editors, *Proceedings of the 2010 ACM SIGPLAN Workshop on Partial Evaluation and Program Manipulation, PEPM 2010*, pages 63–72. ACM, 2010.

10. Th. Coquand. An analysis of girard's paradox. In *Symposium on Logic in Computer Science*, Cambridge, MA, 1986. IEEE Computer Society Press.

11. Th. Coquand and G. Huet. Constructions: A higher order proof system for mechanizing mathematics. In *EUROCAL'85*, volume 203 of *Lecture Notes in Computer Science*, Linz, 1985. Springer-Verlag.

12. Th. Coquand and G. Huet. The Calculus of Constructions. *Information and Computation*, 76(2/3), 1988.

13. Th. Coquand and C. Paulin-Mohring. Inductively defined types. In P. Martin-Löf and G. Mints, editors, *Proceedings of Colog'88*, volume 417 of *Lecture Notes in Computer Science*. Springer-Verlag, 1990.

14. P. Dybjer. Inductive families. *Formal Asp. Comput.*, 6(4):440–465, 1994.

15. G. Gonthier. Formal proof the four-color theorem. *Notices of the AMS*, 55(11):1382–1393, December 2008. `http://www.ams.org/notices/200811/tx081101382p.pdf`.

16. G. Gonthier and A. Mahboubi. An introduction to small scale reflection in coq. *J. Formalized Reasoning*, 3(2):95–152, 2010.

17. X. Leroy. Formal verification of a realistic compiler. *Commun. ACM*, 52(7):107–115, 2009.

18. P. Martin-Löf. *Intuitionistic Type Theory*. Studies in Proof Theory. Bibliopolis, 1984.

19. G. Morrisett and al. The Ynot project. `http://ynot.cs.harvard.edu/`.

20. C. Paulin-Mohring. Inductive Definitions in the System Coq - Rules and Properties. In M. Bezem and J.-F. Groote, editors, *Proceedings of the conference Typed Lambda Calculi a nd Applications*, number 664 in Lecture Notes in Computer Science, 1993. LIP research report 92-49.

21. C. Paulin-Mohring. *Tools for Practical Software Verification, LASER 2011 summerschool, Revised Tutorial Lectures*, chapter Introduction to the Coq proof-assistant for practical software verification, pages 45–95. Number 7682 in Lecture Notes in Computer Science. Springer-Verlag, 2012.

22. L.C. Paulson. A fixedpoint approach to implementing (co)inductive definitions. In Alan Bundy, editor, *Proceedings of the 12th International Conference on Automated Deduction*, volume 814 of *LNAI*, pages 148–161. Springer-Verlag, 1994.

23. F. Pfenning and C. Paulin-Mohring. Inductively defined types in the Calculus of Constructions. In *Proceedings of Mathematical Foundations of Programming Semantics*, volume 442 of *Lecture Notes in Computer Science*. Springer-Verlag, 1990. technical report CMU-CS-89-209.

24. The Coq Development Team. *The Coq Proof Assistant Reference Manual – Version V8.4*, 2012. http://coq.inria.fr.
25. L. Théry and G. Hanrot. Primality proving with elliptic curves. In Klaus Schneider and Jens Brandt, editors, *Theorem Proving in Higher Order Logics, TPHOLs 2007*, volume 4732 of *Lecture Notes in Computer Science*, pages 319–333. Springer, 2007. See also: Certifying Prime Number with the Coq prover `http://coqprime.gforge.inria.fr/`.
26. The Univalent Foundations Program. *Homotopy Type Theory Univalent Foundations of Mathematics*. Institute for Advanced Study, 2013. `http://homotopytypetheory.org/book/`.

Deduction modulo theory

Gilles Dowek

Inria, 23 avenue d'Italie, CS 81321, 75214 Paris Cedex 13, France.
gilles.dowek@inria.fr

1 Introduction

1.1 Weaker vs. stronger systems

Contemporary proof theory goes into several directions at the same time. One of them aims at analyzing proofs, propositions, connectives, etc., that is at decomposing them into more atomic objects. This often leads to design systems that are weaker than Predicate logic, but that have better algebraic or computational properties, and to try to reconstruct part of Predicate logic on top of these systems. Propositional logic, linear logic, deep inference, equational logic, explicit substitution calculi, etc. are examples of such systems. From this point of view, Predicate logic appears more as the ultimate goal of the journey, than as its starting point.

Another direction considers that very little can be expressed in pure Predicate logic and that stronger systems are needed, for instance to express genuine mathematical proofs. Axiomatic theories, modal logics, type theories, etc. are examples of such systems that are more expressive than pure Predicate logic. There, Predicate logic is the starting point of the journey.

Although both points of view coexist in many research projects, these two approaches to proof theory often lead to different systems and different problems. *Deduction modulo theory* is part of the second group, as it focuses on proofs in theories. The concern of integrating theories to proof theory is that of several research groups. See, for instance, [52] and [54] for related approaches.

1.2 Logical vs. theoretical systems

To design a system stronger than pure Predicate logic, several ways are possible. One is to extend Predicate logic with new logical operators, that is to design a logic, the second is to introduce function symbols and predicate symbols within Predicate logic and state axioms expressing the meaning of these symbols, that is to design a theory. The first approach can be illustrated by modal logics, the second by arithmetic or set theory. Simple type theory belongs to both groups as it can be defined either as a logic, in which case it is more often called *higher-order logic*, or as a theory in Predicate logic [32].

Deduction modulo theory is part of the second, theoretical rather than logical, group, as, like Predicate logic, it is a framework in which it is possible to define many theories.

1.3 Axioms vs. reduction rules

But, the main difference between Deduction modulo theory and the axiomatic approach is that a, in Deduction modulo theory, *theory* is not defined as a set of axioms, but as a set of reduction rules.

Indeed, axioms jeopardize most of the properties of proofs of pure Predicate logic. For instance, in pure Predicate logic, a constructive Natural deduction cut free proof always ends with an introduction rule, hence a constructive cut free existential proof always ends with an introduction rule of the existential quantifier. But this result does not extend to axiomatic theories, as a constructive cut free proof in a theory may also end, for instance, with the axiom rule.

In the same way, in automated theorem proving in pure Predicate logic, the search space of the proposition $\bot$ is always finite. But this result does not extend to axiomatic theories, that can generate an infinite search space for the proposition $\bot$.

To overcome these problems, axioms, in Deduction modulo theory, are replaced by sets of reduction rules. For instance, the axioms

$$\forall y \ (0 + y = y)$$

$$\forall x \forall y \ (S(x) + y = S(x + y))$$

are replaced by the reduction rules

$$0 + y \longrightarrow y$$

$$S(x) + y \longrightarrow S(x + y)$$

These reduction rules define a congruence $\equiv$ on propositions, and deduction is performed modulo this congruence. For instance, with the reduction rules above the propositions $2 + 2 = 4$ and $4 = 4$ are congruent, hence any proof of the latter is a proof of the former. If, to define equality, we add the following rules [1], which directly rewrite atomic propositions

$$0 = 0 \longrightarrow \top$$

$$S(x) = 0 \longrightarrow \bot$$

$$0 = S(y) \longrightarrow \bot$$

$$S(x) = S(y) \longrightarrow x = y$$

then the proposition $2 + 2 = 4$ and $\top$ are congruent, and any proof of $\top$—for instance the mere application of the introduction rule for $\top$—is a proof of the proposition $2 + 2 = 4$

$$\frac{}{\vdash 2 + 2 = 4} \ \top\text{-intro}$$

1.4 Deduction vs. computation

In the example above, the proposition $2 + 2 = 4$ is provable because it reduces to $\top$. More generally, all propositions that reduce to $\top$ are provable. But the converse is not true: not all provable propositions reduce to $\top$. Indeed, reducibility to $\top$ is often a decidable property, while provability is not.

On the contrary, the fact that the proposition $2 + 2 = 4$ has a trivial proof, because it reduces to $\top$, shows that the truth of this proposition rests on a mere computation and not on a genuine deduction.

Thus, Deduction modulo theory also permits one to distinguish the computation part from the deduction part within a proof, whereas Predicate logic flattens computation and deduction at the same level.

1.5 The origins of Deduction modulo theory

Deduction modulo theory was first introduced in the area of automated theorem proving.

Indeed, in Resolution, or in other automated theorem proving methods, instead of using equational axioms, for instance the associativity axiom, we often replace standard unification with equational unification, for instance unification modulo associativity [57]. In the same way, in Simple type theory, instead of using the β-conversion axiom, we replace standard unification with equational unification modulo β-equivalence: higher-order unification [2, 48, 49]. The automated theorem proving method obtained this way is called *Equational resolution*.

A way to prove the soundness and completeness of Equational resolution is to introduce a Natural deduction system, or a Sequent calculus system, where propositions are identified modulo associativity, or modulo β-equivalence. Then, this system can be proved to be equivalent to the axiomatic presentation of the theory. Finally, the soundness and completeness of Equational resolution are proved relatively to this system, where every deduction step is performed modulo the congruence.

So Deduction modulo theory comes from automated theorem proving. But it was soon understood that this idea of identifying propositions modulo a congruence was also the idea behind the notion of definitional equality in Martin-Löf's Intuitionistic type theory [53] and that Deduction modulo theory could also be seen as an extension of this notion of definitional equality to Predicate logic.

Another source of inspiration is the extension of Natural deduction with folding and unfolding rules, introduced by Prawitz [58, 23, 43, 40, 24]. In this system, it is not possible to identify an atomic proposition P with a proposition A. But, it is possible to introduce non logical deduction rules

$$\frac{A}{P} \qquad\qquad \frac{P}{A}$$

The relation between the two frameworks is detailed in [28].

2 Proof Systems

The idea of reasoning modulo a theory can be used in different formalisms: Natural deduction, Sequent calculus, λ-calculus, etc. Thus, Deduction modulo theory exists in many flavors.

2.1 Natural Deduction modulo theory

Let us start with constructive Natural deduction. The rules of constructive Natural deduction modulo theory are obtained by transforming the rules of constructive Natural deduction, to allow to use of the congruence. For instance, the rule

$$\frac{\Gamma \vdash A \Rightarrow B \quad \Gamma \vdash A}{\Gamma \vdash B} \Rightarrow\text{-elim}$$

is transformed into

$$\frac{\Gamma \vdash C \quad \Gamma \vdash A}{\Gamma \vdash B} \Rightarrow\text{-elim} \quad \text{if } C \equiv (A \Rightarrow B)$$

where the proposition $A \Rightarrow B$ is replaced by any congruent proposition C. Applying the same transformation to all Natural deduction rules yields the system of Figure 1.

For instance, consider the congruence defined by the *subset reduction rule*

$$x \subseteq y \longrightarrow \forall z \, (z \in x \Rightarrow z \in y)$$

The sequent $\vdash s \subseteq s$ has the proof

$$\frac{\dfrac{\overline{z \in s \vdash z \in s} \text{ axiom}}{\vdash z \in s \Rightarrow z \in s} \Rightarrow\text{-intro}}{\vdash s \subseteq s} \langle z, z \in s \Rightarrow z \in s \rangle \; \forall\text{-intro}$$

Note that when two propositions A and B are provably equivalent, that is when $A \Leftrightarrow B$ is provable, then the proposition A has a proof if and only if the proposition B has a proof, but the propositions A and B need not have the same proofs. In contrast, when two propositions are congruent, that is when $A \equiv B$, then every proof of A is a proof of B and vice versa, thus the propositions A and B have the same proofs.

Sequent calculus modulo theory can be defined in the same way: the rule

$$\frac{\Gamma \vdash A \quad \Gamma, B \vdash \Delta}{\Gamma, A \Rightarrow B \vdash \Delta} \Rightarrow\text{-left}$$

for instance, is transformed into

$$\frac{\Gamma \vdash A \quad \Gamma, B \vdash \Delta}{\Gamma, C \vdash \Delta} \Rightarrow\text{-left} \quad \text{if } C \equiv (A \Rightarrow B)$$

137

$$\frac{}{\Gamma, A \vdash B} \text{ if } A \equiv B \quad \text{axiom}$$

$$\frac{}{\Gamma \vdash A} \text{ if } A \equiv \top \quad \top\text{-intro}$$

$$\frac{\Gamma \vdash B}{\Gamma \vdash A} \text{ if } B \equiv \bot \quad \bot\text{-elim}$$

$$\frac{\Gamma \vdash A \quad \Gamma \vdash B}{\Gamma \vdash C} \text{ if } C \equiv (A \wedge B) \quad \wedge\text{-intro}$$

$$\frac{\Gamma \vdash C}{\Gamma \vdash A} \text{ if } C \equiv (A \wedge B) \quad \wedge\text{-elim}$$

$$\frac{\Gamma \vdash C}{\Gamma \vdash B} \text{ if } C \equiv (A \wedge B) \quad \wedge\text{-elim}$$

$$\frac{\Gamma \vdash A}{\Gamma \vdash C} \text{ if } C \equiv (A \vee B) \quad \vee\text{-intro}$$

$$\frac{\Gamma \vdash D \quad \Gamma, A \vdash C \quad \Gamma, B \vdash C}{\Gamma \vdash C} \text{ if } D \equiv (A \vee B) \quad \vee\text{-elim}$$

$$\frac{\Gamma \vdash B}{\Gamma \vdash C} \text{ if } C \equiv (A \vee B) \quad \vee\text{-intro}$$

$$\frac{\Gamma, A \vdash B}{\Gamma \vdash C} \text{ if } C \equiv (A \Rightarrow B) \quad \Rightarrow\text{-intro}$$

$$\frac{\Gamma \vdash C \quad \Gamma \vdash A}{\Gamma \vdash B} \text{ if } C \equiv (A \Rightarrow B) \quad \Rightarrow\text{-elim}$$

$$\frac{\Gamma \vdash A}{\Gamma \vdash B} \langle x, A \rangle \ \forall\text{-intro} \text{ if } B \equiv (\forall x\ A) \text{ and } x \notin FV(\Gamma)$$

$$\frac{\Gamma \vdash B}{\Gamma \vdash C} \langle x, A, t \rangle \ \forall\text{-elim} \text{ if } B \equiv (\forall x\ A) \text{ and } C \equiv [t/x]A$$

$$\frac{\Gamma \vdash C}{\Gamma \vdash B} \langle x, A, t \rangle \ \exists\text{-intro} \text{ if } B \equiv (\exists x\ A) \text{ and } C \equiv [t/x]A$$

$$\frac{\Gamma \vdash C \quad \Gamma, A \vdash B}{\Gamma \vdash B} \langle x, A \rangle \ \exists\text{-elim} \text{ if } C \equiv (\exists x\ A) \text{ and } x \notin FV(\Gamma B)$$

Fig. 1. Natural Deduction Modulo Theory

where the proposition $A \Rightarrow B$ is replaced by any proposition C such that $C \equiv (A \Rightarrow B)$. And the other rules are transformed alike. See, for instance, [38] for a description of the full system.

Another variant of Natural deduction modulo theory and Sequent calculus modulo theory is *Super-deduction* [62, 18]. In Super-deduction, new deduction rules are computed from the reduction rules. For instance, the subset reduction rule yields the deduction rules

$$\frac{\Gamma, z \in x \vdash z \in y}{\Gamma \vdash x \subseteq y} \, z \notin FV(\Gamma)$$

$$\frac{\Gamma \vdash x \subseteq y \quad \Gamma \vdash z \in x}{\Gamma \vdash z \in y}$$

These rules are closer to the informal mathematical style than, for instance, Natural deduction rules. Indeed, to prove $x \subseteq y$, we often consider a generic element in x and prove that it is in y without using the universal quantifier and the implication of the proposition $\forall z \ (z \in x \Rightarrow z \in y)$. The fact that these derived rules use atomic propositions only also explains why connectives and quantifiers are less often used in informal proofs than in formal ones.

2.2 Polarized deduction modulo theory

In Natural deduction modulo theory and in Sequent calculus modulo theory, the reduction rules are just used to define the congruence $\equiv$. In fact, this congruence does not even need to be defined with reduction rules and it could be any congruence, provided it is decidable and it does not identify non-atomic propositions with different head symbols. But we may also want to stress that computation is oriented and take, in these rules, the condition $C \longrightarrow^* (A \Rightarrow B)$ instead of $C \equiv (A \Rightarrow B)$, meaning that in the sequent $\Gamma, C \vdash \Delta$, the proposition C can only be reduced.

In particular, the axiom rule

$$\frac{}{\Gamma, A \vdash B} \text{ axiom} \quad \text{if } A \equiv B$$

would be restated

$$\frac{}{\Gamma, A \vdash B} \text{ axiom} \quad \text{if } A \longrightarrow^* C \text{ and } B \longrightarrow^* C$$

If the theory contains rewrite rules on terms only, and t and u are two terms such that $t \equiv u$, it is still possible to prove the sequent $P(t) \vdash P(u)$. But when t and u do not have a common reduct, the proof of $P(t) \vdash P(u)$ contains cuts. In other words, in this particular case, the Sequent calculus modulo theory has the cut elimination property if and only if the reduction system is confluent [30] and Newman's algorithm [55]—which permits transforming an equational proof into a valley proof—is a cut-elimination algorithm.

This idea of using a rewrite relation rather than a congruence in the deduction rules can be developed further: the subset reduction rule permits to prove the equivalence

$$x \subseteq y \Leftrightarrow \forall z \ (z \in x \Rightarrow z \in y)$$

Thus, when the atomic proposition P reduces to the proposition A, P and A must be equivalent. For instance, it is not possible to reduce $Equilateral(x)$ to $Isosceles(x)$ because a triangle may be isosceles without being equilateral.

More generally, it is easy to transform an axiom of the form $P \Leftrightarrow A$ into a reduction rule $P \longrightarrow A$, but, although it is possible [17], it is not easy to transform an axiom of the form $P \Rightarrow A$ into a reduction rule. We want to replace such an axiom with a rule that permits reducing P into A when P is a hypothesis, but not when it is a goal.

This leads to an extension of Deduction modulo theory, called *Polarized deduction modulo theory* where reduction rules are classified into positive and negative, the positive rules may apply to the positive occurrences of atomic propositions and the negative ones to the negative occurrences.

For instance, in Polarized sequent calculus modulo theory, the left rule of the implication is stated

$$\frac{\Gamma \vdash A \quad \Gamma, B \vdash \Delta}{\Gamma, C \vdash \Delta} \ \Rightarrow\text{-left} \quad \text{if } C \longrightarrow^*_- (A \Rightarrow B)$$

and its right rule

$$\frac{\Gamma, A \vdash B}{\Gamma \vdash C} \ \Rightarrow\text{-right} \quad \text{if } C \longrightarrow^*_+ (A \Rightarrow B)$$

Polarized deduction modulo theory is the flavor of Deduction modulo theory that is more often used in automated theorem proving.

The first reason is that, in clause based theorem proving, a reduction rule of the form

$$x \in y \cup z \longrightarrow x \in y \lor x \in z$$

can be used to reduce a positive literal in a clause but not a negative one. For instance, the clause $L_1 \lor L_2 \lor a \in b \cup c$ reduces to the clause $L_1 \lor L_2 \lor a \in b \lor a \in c$, but the clause $L_1 \lor L_2 \lor \neg a \in b \cup c$ reduces to the proposition $L_1 \lor L_2 \lor \neg(a \in b \lor a \in c)$ that is not a clause. In contrast, if we replace this reduction rule by the polarized rules

$$x \in y \cup z \longrightarrow_- x \in y \lor x \in z$$

$$x \in y \cup z \longrightarrow_+ x \in y$$

$$x \in y \cup z \longrightarrow_+ x \in z$$

then the clause $L_1 \lor L_2 \lor \neg a \in b \cup c$ reduces to the clauses $L_1 \lor L_2 \lor \neg a \in b$ and to $L_1 \lor L_2 \lor \neg a \in c$. More generally, any reduction system can be transformed this way to a clausal one [42].

The second reason is that any consistent set of axioms can be transformed into a Polarized reduction system that is classically equivalent [29, 14] and some sets of axioms can be transformed into a Polarized reduction system that is constructively equivalent [11].

Interestingly, this result has been proved with applications to automated theorem proving in mind, it uses automated theorem proving methods, but it is a purely proof-theoretical result.

2.3 Expressing theories in Deduction modulo theory

The early work on expressing theories in Deduction modulo theory was focused on specific theories: Simple type theory [34], Arithmetic [39, 1], Set theory [37], etc.

Then, as already said, systematic ways of transforming sets of axioms into sets of reduction rules have been investigated [29, 14, 11].

2.4 The $\lambda\Pi$-calculus modulo theory

The early developments of Deduction modulo theory were independent of the proofs-as-algorithms paradigm, also known as the Brouwer-Heyting-Kolmogorov interpretation, that is the idea that a proof of $A \Rightarrow B$, for instance, is an algorithm transforming proofs of A into proofs of B. In Deduction modulo theory, like in Predicate logic, terms, propositions, and proofs belong to three different languages, and proofs are not terms. But we have mentioned that one of the origins of Deduction modulo theory was the definitional equality of Martin-Löf's Intuitionistic type theory. This suggests that this idea of identifying congruent propositions can also be useful in systems based on the proofs-as-algorithms paradigm.

The simplest system to express proofs of Predicate logic as algorithms is the λ-calculus with dependent types [47], also know as the $\lambda\Pi$-calculus. This leads to the development of an extension of the $\lambda\Pi$-calculus, called the $\lambda\Pi$-calculus modulo theory [22]. This system is closely related to Martin-Löf's logical framework [56].

Any theory that can be expressed in minimal Deduction modulo theory, that is in the restriction of Deduction modulo theory, where the only logical operators are the implication and the universal quantifier, can be expressed in the $\lambda\Pi$-calculus modulo theory. In particular Simple type theory can be expressed in the $\lambda\Pi$-calculus modulo theory. An interesting point here is that the Calculus of Constructions [20] has been designed to express proofs of Simple type theory as algorithms. It happens that $\lambda\Pi$-calculus modulo theory also can express those proofs as algorithms. This suggests that the Calculus of Constructions itself could be expressed in the $\lambda\Pi$-calculus modulo theory, and this is indeed the case [22]. The embedding of the Calculus of Constructions into the $\lambda\Pi$-calculus modulo theory follows closely the expression of Simple type theory in Deduction modulo theory.

It happens *a posteriori* that this embedding of the Calculus of Constructions into the $\lambda\Pi$-calculus modulo theory can be seen as an extension of the $\lambda\Pi$-calculus with an impredicative universe *à la* Tarski [3] and thus that there is a strong link between the expression of Simple type theory in Predicate logic and the notion of universe *à la* Tarski.

3 Properties

3.1 Models

The usual models of classical Predicate logic, valued in $\{0, 1\}$, can be used for Deduction modulo theory. A congruence $\equiv$ is said to be valid in a model when $A \equiv B$ implies $[\![A]\!]_\phi = [\![B]\!]_\phi$ for all valuations ϕ, and a soundness and completeness theorem can be proved using standard methods.

Like for Predicate logic, the set of truth values $\{0, 1\}$ can be extended to any Boolean algebra, allowing to prove a stronger completeness theorem: given a theory, there exists a model such that the propositions valid in this model are exactly the propositions provable in this theory.

Boolean algebras can be extended to Heyting algebras to define a sound and complete semantics for constructive logic.

However, in all these models—valued in $\{0, 1\}$, in Boolean algebras and in Heyting algebras—, two provably equivalent propositions always have the same truth value: if $A \Leftrightarrow B$ is valid, then $A \Rightarrow B$ and $B \Rightarrow A$ are valid, hence $[\![A]\!]_\phi \leq [\![B]\!]_\phi$ and $[\![B]\!]_\phi \leq [\![A]\!]_\phi$ and by antisymmetry $[\![A]\!]_\phi = [\![B]\!]_\phi$. Thus, there is no way to make a difference, in the model, between provable equivalence and congruence: whether A and B are just equiprovable or have the same proofs, they have the same truth value.

A way to overcome this is to extend Boolean algebras and Heyting algebras by dropping the antisymmetry condition on the relation $\leq$. This relation is then a pre-order and the algebras defined this way can be called *pre-Boolean* algebras [10] and *pre-Heying* algebras [31]. The soundness theorem is proved exactly the same way—antisymmetry is never used in this proof—, and the completeness is simpler as the class of models is larger. This corresponds to the intuition that the relation $\leq$, defined by $A \leq B$ if $A \Rightarrow B$ is provable, is reflexive and transitive, but not antisymmetric.

This way, two provably equivalent propositions may be interpreted by distinct truth values, unlike two congruent propositions that must be interpreted by the same one, and it is possible to define models where a proposition A is interpreted by the set of its proofs.

When a theory has a model valued in some pre-Heyting algebra it is consistent, when it has a model valued in all pre-Heyting algebras it is said to be *super-consistent*.

3.2 Cut-elimination

Proof-reduction is defined in Deduction modulo theory in the same way as in Predicate logic, but the difference is that it does not always terminate. Indeed, if we define a theory with the reduction rule $P \longrightarrow (P \Rightarrow Q)$ the sequent $\vdash Q$

has the following proof

$$
\cfrac{
  \cfrac{
    \cfrac{\overline{P \vdash P \Rightarrow Q}\;\text{axiom} \quad \overline{P \vdash P}\;\text{axiom}}{P \vdash Q}\;\Rightarrow\text{-elim}
  }{\vdash P \Rightarrow Q}\;\Rightarrow\text{-intro}
  \qquad
  \cfrac{
    \cfrac{\overline{P \vdash P \Rightarrow Q}\;\text{axiom} \quad \overline{P \vdash P}\;\text{axiom}}{P \vdash Q}\;\Rightarrow\text{-elim}
  }{\vdash P}\;\Rightarrow\text{-intro}
}{\vdash Q}\;\Rightarrow\text{-elim}
$$

that contains a cut and that reduces to itself.

Moreover, it is possible to prove that all cut free, that is irreducible, proofs end with an introduction rule, thus not only this proof does not terminate, but the sequent $\vdash Q$ has no cut free proof.

And a similar example can be built with a terminating reduction system [38].

Not only some theories have the cut-elimination property and some others do not, but this property is even undecidable [17, 46].

Thus, unlike for axiomatic theories, the notion of proof-reduction can be defined in a generic, theory independent, way, and the properties of cut free proofs, such as the property that the last rule of a cut free proof is an introduction rule can be proved in a generic way. But, the proof-termination theorem itself must be proved for each theory.

Using a method introduced to prove the termination of proof reduction in Simple type theory [41], we can prove that proof-reduction terminates in some theory, if a reducibility candidate $[\![A]\!]$ can be associated to each proposition A, in such a way that two congruent propositions are associated with the same reducibility candidate [38]

$$A \equiv B \text{ implies } [\![A]\!] = [\![B]\!]$$

This association of a reducibility candidate to each proposition is thus a model valued in the algebra of the reducibility candidates and the condition that two congruent propositions are associated with the same reducibility candidate is the validity of this congruence in this model.

This way, we get that if a theory has a model valued in the algebra of reducibility candidates, then proof-reduction strongly terminate.

The algebra of reducibility candidates is a pre-Heyting algebra—but not a Heyting algebra—thus we also get that proof-reduction terminates in super-consistent theories.

This semantic view on termination of proof reduction theorems also permits to relate these termination proofs to the so called *semantic* cut-elimination proofs that proceed by proving a completeness result for cut free provability. First, without proving the termination of proof-reduction, it is possible to prove directly that, in a super-consistent theory, each provable proposition has a cut free proof [36, 10]. This completeness proof does not use the pre-Heyting algebra of reducibility candidates but a simpler one.

Then, in some non super-consistent theories, proof reduction does not terminate, but each provable proposition has a cut free proof [44]. An example is obtained by replacing the proposition Q by $\top$ in the example above. This proof still fails to terminate but the sequent $\vdash \top$ has another proof, that is cut free. Such cut-elimination theorems can only be proved via a completeness theorem and, when they are proved constructively, the constructive content of these proofs is a proof-transformation algorithm, that need not be related to proof-reduction.

Finally, some theories do not have the cut elimination property, but they can sometimes be extended to theories that have this property by adding derivable reduction rules [17, 15]. This saturation process can be compared to Knuth-Bendix method [51]—remember that confluence is a special case of cut-elimination—that does not prove that all reduction systems are confluent, but that, in some cases, it is possible to extend a reduction system with derivable rules, to make it confluent.

3.3 Automated theorem proving methods

Deduction modulo theory has been introduced to design and study automated theorem proving methods. The first method introduced was a variant of Resolution [35] that was too complicated because rules were not polarized. Thus, clauses could rewrite to non clausal propositions that needed to be dynamically transformed into clausal form. Polarization permitted to simplify the method [33] and also to understand better its relation to other methods. This method is complete if and only if the theory has the cut-elimination property [45].

Imagine we have a clause

$$L_1 \lor L_2 \lor a \in b \cup c$$

and a negative reduction rule

$$x \in y \cup z \longrightarrow_- x \in y \lor x \in z$$

then applying this rule to this clause yields the clause

$$L_1 \lor L_2 \lor a \in b \lor a \in c$$

But instead of this reduction rule, we could have taken a clause

$$\neg x \in y \cup z \lor x \in y \lor x \in z$$

and Resolution, applied to the literal $a \in b \cup c$ and the underlined literal in the new clause, would have yielded the same result. Thus, there is no need to extend Resolution to handle reduction rules, but reduction rules can just be seen as special clauses, called *one-way clauses*. The Resolution rule cannot be applied to two one-way clauses and when it is applied to a one-way clause and an ordinary one, only the literal corresponding to the left-hand side of the reduction rule can be used. Thus, Polarized resolution modulo theory is just another variant of Equational resolution with clause restrictions—like the Set of support [63] and the Semantic resolution [61] strategies—and literal restrictions—like Ordered resolution.

But, unlike other variants of Resolution, its completeness is equivalent to a cut-elimination theorem. Thus, it permits to handle theories, such as Simple type theory, that cannot be handled, for instance, with Ordered resolution, as the completeness of Polarized resolution modulo the rules of Simple type theory implies cut elimination for Simple type theory and, unlike that of Ordered resolution, it cannot be proved in Simple type theory [16].

A side effect of this work is to show that, surprisingly, clause restriction strategies—such as the Set of support or the Semantic resolution strategy—and literal restriction strategies—such as Ordered resolution—can be combined, provided we do not consider theories that are just consistent, but theories that also have the cut elimination property.

These remarks also showed the way to combine this method with other selection strategies in Resolution. In particular, it has been shown that this restriction is compatible with Ordered resolution [12].

Besides Resolution, other proof-search methods have been investigated, in particular direct search in cut free sequent calculus modulo theory, also known as the *tableaux* method [9].

4 Implementations

The early work on Deduction modulo theory only led to experimental implementations. But more mature systems have been developed in the recent years.

4.1 Dedukti

Dedukti [6, 8, 59] is an implementation of the $\lambda \Pi$-calculus modulo theory. It is thus based on the proofs-as-algorithms paradigm and proof-checking is reduced to type-checking. But type-checking itself may require an arbitrary amount of computation to check the congruence of two propositions.

Dedukti is a parametric system: by changing the reduction rules, we change the theory in which the proofs are checked. Thus Dedukti is a logical framework [47]. As the proofs of many different systems can be expressed in this framework Dedukti is mostly used to check proofs developed in other systems—hence its name: "to deduce" in Esperanto—: HOL [4], Focalize [19], Coq [7, 3], etc. as well as proofs produced by automated theorem proving systems, such as iProver, Zenon, iProver modulo, and Zenon modulo. The current goal of the project is to be able to interface proofs developed in different systems, and defining a standard for proofs in various theories, much the same way standards are defined, for instance, for SMT solvers [5, 60].

4.2 iProver modulo, Super Zenon and Zenon modulo

iProver modulo [13] is an implementation of Ordered polarized resolution modulo theory. It is developed as an extension of iProver. It has shown convincing experimental results compared to the axiomatic approach. A tool automatically orienting axioms into rewriting systems usable by iProver Modulo is also available.

Super Zenon [50] is an implementation of Tableaux modulo theory specifically designed for a variant of Class theory—Second order logic—called *B set theory*, and using Super-deduction instead of the original Deduction modulo theory.

Zenon modulo [25, 26] is a generic implementation of the Tableaux modulo theory method. It comes with a heuristic that turns axioms into rewrite rules before performing proof-search, and also with a new hand-tailored expression of B set theory as a set of rewrite rules.

5 Trends and Open Problems

In recent years, the effort in Deduction modulo theory has been put on the development of implementations. In particular, we do not know how far we can go in interfacing proof systems using a logical framework such as Dedukti. We also need to investigate how having user defined reduction rules can impact tactic based proof development.

In automated theorem proving we do not understand yet how to mix Resolution modulo theory with equality specific methods such as superposition.

On the more proof-theoretical side, we know that super-consistency is a sufficient condition for the strong termination of proof reduction but we do not know if it is a necessary condition. As suggested in [21], the notion of super-consistency may require some adjustment so that we can prove that it is a necessary and sufficient condition for proof termination. Finally, some extension of Deduction modulo theory allow congruences that identify non-atomic propositions with different head-symbols [27], in particular isomorphic types such as $A \Rightarrow (B \wedge C)$ and $(A \Rightarrow B) \wedge (A \Rightarrow C)$, but we do not know yet how far we can go in this direction.

References

1. L. Allali. Algorithmic equality in Heyting arithmetic modulo. *Higher Order Rewriting*, 2007.
2. P.B. Andrews. Resolution in type theory. *The Journal of Symbolic Logic*, 36, 1971, pp. 414-432.
3. A. Assaf. A Calculus of Constructions with explicit subtyping. Manuscript, 2014.
4. A. Assaf. *Traduction de HOL en Dedukti*. Master thesis, 2012.
5. F. Besson, P. Fontaine, and L. Théry, A Flexible Proof Format for SMT: a Proposal *Proof Exchange for Theorem Proving*, 2011.
6. M. Boespflug. *Conception d'un noyau de vérification de preuves pour le lambda-Pi-calcul modulo*. Doctoral thesis, École polytechnique, 2011.
7. M. Boespflug and G. Burel. CoqInE: translating the calculus of inductive constructions into the lambda-pi-calculus modulo. *Proof Exchange for Theorem Proving*, CEUR Workshop Proceedings 878, 2012, pp. 44-50.
8. M. Boespflug, Q. Carbonneaux, and O. Hermant. The $\lambda\Pi$-calculus Modulo as a Universal Proof Language. *Proof Exchange for Theorem Proving*, CEUR Workshop Proceedings 878, 2012, pp. 28-43.

146

9. R. Bonichon and O. Hermant. A semantic completeness proof for TaMeD. *LPAR*, Lecture Notes in Computer Science 4246, Springer, 2006, pp. 167-181.

10. A. Brunel, O. Hermant, and C. Houtmann. Orthogonality and Boolean Algebras for Deduction Modulo. *Typed Lambda Calculus and Applications*, Lecture Notes in Computer Science 6990, Springer, 2011, pp. 76-90.

11. G. Burel. Automating theories in intuitionistic logic. S. Ghilardi and R. Sebastiani (eds.), FroCoS, Lecture Notes in Artificial Intelligence 5749, Springer, 2009, pp. 181-197.

12. G. Burel. Embedding deduction modulo into a prover. A. Dawar and H. Veith (eds.), *CSL*, Lecture Notes in Computer Science 6247, Springer, 2010, pp. 155-169.

13. G. Burel. Experimenting with deduction modulo. V. Sofronie-Stokkermans and N. Bjørner (eds.), *CADE*, Lecture Notes in Computer Science 6803, Springer, 2011, pp. 162-176.

14. G. Burel. From Axioms to Rewriting Rules. Manuscript.

15. G. Burel. Cut Admissibility by Saturation. *RTA-TLCA*, Lecture Notes in Computer Science 8560, Springer, 2014.

16. G. Burel and G. Dowek. How can we prove that a proof search method is not an instance of another? *Fourth International Workshop on Logical Frameworks and Meta-Languages: Theory and Practice.* ACM International Conference Proceeding Series, 2009.

17. G. Burel and C. Kirchner. Regaining cut admissibility in deduction modulo using abstract completion. *Information and Computation*, 208(2), 2010, pp. 140-164.

18. P. Brauner, C. Houtmann, and C. Kirchner. Superdeduction at work. *Rewriting, Computation and Proof, Essays dedicated to Jean-Pierre Jouannaud on the occasion of his 60th birthday*, Lectures Notes in Computer Science 4600, Springer, 2007, pp. 132-166.

19. R. Cauderlier. *Traits orients objet en $\lambda\Pi$-calcul modulo : Compilation de FoCaLize vers Dedukti.* Master thesis, 2012.

20. T. Coquand and G. Huet. The calculus of constructions. *Information and Computation*, 76, 1988, pp. 95-120.

21. D. Cousineau. On completeness of reducibility candidates as a semantics of strong normalization. *Logical Methods in Computer Science*, 8(1), 2012.

22. D. Cousineau and G. Dowek. Embedding pure type systems in the lambda-Pi-calculus modulo. S. Ronchi Della Rocca, *Typed Lambda Calculi and Applications*, Lecture Notes in Computer Science 4583, Springer, 2007, pp. 102-117.

23. M. Crabbé. Non-normalisation de la théorie de Zermelo. Manuscript, 1974.

24. M. Crabbé. Stratification and cut-elimination. *The Journal of Symbolic Logic*, 56(1), 1991, pp. 213-226.

25. D. Delahaye, D. Doligez, F. Gilbert, P. Halmagrand, and O. Hermant, Zenon Modulo: When Achilles Outruns the Tortoise using Deduction Modulo. *Logic for Programming, Artificial Intelligence, and Reasoning.* Lecture Notes in Computer Science 8312, Springer, 2013, pp. 274-290.

26. D. Delahaye, D. Doligez, F. Gilbert, P. Halmagrand, O. Hermant. Proof Certication in Zenon Modulo: When Achilles Uses Deduction Modulo to Outrun the Tortoise with Shorter Steps. *International Workshop on the Implementation of Logics*, 2013.

27. A. Díaz-Caro and G. Dowek. Simply Typed Lambda-Calculus Modulo Type Isomorphism. Manuscript, 2014.

28. G. Dowek. About folding-unfolding cuts and cuts modulo. *Journal of Logic and Computation* 11(3), 2001, pp. 419-429.

29. G. Dowek. What is a theory? H. Alt, A. Ferreira (Eds.), *Symposium on Theoretical Aspects of Computer Science*, Lecture Notes in Computer Science 2285, Springer, 2002, pp. 50-64.

30. G. Dowek. Confluence as a cut elimination property. R. Nieuwenhuis (Ed.), *Rewriting Technique and Applications*, Lecture Notes in Computer Science 2706, Springer, 2003, pp 2-13.

31. G. Dowek. Truth values algebras and proof normalization. *TYPES 2006*, Lectures Notes in Computer Science 4502, Springer, 2007.

32. G. Dowek, Skolemization in Simple Type Theory: the Logical and the Theoretical Points of View, C. Benzmller, C. Brown, J. Siekmann and R. Statman (eds.), *Festschrift in Honour of Peter B. Andrews on his 70th Birthday*. College Publications, 2008.

33. G. Dowek. Polarized resolution modulo. *IFIP Theoretical Computer Science*, 2010.

34. G. Dowek, Th. Hardin, and C.Kirchner. HOL-lambda-sigma: an intentional first-order expression of higher-order logic. *Mathematical Structures in Computer Science*, 11, 2001, pp. 1-25.

35. G. Dowek, Th. Hardin, and C. Kirchner. Theorem proving modulo. *Journal of Automated Reasoning*, 31(1), 2003, pp. 33-72.

36. G. Dowek, O. Hermant. A Simple Proof that Super-Consistency Implies Cut Elimination. *Notre Dame Journal of Formal Logic*, 53(4), 2012, pp. 439-456.

37. G. Dowek and A. Miquel. Cut elimination for Zermelo's set theory. Manuscript.

38. G. Dowek and B. Werner. Proof normalization modulo. *The Journal of Symbolic Logic*, 68(4), 2003, pp. 1289-1316.

39. G. Dowek and B. Werner. Arithmetic as a theory modulo. J. Giesel (Ed.), *Term rewriting and applications*, Lecture Notes in Computer Science 3467, Springer, 2005, pp. 423-437.

40. J. Ekman. *Normal proofs in set theory*. Doctoral thesis, Chalmers university of technology and University of Göteborg, 1994.

41. J.-Y. Girard. Une extension de l'interprétation de Gödel à l'analyse et son application à l'élimination des coupures dans l'analyse et la théorie des types. J.E. Fenstad (Ed.) *Second Scandinavian Logic Symposium*, North-Holland, 1970.

42. Jianhua Gao. Clausal Presentation of Theories in Deduction Modulo. *Journal of Computer Science and Technology*, 28(6), 2013, pp. 1085-1096. DOI: 10.1007/s11390-013-1399-0.

43. L. Hallnäs. *On normalization of proofs in set theory*. Doctoral thesis, University of Stockholm, 1983.

44. O. Hermant. Semantic cut elimination in the Intuitionnistic Sequent Calculus. *Typed Lambda Calculus and Applications*, Lecture Notes in Computer Science 3461, Springer, 2005, pp. 221-233.

45. O. Hermant. Resolution is cut-free. *Journal of Automated Reasoning*, 44(3), 2010, pp. 245-276.

46. O. Hermant. Personnal communication.

47. R. Harper, F. Honsell, G. Plotkin. A Framework for Defining Logics. *Proceedings of Logic in Computer Science*, 1987, pp. 194-204.

48. G. Huet. A mechanisation of Type Theory. *Third International Joint Conference on Artificial Intelligence*, 1973, pp. 139-146.

49. G. Huet. A Unification Algorithm for Typed λ-calculus. *Theoretical Computer Science*, 1, 1975, pp. 27-57.

50. M. Jacquel, K. Berkani, D. Delahaye, and C. Dubois. Tableaux Modulo Theories Using Superdeduction - An Application to the Verification of B Proof Rules with the Zenon Automated Theorem Prover. IJCAR, 2012, pp. 332-338.

51. D.E. Knuth and P.B. Bendix. Simple word problems in universal algebras. J. Leech (Ed.), *Computational Problems in Abstract Algebra*, Pergamon Press, 1970, pp. 263-297.

52. S. Negri and J. Von Plato. Cut elimination in the presence of axioms. *Bulletin of Symbolic Logic*, 4(4), 1998, pp. 418-435.

53. P. Martin-Löf. *Intuitionistic type theory*. Bibliopolis, 1984.

54. A. Naibo. *Le statut dynamique des axiomes: Des preuves aux modèles*. Doctoral thesis, 2013.

55. M.H.A. Newman. On theories with a combinatorial definition of "equivalence". *Annals of Mathematics*, 43, 2, 1942, pp. 223-243.

56. B. Nordström, K. Petersson, and J.M. Smith. Martin-Löf's type theory. S. Abramsky, D. Gabbay,and T. Maibaum (eds.) *Handbook of Logic in Computer Science*, Clarendon Press, 2000, pp. 1-37.

57. G. Plotkin. Building-in equational theories. *Machine Intelligence*, 7, 1972, pp. 73-90.

58. D. Prawitz. *Natural Deduction, a Proof-theoretical Study*. 1965.

59. R.Saillard. Towards explicit rewrite rules in the $\lambda\Pi$-calculus modulo. *International Workshop on the Implementation of Logics*, 2013.

60. A. Stump, D. Oe, A. Reynolds, L. Hadarean, and C. Tinelli. SMT Proof Checking Using a Logical Framework. *Formal Methods in System Design* 42 (1), 91-118

61. J.R. Slagle. Automatic theorem proving with renamable and semantic resolution. *J. ACM*, 14, 1967, pp. 687-697.

62. B. Wack. *Typage et déduction dans le calcul de réécriture*. Doctoral Thesis, Université Henri Poincaré Nancy 1, 2005.

63. L. Wos, G.A. Robinson, D.F. Carson. Efficiency and completeness of the set of support strategy in theorem proving. *J. ACM*, 12, 1965, pp. 536-541.

Foundational Proof Certificates

Dale Miller

INRIA-Saclay and LIX/École Polytechnique
dale.miller@inria.fr

1 Introduction

Consider a world where exporting proof evidence into a well defined, universal, and permanent format is taken as "feature zero" for computational logic systems. In such a world, provers will communicate and share theorems and proofs; libraries will archive and organize proofs; and marketplaces of proofs would be open to any prover that admits checkable proof objects. In that world, proof checkers will be the new gatekeepers: they will be entrusted with the task of checking that claimed proof evidence elaborates into a formal proof.

Logicians and proof theorists have worked on defining notions of proof that are not based on technology and do not have version numbers attached to them. There are many such proof systems in the literature: Hilbert-Frege proofs, Gentzen's sequent calculus proofs, Prawitz's natural deduction proofs, etc. Each of these proof systems have been given precise syntax and meaning. While such well studied proof descriptions exist, a quick review of the current state of automated and interactive theorem provers reveals that provers seldom output their "proof evidence" using such proof systems. While there is a lot of interest in having provers share and trust each other's proofs (see, for example, [3, 10, 28]) most of that work has been based on building bridges between two specific provers: a change in the version number of one prover can cause that bridge to collapse.

The ProofCert project [22] has as one of its goals the development of a flexible framework for defining the semantics of a wide range of proof evidence in such a way that provers would define the meaning of their own proof evidence and trusted proof checkers would be able to interpret that meaning and check its formal correctness. To achieve this goal, we must first be able to separate proof evidence from its provenance and then provide a formal and clear framework for defining the semantics of proof evidence. The ProofCert project is focused on the problem of checking *formal* proof: there is no assumption made that such formal proofs are actually readable by humans.

2 Defining the semantics of proof evidence

The wide range of provers in use today represent their "proof evidence" in many different ways. Such evidence might be resolution refutations, sets of links between atoms in a formula, natural deduction proofs, typed λ-terms, or proof scripts. If we insist that provers output their proof evidence as a document that could be transmitted and formally checked, we immediately face the problem that there will necessarily be many different languages and structures describing proof evidence. How can we deal with such a plurality of proof languages?

Similar situations have, of course, appeared and been addressed within computer science. For example, consider the problem of defining the static and dynamic semantics of programming languages. Sometimes, a particular implementation of a given programming language actually defines that language's syntax and semantics. Such an ad hoc notion of language definition has largely been replaced by the use of formal frameworks where syntax and semantics are defined in a universal and permanent (*i.e.,*, technology independent) fashion. For example, today, it is common that a programming language's syntax is given using the formal framework of grammars. Using grammars, one describes "declaratively" how to generate the set of legal programming language expressions. Advantages of such a system are numerous: anyone can now implement their own parser while knowing the specification of what they need to implement. Furthermore, people can now attempt to automate the entire process of producing parsers from grammars: in this way, the correctness of many parsers can be reduced to the problem of establishing the correctness of the parser generator. The use of such a framework does come with some costs. For example, context free grammars can be ambiguous and it is undecidable in general to tell if a given such grammar is ambiguous. Also, parser generators usually support restricted sets of grammars (*e.g.*, LALR(1), LR(k)) and the syntax of a given programming language might have to be simplified to conform to various restrictions imposed by generators.

Similar observations also hold for the problem of defining the dynamic semantics of a programming language. It is common place (at least in the research community) to formally define the semantics of a programming language using the "declarative" techniques found in operational semantics [16, 31]. Using such semantic specifications it is possible to define a programming language precisely enough that various compilers and interpreters can be build for the same language [27] and for formal theorems to be proved about them.

In the rest of this paper, we shall describe the *foundational proof certificate* framework that can be used to define the semantics of a wide range of proof evidence.

3   The chemistry of inference

The foundational approach to proof certificates is a framework where large inference rules are built from small inference rules. In particular, we first identify the "atoms" of inference and the "rules of chemistry" that then allow us to build the "molecules" of inference.

3.1 The atoms: sequent calculus inference rules

The smallest elements of inference that we consider come from Gentzen's sequent calculus [11]: we shall assume that the reader is familiar with the basics of sequent calculus.

While the foundational proof certificate framework can be described for both classical and intuitionistic first order logics, we restrict our attention here to just classical logic in order to simplify our presentation. Given this simplification, we can also assume that formulas are placed into negation normal form (*i.e.*, negations have only atomic scope) and if B is not atomic, the expression $\neg B$ is meant to denote the formula that results from using de Morgan dualities to push the outermost negation in over all connectives. We can also limit ourselves to using only one-sided sequents, written $\vdash \Delta$. Here, Δ is a collection of formulas: Gentzen used lists of formulas exclusively but we shall use multisets instead.

The inference rules of Gentzen's sequent calculus can be divided into three groups: *structural rules*, *identity rules*, and *introduction rules*. We shall consider these rules as they are applied to one-sided sequents.

The structural rules are the familiar rules of *exchange*, *weakening*, and *contractions*. We shall not use the exchange rule here since it is meaningless when used with multisets of formulas. Ultimately, weakening and contraction will not be separate rules but will be built into other rules.

There are two identity rules, namely, the *initial rule* and the *cut rule*:

$$\frac{\vdash \Delta_1, B \quad \vdash \Delta_2, \neg B}{\vdash \Delta_1, \Delta_2} \; cut \qquad \frac{}{\vdash B, \neg B, \Delta} \; init$$

These rules are collectively called "identity" rules since they both involve checking that a formula is identical to the negation normal form of another formula. Note that the weakening rule is built into the *init* rule. Gentzen proved that all instances of these identity rules can be eliminated except for instances of the *init* rule where B is an atomic formula.

The introduction rules give meaning to the logical connectives. In the version of classical logic we are considering, the only logical connectives are $\wedge$, $\vee$, $\forall$, and $\exists$. We shall take the following familiar rules for the quantifiers:

$$\frac{\vdash \Delta, [y/x]B}{\vdash \Delta, \forall x.B} \qquad \text{and} \qquad \frac{\vdash \Delta, [t/x]B}{\vdash \Delta, \exists x.B} \; .$$

For the propositional connectives, we find different possibilities. For example, the one-sided sequent calculus rules most closely related to Gentzen's two-sided sequent calculus rules would be

$$\frac{\vdash \Delta, B \quad \vdash \Delta, C}{\vdash \Delta, B \wedge C} \; \wedge\text{-}I_a \qquad \frac{\vdash \Delta, B}{\vdash \Delta, B \vee C} \; \vee\text{-}I_a \qquad \frac{\vdash \Delta, C}{\vdash \Delta, B \vee C} \; \vee\text{-}I_a.$$

Other natural possibilities exist:

$$\frac{\vdash \Delta_1, B \quad \vdash \Delta_2, C}{\vdash \Delta_1, \Delta_2, B \wedge C} \; \wedge\text{-}I_m \qquad \frac{\vdash \Delta, B, C}{\vdash \Delta, B \vee C} \; \vee\text{-}I_m.$$

152

As we have learned from linear logic [12], the inference rules for conjunction and disjunction (and their units) can come in two forms: the additive rules (displayed above with the subscript a) and the multiplicative rules (displayed above with the subscript m).

It is the case, of course, that when both contraction and weakening are available, the additive and the multiplicative versions of these rules are interchangeable: as a result, most sequent calculus systems select one version of these rules only. For example, many papers dealing with theorem proving in classical logic commonly use the $\wedge\text{-}I_a$ and $\vee\text{-}I_m$ rules since they are invertible.

In our case, we are striving to collect a good set of atomic inferences and we are helped if we can allow for having both additive and multiplicative versions of these inference rules. We shall use the following technique to disambiguate when these rules are applied. First, we introduce two versions of the conjunction $\wedge^+$, $\wedge^-$ and two versions of the disjunction $\vee^+$, $\vee^-$. Second, we write the additive and multiplicative inference rules as

$$\frac{\vdash \Delta, B \qquad \vdash \Delta, C}{\vdash \Delta, B \wedge^- C} \qquad \frac{\vdash \Delta, B}{\vdash \Delta, B \vee^+ C} \qquad \frac{\vdash \Delta, C}{\vdash \Delta, B \vee^+ C}$$

$$\frac{\vdash \Delta_1, B \qquad \vdash \Delta_2, C}{\vdash \Delta_1, \Delta_2, B \wedge^+ C} \qquad \frac{\vdash \Delta, B, C}{\vdash \Delta, B \vee^- C}$$

Note that the rules for the negative connectives $\wedge^-$, $\vee^-$ are invertible. (For the sake of completeness, we introduce the polarized forms for the true and false constants t^-, t^+, f^-, f^+ in the next section.) Third, if B is a first-order formula (an "unpolarized" formula), we shall write $\hat{B}$ to denote any ("polarized") formula that results from replacing all occurrences of $\wedge$ and $\vee$ in B with one of the signed versions of the corresponding connective. If B contains n occurrences of either conjunction and disjunction, then $\hat{B}$ ranges overs the 2^n polarized forms of B.

3.2 The chemistry of inference: focused proof systems

While the sequent calculus rules capture tiny steps in deduction, they are poorly equipped to capture larger scale notions of inference. What is needed is a means of organizing these small rules into larger and more familiar rules. For example, the work on *uniform proofs* in the late 1980's [25, 26] provided a restriction on sequent calculus proofs that allowed such proofs to capture the alternating phases of goal-reduction and backchaining that take place within logic programming proof search. While the technique of uniform proofs provided a description of the simple structure of inference in logic programming, it was not flexible enough to capture many other forms of inference in other computational logic systems. As Andreoli [1] showed, when one moves the alternating phase structure of uniform proofs to linear logic, a much more flexible means for structuring proofs arises. Andreoli called his new proof system a *focused* proof system. Comprehensive focused proof systems for classical and intuitionistic logic were later introduced by Liang and Miller [17, 18]. We shall now present more details of LKF [18], a focused proof system for classical logic.

$$\frac{}{\vdash \Theta \Downarrow t^+} \qquad \frac{\vdash \Theta \Downarrow B_1 \quad \vdash \Theta \Downarrow B_2}{\vdash \Theta \Downarrow B_1 \wedge^+ B_2} \qquad \frac{\vdash \Theta \Downarrow B_i \quad i \in \{1,2\}}{\vdash \Theta \Downarrow B_1 \vee^+ B_2} \qquad \frac{\vdash \Theta \Downarrow [t/x]B}{\vdash \Theta \Downarrow \exists x.B}$$

$$\frac{\vdash \Theta \Uparrow \Gamma}{\vdash \Theta \Uparrow f^-, \Gamma} \qquad \frac{\vdash \Theta \Uparrow A, B, \Gamma}{\vdash \Theta \Uparrow A \vee^- B, \Gamma} \qquad \frac{}{\vdash \Theta \Uparrow t^-, \Gamma} \qquad \frac{\vdash \Theta \Uparrow A, \Gamma \quad \vdash \Theta \Uparrow B, \Gamma}{\vdash \Theta \Uparrow A \wedge^- B, \Gamma}$$

$$\frac{\vdash \Theta \Uparrow [y/x]B, \Gamma}{\vdash \Theta \Uparrow \forall x.B, \Gamma} \qquad \frac{\vdash \Theta \Uparrow B \quad \vdash \Theta \Uparrow \neg B}{\vdash \Theta \Uparrow \cdot} \; cut \qquad \frac{}{\vdash \neg P_a, \Theta \Downarrow P_a} \; init$$

$$\frac{\vdash \Theta, C \Uparrow \Gamma}{\vdash \Theta \Uparrow C, \Gamma} \; store \qquad \frac{\vdash \Theta \Uparrow N}{\vdash \Theta \Downarrow N} \; release \qquad \frac{\vdash P, \Theta \Downarrow P}{\vdash P, \Theta \Uparrow \cdot} \; decide$$

Here, P is a positive formula; N is a negative formula; P_a is a (positive) atom; and C is a positive formula or negated atom. Also, y is a variable that is not free in any formula in the conclusion sequent of the $\forall$ introduction rule.

Fig. 1. LKF rules

We shall call a polarized formula *positive* if it is either atomic, or its top-level connective is $\wedge^+$, $\vee^+$, t^+, f^+, or $\exists$. Similarly, a polarized formula is *negative* if it is either a negated atom, or its top-level connectives is $\wedge^-$, $\vee^-$, t^-, f^-, or $\forall$.

The inference rules for *LKF* are given in Figure 1. Note that these inference rules involve sequents of the form $\vdash \Theta \Uparrow \Gamma$ and $\vdash \Theta \Downarrow B$ where Θ is a multiset of formulas, Γ is a list of formulas, and B is a formula (all formulas are polarized formulas). Such sequents can be approximated as the one-sided sequents $\vdash \Theta, \Gamma$ and $\vdash \Theta, B$, respectively. Furthermore, introduction rules are applied to either the first element of the list Γ in the $\Uparrow$ sequent or the formula B in the $\Downarrow$ sequent. This occurrence of the formula B is called the *focus* of that sequent. Proofs in *LKF* are built using two kinds of alternating *phases*. The *negative* phase is composed of invertible inference rules and only involves $\Uparrow$-sequents in the conclusion and premise. The other phase is the *positive* phase: here, applications of such inference rules often require choices. In particular, the introduction rule for the disjunction requires selecting either the left or right disjunct and the introduction rule for the existential quantifier requires selecting a term for instantiating the quantifier. The initial rule can terminate a positive phase and the cut rule can restart a negative phase. Finally, there are three structural rules in *LKF*. The *store* rule recognizes that the first formula to the right of the $\Uparrow$ is either a negative atom or a positive formula: such a formula does not have an invertible inference rule and, hence, its treatment is delayed by storing it on the left. The *release* rule is used when the formula under focus (*i.e.*, the formula to the right of the $\Downarrow$) is no longer positive: at such a moment, the phase changes to the negative phase. Finally, the *decide* rule is used at the end of the negative phase to start a positive phase by selecting a previously stored positive formula as the new focus. Note that the contraction rule is built into the decide rule and that the context Θ is treated additively even by the multiplicative connective $\wedge^+$.

It is proved in [18] that if B is a classical theorem and $\hat{B}$ is any polarization of B, then $\vdash \cdot \Uparrow \hat{B}$ has an *LKF* proof. Conversely, if $\vdash \cdot \Uparrow \hat{B}$ is provable in *LKF* then

B is a theorem. Thus the different polarizations do not change *provability* but can radically change the structure of proofs. A simple induction on the structure of an *LKF* proof of $\vdash \cdot \Uparrow B$ (for some polarized formula B) reveals that every formula that occurs to the left of $\Uparrow$ or $\Downarrow$ in one of its sequents is either a negated atom or a positive formula. Also, it is immediate that the only occurrence of a contraction rule is within the decide rule: thus, only the positive formulas are contracted. Since there is flexibility in how formulas are polarized, the choice of polarization can, at times, lead to greatly reduced opportunities for contraction. When one is able to eliminate or constrain contractions, naive proof search can sometimes become a decision procedure.

It is the negative and positive phases that are the *macro* or *synthetic* inference rules: these are the molecules of inference and are built from the atoms of inference. Note that phases are organized as follows:

$$\cdots \frac{\frac{\vdash \Theta_{i1} \Uparrow \cdot \quad \cdots \quad \vdash \Theta_{ij} \Uparrow \cdot}{\frac{\vdash \Theta \Uparrow N_i}{\vdash \Theta \Downarrow N_i}\ release} \quad \cdots}{\frac{\vdash \Theta \Downarrow P}{\vdash \Theta \Uparrow \cdot}\ decide}$$

Specifically, a positive phase has as its root a decide rule and as leaves the conclusions of release rules, while the negative phase ends with a sequent that is the conclusion of a decide rule. Together, a negative phase above a positive phase (a pair sometimes called a *bi-pole*) has a conclusion of the form $\vdash \Theta \Uparrow \cdot$ and premises of the form $\vdash \Theta' \Uparrow \cdot$ where Θ is a sub-multiset of Θ'. Thus, the sequence of decide and release rules determine boundaries between phases. A phase may also end, of course, with the introduction rules for t^- and t^+ and the initial rule.

3.3   The engineering of inference rules

To illustrate the possibilities allowed by *LKF*, we now consider a couple of different approaches to building *LKF* proofs for a propositional tautology B.

One approach involves picking $\hat{B}$ to be the polarized version of this formula that contains only the negative connectives $t^-, f^-, \wedge^-, \vee^-$. In this case, the only *LKF* proofs of $\vdash \cdot \Uparrow \hat{B}$ have exactly one negative phase: it has as a conclusion $\vdash \cdot \Uparrow \hat{B}$ and has a possibly exponential number (in the size of B) of premises of the form $\vdash L_1, \ldots, L_n \Uparrow \cdot$, where $L_1, \ldots, L_n$ are literals (atoms or negated atoms). Note that the negative phase is completely determinate (computed functionally from $\hat{B}$). The premises of this negative phase are of the form $\vdash L_1, \ldots, L_n \Uparrow \cdot$. The only way to prove such a sequent in *LKF* (without cut) is to use one occurrence each of the decide and init rules.

Note that this use of *LKF* and negative polarities leads to the following *protocol* for proving a tautology: first use negative rules to compute essentially the conjunctive normal form of B and then show that there are complementary pairs in each of the remaining premises. Determining these complementary pairs could be done by having some external source of information (like an oracle) provide the right answer or by a search. Such a search would be rather shallow and easily performed in a complete fashion. Thus, such a protocol could be used to define a simple kind of *proof certificate*: the certificate would announce that the negative connectives are to be used and that the complementary pair of literals in each premise is either explicitly listed in that certificate or that the certificate checker should do the search for complementary literals. If the certificate provides the complementary pairs, then the certificate would be exponentially large (based on the size of B) or it would be constant sized. In either case, the checking time for this certificate would be exponential since checking involves computing the conjunctive normal form of B.

Consider the appropriateness of such an approach for showing that the formula $B = (p \vee (C \vee \neg p))$ is a tautology: here p is a propositional constant and C is a possibly large propositional formula. Clearly, this formula is tautologous. While using the protocol above to prove this formula would work, it is easy to describe a more direct proof, one where we would like to insert some "clever" information into the proof building stage. To do this, we use the positive connectives t^+, f^+, $\wedge^+$, $\vee^+$. The "clever" choices are injected twice into the proof with the mark †. The subformula C is avoided in this proof.

$$
\cfrac{
  \cfrac{
    \cfrac{
      \cfrac{
        \cfrac{
          \cfrac{
            \cfrac{}{\vdash B, \neg p \Downarrow p} \; init
          }{\vdash \hat{B}, \neg p \Downarrow (p \vee^+ \hat{C}) \vee^+ \neg p} \; \dagger
        }{\vdash \hat{B}, \neg p \Uparrow \cdot} \; decide
      }{\vdash \hat{B} \Uparrow \neg p} \; store
    }{\vdash \hat{B} \Downarrow \neg p} \; release
  }{\vdash \hat{B} \Downarrow (p \vee^+ \hat{C}) \vee^+ \neg p} \; \dagger
}{\vdash \hat{B} \Downarrow \cdot} \; decide
$$

Clearly, different polarities can lead to rather different disciplines for organizing proofs. The negative phase does not listen to any outside oracles: instead, it simply performs a (determinate) computation that carries a concluding sequent to a list of premises. On the other hand, the positive phase consumes information—such as which branch of a disjunction or which instance of an existential quantifier to consider. That information can be supplied, in principle, from either an oracle (by reading a proof certificate) or a non-deterministic search. We now formalize more carefully how to integrate a proof certificate with focused proof construction.

3.4   Clerks, experts, certificates, and indexes

In order to translate the information in a given proof certificate into instructions to drive the kernel's (that is, *LKF*'s) inference rules, we use the notion of *clerks and experts* [7,8]. An analogy can help motivate our proposed design. Imagine an accounting office that needs to check that a certain collection of financial

$$\frac{true_e(\Xi)}{\Xi \vdash \Theta \Downarrow t^+} \qquad \frac{\Xi_1 \vdash \Theta \Downarrow B_1 \quad \Xi_2 \vdash \Theta \Downarrow B_2 \quad \wedge_e(\Xi, \Xi_1, \Xi_2)}{\Xi \vdash \Theta \Downarrow B_1 \wedge^+ B_2}$$

$$\frac{\Xi' \vdash \Theta \Downarrow B_i \quad i \in \{1, 2\} \quad \vee_e(\Xi, \Xi', i)}{\Xi \vdash \Theta \Downarrow B_1 \vee^+ B_2} \qquad \frac{\Xi' \vdash \Theta \Downarrow [t/x]B \quad \exists_e(\Xi, \Xi', t)}{\Xi \vdash \Theta \Downarrow \exists x.B}$$

$$\frac{\Xi' \vdash \Theta \Uparrow \Gamma \quad f_c(\Xi, \Xi')}{\Xi \vdash \Theta \Uparrow f^-, \Gamma} \qquad \frac{\Xi' \vdash \Theta \Uparrow A, B, \Gamma \quad \vee_c(\Xi, \Xi')}{\Xi \vdash \Theta \Uparrow A \vee^- B, \Gamma}$$

$$\frac{}{\Xi \vdash \Theta \Uparrow t^-, \Gamma} \qquad \frac{\Xi_1 \vdash \Theta \Uparrow A, \Gamma \quad \Xi_2 \vdash \Theta \Uparrow B, \Gamma \quad \wedge_c(\Xi, \Xi_1, \Xi_2)}{\Xi \vdash \Theta \Uparrow A \wedge^- B, \Gamma}$$

$$\frac{\Xi' \vdash \Theta \Uparrow [y/x]B, \Gamma \quad \forall_c(\Xi, \Xi') \quad y \text{ not free in } \Xi, \Theta, \Gamma, B}{\Xi \vdash \Theta \Uparrow \forall x.B, \Gamma}$$

$$\frac{\Xi' \vdash \Theta, \langle l, C \rangle \Uparrow \Gamma \quad store_c(\Xi, C, \Xi', l)}{\Xi \vdash \Theta \Uparrow C, \Gamma} \; store$$

$$\frac{\Xi_1 \vdash \Theta \Uparrow B \quad \Xi_2 \vdash \Theta \Uparrow \neg B \quad cut_e(\Xi, \Theta, \Xi_1, \Xi_2, B)}{\Xi \vdash \Theta \Uparrow \cdot} \; cut$$

$$\frac{\Xi' \vdash \Theta \Uparrow N \quad release_e(\Xi, \Xi')}{\Xi \vdash \Theta \Downarrow N} \; release \qquad \frac{init_e(\Xi, \Theta, l) \quad \langle l, \neg P_a \rangle \in \Theta}{\Xi \vdash \Theta \Downarrow P_a} \; init$$

$$\frac{\Xi' \vdash \Theta \Downarrow P \quad decide_e(\Xi, \Theta, \Xi', l) \quad \langle l, P \rangle \in \Theta \quad positive(P)}{\Xi \vdash \Theta \Uparrow \cdot} \; decide$$

Fig. 2. LKF^a: LKF augmented with premises that invoke clerks and experts.

documents represents a legal transaction. The office workers called experts are given the responsibility of looking into the collection and extracting information: they must *decide* into which series of transactions to dig and they need to know when to *release* their findings for storage and later reconsideration. On the other hand, the clerks are responsible for taking information released by the experts and performing some computations on them, including their *indexing* and *storing*. Of course, the division of labor between experts and clerks arises from the different characteristics of the positive and negative phases of proof structures in *LKF*.

The inference rules in Figure 2 define LKF^a, the *augmented LKF* proof system. The augmentation adds three kinds of objects to *LKF*. The first is the actual proof certificate as a term: the syntactic variable Ξ ranges over such certificates. The second addition is the extra premises to all inference rules. The positive inference rules have calls to *experts*: these are predicates that know how to extract information from proof certificates in order to supply information required by a positive inference rule. Thus, the expert for the existential induction rule, $\exists_e(\Xi, \Xi', t)$ is supposed to hold when the certificate Ξ indicates that the term t

can be used to instantiate the corresponding $\exists$ quantifier: once that information has been extracted, the remaining certificate is Ξ'.

The third item that we need to add to the LKF^a proof system was hinted at with the office analogy above: when a clerk releases some information to be considered later, that item must be stored. Storing must of course support "recall" (embodied by the decide inference rule). To do such store and recall flexibly, we shall allow the office workers to agree on an actual indexing scheme for stored formulas. Such index schemes can be various. For example, we have found all the following indexes useful in different styles of proof evidence: the formula itself, a de Bruijn number, a formula occurrence, a link name in a proof net, a line number in a Frege proof, and a clause number in a resolution refutation. Note that in LKF^a, the context Θ does not denote a multiset of formulas but rather a multiset of pairs $\langle l, C \rangle$ where l is an index and C is a formula. It is the responsibility of the store clerk to compute an index for the formula that is moved from the right to the left of the $\Uparrow$. Similarly, the decide rule selects a formula from the Θ context by providing an index.

It now is clear what a proof certificate must contain. First, it must describe both the datatypes used to build certificates and indexes. Second, it must provide a method of polarizing a formula B into $\hat{B}$. Third, it must provide the specification of the various clerks and experts. These can be described either as inference rules themselves or, equivalently, as simple logic programs. Note that the rules in LKF^a are always sound, no matter how one specifies the clerks and experts. This soundness is an important feature for a checker that must be trusted even though significant specifications (and code) are supplied from outside the kernel.

3.5 Examples of clerks and experts

We present here two examples of how one can specify clerks and experts. Given our presentation to this point, it will probably be difficult to understand the details of how these specifications work. More details can be found in [8] from where these examples are taken. Here, we will limit ourselves to some observations. We shall also use λProlog syntax to provide the specifications of clerks and experts: inference rules could have also been used but the λProlog syntax is more compact (see also the discussion in Section 4).

$$\texttt{cnf : cert} \qquad\qquad \texttt{idx : form -> index}$$

$$\forall C.\ store_c(\texttt{cnf}, C, \texttt{cnf}, \texttt{idx}(C)). \qquad \wedge_c(\texttt{cnf}, \texttt{cnf}, \texttt{cnf}).$$
$$\forall \Theta \forall l.\ init_e(\texttt{cnf}, \Theta, l). \qquad\qquad \vee_c(\texttt{cnf}, \texttt{cnf}).$$
$$\forall \Theta \forall l.\ decide_e(\texttt{cnf}, \Theta, \texttt{cnf}, l). \qquad f_c(\texttt{cnf}, \texttt{cnf}).$$
$$release_e(\texttt{cnf}, \texttt{cnf}).$$

Fig. 3. A checker based on a simple decision procedure

A decision procedure as a proof certificate. Figure 3 presents a concrete specification of the decision procedure for propositional formulas described in Section 3.3. Figure 3 first lists the constructors for certificates (type `cert`) and indexes (type `idx`). In this case, there is a unique inhabitant of type `cert` and that is just the name of this decision procedure (that is, this part of the certificate contains no information). Similarly, the only way to build an index is to use the formula itself (thereby trivializing the indexing mechanism here). Since we are assuming that conjunction and disjunction are polarized negatively, only clerks are associated to introduction rules. Finally, the expert predicates for the initial rule and the decide rule are not acting like experts at all: they formally allow any context Θ and any index l to be related to the `cnf` certificate. Such behavior is fine, however, since the rules in Figure 2 make additional checks on Θ and l and these checks actually discover complementary pairs of literals.

This certificate illustrates an important aspect of our proposal for FPC: some detail from a proof can, in principle, be elided and this may not cause a problem for proof checking. In the case of this certificate format, there might be several proofs of a sequent containing just literals, since it might contain many different complementary pairs. One could rewrite this certificate format to explicitly contain a *mating*, that is a set of pairs of complementary literals that spans all such clauses [2]. Such a mating is, of course, possibly exponentially large with respect to the tautology being checked. But if we allow for some search, we can do some "proof reconstruction" that involves searching for complementary pairs. Allowing such reconstruction makes it possible for this FPC to have a constant size instead of the possibly exponential size for recording an explicit mating.

Resolution refutations as proof certificates. Figure 4 lists the constructors and the clerks and experts that can be used to specify the semantics of a simple form of resolution refutation. There are two key parts of this checker. First, if two clauses C_1 and C_2 resolve to yield clause C_0, then there is an *LKF* proof of $\vdash \neg C_1, \neg C_2 \Uparrow C_0$ that has decide depth 3 or less (the decide depth of a proof is the maximum number of decide rules along a path in that proof). The first set of specifications in Figure 4 describe how the clerks and experts can be specified to check for the existence of such small proofs. Second, clauses are indexed by natural numbers and the resolution refutation is a list of triples. Each of these triples are checked using the specification above to confirm that it is a valid binary resolution and then the cut-rule is used to integrate the resolvent into the other clauses. The second set of clauses specify clerks and experts that direct the LKF^a kernel to trace out a proof whose backbone is a series of cuts all of whose left premises are the small proofs that are responsible for checking claimed binary resolutions. For more explanation about this certificate, see [8, 23].

4   A reference checker

Logic programming can sometimes be used to convert a declarative specification into a prototype implementation. For example, versions of Prolog can be used to

```
idx : int -> index              lit : form -> index
dl  : list int -> cert         ddone : cert
```

$\forall L. \vee_c(\mathtt{dl}(L), \mathtt{dl}(L)).$ $\forall L. \mathit{true}_e(\mathtt{dl}(L)).$

$\forall L. f_c(\mathtt{dl}(L), \mathtt{dl}(L)).$ $\forall L. \forall_c(\mathtt{dl}(L), \mathtt{dl}(L)).$

$\forall C \forall L. \mathit{store}_c(\mathtt{dl}(L), C, \mathtt{dl}(L), \mathtt{lit}(C)).$ $\forall L. \exists_e(\mathtt{dl}(L), \mathtt{dl}(L), T).$

$\forall L \forall P \forall \Theta. \mathit{decide}_e(\mathtt{dl}(L), \Theta, \mathtt{ddone}, \mathtt{lit}(P)).$ $\forall L. \wedge_e(\mathtt{dl}(L), \mathtt{dl}(L), \mathtt{dl}(L)).$

$\forall I \forall \Theta. \mathit{decide}_e(\mathtt{dl}([I]), \Theta, \mathtt{dl}([]), \mathtt{idx}(I)).$ $\forall l \forall \Theta. \mathit{init}_e(\mathtt{ddone}, \Theta, l).$

$\forall I \forall J \forall \Theta. \mathit{decide}_e(\mathtt{dl}([I,J]), \Theta, \mathtt{dl}([J]), \mathtt{idx}(I)).$ $\forall l \forall L \forall \Theta. \mathit{init}_e(\mathtt{dl}(L), \Theta, l).$

$\forall I \forall J \forall \Theta. \mathit{decide}_e(\mathtt{dl}([J,I]), \Theta, \mathtt{dl}([J]), \mathtt{idx}(I)).$ $\forall L. \mathit{release}_e(\mathtt{dl}(L), \mathtt{dl}(L)).$

```
rdone : cert      rlist  :       list (int * int * int) -> cert
                  rlisti : int -> list (int * int * int) -> cert
```

$\forall R. f_c(\mathtt{rlist}(R), \mathtt{rlist}(R)).$

$\forall C \forall l \forall R. \mathit{store}_c(\mathtt{rlisti}(l, R), C, \mathtt{rlist}(R), \mathtt{idx}(l)).$

$\mathit{true}_e(\mathtt{rdone}).$

$\forall I \forall \Theta. \mathit{decide}_e(\mathtt{rlist}([]), \Theta, \mathtt{rdone}, \mathtt{idx}(I)) \;\text{:-}\; \langle \mathtt{idx}(I), \mathit{true} \rangle \in \Theta.$

$\forall I, J, K, R, C, N, \Theta. \mathit{cut}_e(\mathtt{rlist}([\langle I, J, K \rangle | R]), \Theta, \mathtt{dl}([I,J]), \mathtt{rlisti}(K, R), N) \;\text{:-}$
$\langle \mathtt{idx}(K), C \rangle \in \Theta, \; \mathtt{negate}(C, N).$

Fig. 4. Resolution certificate definition in two parts

directly convert certain grammar specifications into parsers [29]. Logic programming languages such as Prolog, λProlog, and Elf have also been successfully used to provide direct implementations of the operational semantic specifications of programming languages [5, 21, 20].

Given that the kernel of an FPC checker is specified by inference rules (such as those in Figure 2) and that some forms of proof reconstruction should be supported during proof checking, a natural programming language for this reference checker is a logic programming language. The λProlog programming language [24] is a particularly good choice since it contains typing, abstract datatypes, and higher-order programming in a style similar to ML—the first programming language designed for implementing proof checkers [13]. λProlog goes beyond ML by providing a logically clean notion of binding and (object-level) substitution. Furthermore, λProlog implements both unification and backtracking search, two features critical for implementing proof reconstruction. These two features allow proof certificates to have the option of eliding some proof evidence in the hope that the proof checker can reconstruct the missing details. Allowing a trade-off between certificate size and checking (and proof reconstruction) time is an important feature for designing flexible proof certificate formats [8]. For example, this trade-off makes it possible for the cnf certificate format to not explicitly describe which literals are linked to which literals within a clause: without the explicit information available, the logic programming implementation will do a simple and bounded search to find such linkable literals.

5 Related and future work

Dependent typed λ-calculi have been proposed as frameworks for specifying proof systems in a range of settings. The LF system [14] showed how natural deduction systems for intuitionistic logic could be given elegant and compact specifications. A logic programming language Elf [30] has also been built on top of LF: checking a proof in LF can then involve proving a goal in such a logic programming language. Unification and backtracking search are trusted components for Elf and partial proof reconstruction is possible in that setting.

The LF system (also called the $\lambda\Pi$-calculus) has been extended to allow for *deduction modulo* [9]: in the resulting system, implemented as Dedukti [33], functional computation replaces the proof search style computations within Elf. The LF system has also been extended with side conditions [32] and with external predicates [15] in order to make that proof representation more expressive.

Actually, the *LKF* based kernel that we have described in this paper is not the most expressive. We have experimented with writing kernels for both intuitionistic and classical logics; these are based on linear logic principles as described by the LKU focused proof system [19]. Using techniques developed by Chaudhuri [6], it should be possible to get a completely functioning *LKF* kernel from implementing just an *LJF* kernel.

Besides writing specifications of a number of other forms of proof evidence, we also plan to develop a similar approach to proof involving inductive and co-inductive definitions. With that extension, we should be able to check proof evidence coming from model checkers and inductive theorem provers. We are still doing the research to develop appropriate focusing systems for, essentially, classical and intuitionistic versions of arithmetic: the linear logic theory of fixed points developed by Baelde [4] is our current starting point.

Acknowledgments. The work presented in this paper has been done jointly with Zakaria Chihani and Fabien Renaud and has been funded by the ERC Advanced Grant ProofCert. I thank the anonymous reviewers for their comments on an earlier draft of this paper.

References

1. Jean-Marc Andreoli. Logic programming with focusing proofs in linear logic. *J. of Logic and Computation*, 2(3):297–347, 1992.
2. Peter B. Andrews. Theorem proving via general matings. *J. ACM*, 28(2):193–214, 1981.
3. Michaël Armand, Germain Faure, Benjamin Grégoire, Chantal Keller, Laurent Théry, and Benjamin Werner. A modular integration of SAT/SMT solvers to coq through proof witnesses. In J.-P. Jouannaud and Z. Shao, editors, *Certified Programs and Proofs (CPP 2011)*, *LNCS* 7086, pages 135–150, 2011.

4. David Baelde. Least and greatest fixed points in linear logic. *ACM Trans. on Computational Logic*, 13(1), April 2012.

5. P. Borras, D. Clément, Th. Despeyroux, J. Incerpi, G. Kahn, B. Lang, and V. Pascual. Centaur: the system. In *Third Annual Symposium on Software Development Environments (SDE3)*, pages 14–24, Boston, 1988.

6. Kaustuv Chaudhuri. Classical and intuitionistic subexponential logics are equally expressive. In Anuj Dawar and Helmut Veith, editors, *CSL 2010: Computer Science Logic*, LNCS 6247, pages 185–199, Brno, Czech Republic, August 2010. Springer.

7. Zakaria Chihani, Dale Miller, and Fabien Renaud. Checking foundational proof certificates for first-order logic (extended abstract). In J. C. Blanchette and J. Urban, editors, *Third International Workshop on Proof Exchange for Theorem Proving (PxTP 2013)*, volume 14 of *EPiC Series*, pages 58–66. EasyChair, 2013.

8. Zakaria Chihani, Dale Miller, and Fabien Renaud. Foundational proof certificates in first-order logic. In Maria Paola Bonacina, editor, *CADE 24: Conference on Automated Deduction 2013*, number 7898 in LNAI, pages 162–177, 2013.

9. Denis Cousineau and Gilles Dowek. Embedding pure type systems in the lambda-pi-calculus modulo. In Simona Ronchi Della Rocca, editor, *Typed Lambda Calculi and Applications, 8th International Conference, TLCA 2007, Paris, France, June 26-28, 2007, Proceedings*, LNCS 4583, pages 102–117. Springer, 2007.

10. Pascal Fontaine, Jean-Yves Marion, Stephan Merz, Leonor Prensa Nieto, and Alwen Fernanto Tiu. Expressiveness + automation + soundness: Towards combining SMT solvers and interactive proof assistants. In Holger Hermanns and Jens Palsberg, editors, *TACAS: Tools and Algorithms for the Construction and Analysis of Systems, 12th International Conference*, LNCS 3920, pages 167–181. Springer, 2006.

11. Gerhard Gentzen. Investigations into logical deduction. In M. E. Szabo, editor, *The Collected Papers of Gerhard Gentzen*, pages 68–131. North-Holland, Amsterdam, 1969. Translation of articles that appeared in 1934-35.

12. Jean-Yves Girard. Linear logic. *Theoretical Computer Science*, 50:1–102, 1987.

13. Michael J. Gordon, Arthur J. Milner, and Christopher P. Wadsworth. *Edinburgh LCF: A Mechanised Logic of Computation*, LNCS 78. Springer, 1979.

14. Robert Harper, Furio Honsell, and Gordon Plotkin. A framework for defining logics. *Journal of the ACM*, 40(1):143–184, 1993.

15. Furio Honsell, Marina Lenisa, Luigi Liquori, Petar Maksimovic, and Ivan Scagnetto. LFP: a logical framework with external predicates. In *LFMTP'12: Proceedings of the Seventh International Workshop on Logical Frameworks and Meta-languages, Theory and Practice*, pages 13–22. ACM New York, 2012.

16. Gilles Kahn. Natural semantics. In *Proceedings of the Symposium on Theoretical Aspects of Computer Science*, LNCS 247, pages 22–39. Springer, March 1987.

17. Chuck Liang and Dale Miller. Focusing and polarization in intuitionistic logic. In J. Duparc and T. A. Henzinger, editors, *CSL 2007: Computer Science Logic*, LNCS 4646, pages 451–465. Springer, 2007.

18. Chuck Liang and Dale Miller. Focusing and polarization in linear, intuitionistic, and classical logics. *Theoretical Computer Science*, 410(46):4747–4768, 2009.

19. Chuck Liang and Dale Miller. A focused approach to combining logics. *Annals of Pure and Applied Logic*, 162(9):679–697, 2011.

20. Spiro Michaylov and Frank Pfenning. Natural semantics and some of its metatheory in Elf. In Lars Hallnäs, editor, *Extensions of Logic Programming*. Springer LNCS, 1992.

21. Dale Miller. Formalizing operational semantic specifications in logic. *Concurrency Column of the Bulletin of the EATCS*, October 2008.

22. Dale Miller. Proofcert: Broad spectrum proof certificates. An ERC Advanced Grant funded for the five years 2012-2016, February 2011.

23. Dale Miller. A proposal for broad spectrum proof certificates. In J.-P. Jouannaud and Z. Shao, editors, *CPP: First International Conference on Certified Programs and Proofs*, *LNCS* 7086, pages 54–69, 2011.

24. Dale Miller and Gopalan Nadathur. *Programming with Higher-Order Logic*. Cambridge University Press, June 2012.

25. Dale Miller, Gopalan Nadathur, Frank Pfenning, and Andre Scedrov. Uniform proofs as a foundation for logic programming. *Annals of Pure and Applied Logic*, 51:125–157, 1991.

26. Dale Miller, Gopalan Nadathur, and Andre Scedrov. Hereditary Harrop formulas and uniform proof systems. In David Gries, editor, *2nd Symp. on Logic in Computer Science*, pages 98–105, Ithaca, NY, June 1987.

27. Robin Milner, Mads Tofte, and Robert Harper. *The Definition of Standard ML*. MIT Press, 1990.

28. Lawrence C. Paulson and Jasmin Christian Blanchette. Three years of experience with Sledgehammer, a practical link between automatic and interactive theorem provers. In *IWIL-LPAR*, pages 1–11, 2010.

29. Fernando C. N. Pereira and David H. D. Warren. Definite clauses for language analysis. *Artificial Intelligence*, 13:231–278, 1980.

30. Frank Pfenning. Elf: A language for logic definition and verified metaprogramming. In *4th Symp. on Logic in Computer Science*, pages 313–321, Monterey, CA, June 1989.

31. Gordon Plotkin. A structural approach to operational semantics. DAIMI FN-19, Aarhus University, Aarhus, Denmark, September 1981.

32. Aaron Stump. Proof checking technology for satisfiability modulo theories. In A. Abel and C. Urban, editors, *Logical Frameworks and Meta-Languages: Theory and Practice*, 2008.

33. The Dedukti team. The Dedukti system and homepage. `https://www.rocq.inria.fr/deducteam/Dedukti/index.html`, 2013.

Deep Inference

Alessio Guglielmi

University of Bath
A.Guglielmi@Bath.Ac.UK

1 Introduction

Deep inference could succinctly be described as an extreme form of linear logic [12]. It is a methodology for designing proof formalisms that generalise Gentzen formalisms, *i.e.* the sequent calculus and natural deduction [11]. In a sense, deep inference is obtained by applying some of the main concepts behind linear logic to the formalisms, *i.e.*, to the rules by which proof systems are designed. By doing so, we obtain a *better* proof theory than the traditional one due to Gentzen. In fact, in deep inference we can provide proof systems for more logics, in a more regular and modular way, with smaller proofs, less syntax, less bureaucracy and we have a chance to make substantial progress towards a solution to the century-old problem of the *identity of proofs*. The first manuscript on deep inference appeared in 1999 and the first refereed papers in 2001 [6, 19]. So far, two formalisms have been designed and developed in deep inference: the *calculus of structures* [15] and *open deduction* [17]. A third one, *nested sequents* [5], introduces deep inference features into a more traditional Gentzen formalism.

Essentially, deep inference tries to understand proof composition and proof normalisation (in a very liberal sense including cut elimination [11]) in the most logic-agnostic way. Thanks to it we obtain a deeper understanding of the nature of normalisation. It seems that normalisation is a primitive, simple phenomenon that manifests itself in more or less complicated ways that depend more on the choice of representation for proofs rather than their true mathematical nature. By dropping syntactic constraints, as we do in deep inference compared to Gentzen, we get closer to the semantic nature of proof and proof normalisation.

As I said, the early inspiration for deep inference comes from linear logic. Linear logic, among other ideas, supports the notion that logic has a *geometric* nature, and that a more perspicuous analysis of proofs is possible if we uncover their geometric shape, hidden behind their syntax. We can give technical meaning to this notion by looking for *linearity* in proofs. In the computing world, linearity can be interpreted as a way to deal with *quantity* or *resource*. The significance of linear logic for computer science has stimulated a remarkable amount of research, that continues to these days, and that ranges from the most theoretical investigations in categorical semantics to the implementation of languages and compilers and the verification of software.

Linear logic expresses locality by relying on Gentzen's formalisms. However, these had been developed for classical mathematical logic, for which linearity is not a primitive, natural notion. While attempting to relate *process algebras* (which are foundational models of concurrent computation) to linear logic, I realised that Gentzen's formalisms were inherently inadequate to express the most primitive notion of composition in computer science: *sequential composition*. This is indeed linear, but of a different kind of linearity from that naturally supported by linear logic.

I realised then that the linear logic ideas were to be carried all the way through and that the formalisms themselves had to be 'linearised'. Technically, this turned out to be possible by dropping one of the assumptions that Gentzen implicitly used, namely that the (geometric) shape of proofs is directly related to the shape of formulae that they prove. In deep inference, we do not make this assumption, and we get proofs whose shape is much more liberally determined than in Gentzen's formalisms. As an immediate consequence, we were able to capture process-algebras sequential composition [7], but we soon realised that the new formalism was offering unprecedented opportunities for both a more satisfying general theory of proofs and for more applications in computer science.

2 Proof System(s)

The difference between Gentzen formalisms and deep inference ones is that in deep inference we compose proofs by the same connectives of formulae: if

$$\Phi = \begin{matrix} A \\ \| \\ B \end{matrix} \quad \text{and} \quad \Psi = \begin{matrix} C \\ \| \\ D \end{matrix}$$

are two proofs with, respectively, premisses A and C and conclusions B and D, then

$$\Phi \wedge \Psi = \begin{matrix} A \wedge C \\ \| \\ B \wedge D \end{matrix} \quad \text{and} \quad \Phi \vee \Psi = \begin{matrix} A \vee C \\ \| \\ B \vee D \end{matrix}$$

are valid proofs with, respectively, premisses $A \wedge C$ and $A \vee C$, and conclusions $B \wedge D$ and $B \vee D$. Significantly, while $\Phi \wedge \Psi$ can be represented in Gentzen, $\Phi \vee \Psi$ cannot. That is basically the definition of deep inference and it holds for every language, not just propositional classical logic.

As an example, I will show the standard deep inference system for propositional logic. System SKS is a proof system defined by the following *structural* inference rules (where a and $\bar{a}$ are dual atoms)

$$\text{i}\downarrow \frac{\text{t}}{a \vee \bar{a}} \qquad \text{w}\downarrow \frac{\text{f}}{a} \qquad \text{c}\downarrow \frac{a \vee a}{a}$$

$$\textit{identity} \qquad\qquad \textit{weakening} \qquad\qquad \textit{contraction}$$

,

$$\text{i}\uparrow \frac{a \wedge \bar{a}}{\text{f}} \qquad \text{w}\uparrow \frac{a}{\text{t}} \qquad \text{c}\uparrow \frac{a}{a \wedge a}$$

$$\textit{cut} \qquad\qquad \textit{coweakening} \qquad\qquad \textit{cocontraction}$$

165

and by the following two *logical* inference rules:

$$s\,\frac{A \wedge [B \vee C]}{(A \wedge B) \vee C} \qquad\qquad m\,\frac{(A \wedge B) \vee (C \wedge D)}{[A \vee C] \wedge [B \vee D]} \quad.$$

$$\textit{switch} \qquad\qquad\qquad \textit{medial}$$

A *cut-free* derivation is a derivation where i↑ is not used, *i.e.*, a derivation in SKS \ {i↑}. In addition to these rules, there is a rule

$$=\frac{C}{D} \quad,$$

such that C and D are opposite sides in one of the following equations:

$$A \vee B = B \vee A \qquad\qquad\qquad A \vee \mathsf{f} = A$$
$$A \wedge B = B \wedge A \qquad\qquad\qquad A \wedge \mathsf{t} = A$$
$$[A \vee B] \vee C = A \vee [B \vee C] \qquad\qquad \mathsf{t} \vee \mathsf{t} = \mathsf{t}$$
$$(A \wedge B) \wedge C = A \wedge (B \wedge C) \qquad\qquad \mathsf{f} \wedge \mathsf{f} = \mathsf{f}$$

We do not always show the instances of rule =, and when we do show them, we gather several contiguous instances into one.

For example, this is a valid derivation:

$$= \frac{[a \vee b] \wedge a}{\overset{\|}{([a \vee b] \wedge a) \wedge ([a \vee b] \wedge a)}} = \;\; m\,\frac{\left[\mathsf{c}{\uparrow}\dfrac{a}{a \wedge a} \vee \mathsf{c}{\uparrow}\dfrac{b}{b \wedge b} \right]}{[a \vee b] \wedge [a \vee b]} \wedge \mathsf{c}{\uparrow}\dfrac{a}{a \wedge a} \quad.$$

This derivation illustrates a general principle in deep inference: structural rules on generic formulae (in this case a cocontraction) can be replaced by corresponding structural rules on atoms (in this case c↑).

It is interesting to note that the inference rules for classical logic, as well as the great majority of rules for any logic, all derive from a common template which has been distilled from the semantics of a purely linear logic in the first deep inference paper [15]. Since this phenomenon is very surprising, especially for structural rules such as weakening and contraction, we believe that we might be dealing with a rather deep aspect of logic and we are currently investigating it (see [13, 14] for an informal exposition of the idea).

3 Proof-Theoretical Properties

Locality and linearity are foundational concepts for deep inference, in the same spirit as they are for linear logic. Going for locality and linearity basically means going for *complexity bounded by a constant*. This last idea introduces geometry into the picture, because bounded complexity leads us to *equivalence modulo continuous deformation.* In a few words, the simple and natural definition of deep inference that we have seen above captures these ideas about linearity, locality and geometry, and can consequently be exploited in many ways, and notably:

- to recover a De Morgan premiss-conclusion symmetry that is lost in Gentzen [3];
- to obtain new notions of normalisation in addition to cut elimination [18, 16];
- to shorten analytic proofs by exponential factors compared to Gentzen [8, 10];
- to obtain quasipolynomial-time normalisation for propositional logic [9];
- to express logics that cannot be expressed in Gentzen [32, 5];
- to make the proof theory of a vast range of logics regular and modular [5];
- to get proof systems whose inference rules are local, which is usually impossible in Gentzen [29];
- to inspire a new generation of proof nets and semantics of proofs [30];
- to investigate the nature of cut elimination [16, 20];
- to type optimised versions of the λ-calculus that are not typeable in Gentzen [21, 22];
- to model process algebras [7, 23, 24, 26, 27];
- to model quantum causal evolution [2] ...
- ... and much more.

One of the open questions is whether deep inference might have a positive influence on the proof-search-as-computation paradigm and possibly on focusing. This subject has been so far almost unexplored, but some preliminary work looks very promising [25].

The above references have been selected among those that provide surveys when possible. There is no time in this tutorial to cover all those aspects, therefore I refer the reader to this web page, which contains up-to-date information about deep inference:

$$\text{http://alessio.guglielmi.name/res/cos} \quad .$$

The core topic of every proof-theoretic investigation, namely normalisation, deserves a special mention. Traditionally, normalisation is at the core of proof theory, and this is of course the same for deep inference. Normalisation in deep inference is not much different, in principle, from normalisation in Gentzen theory. In practice, however, the more liberal proof composition mechanism of deep inference completely invalidates the techniques (and the intuition) behind cut elimination procedures in Gentzen systems. Much of the effort of these 15 years of research on deep inference went into recovering a normalisation theory. One of the main ideas is in [15] where we show a technique called *splitting*, which at present is the most general method we know for eliminating cuts in deep inference. Splitting is too technical to be described here. The best reference for those who know Gentzen's theory is probably [4], where splitting is applied to classical predicate logic.

167

On the other hand, we now have techniques that are not as widely applicable but that are of a completely different nature from splitting, which is combinatorial. A surprising, relatively recent result consists in exploiting deep inference's locality to obtain the first purely geometric normalisation procedure, by a topological device that we call *atomic flows* [16, 18]. This means that, at least for classical logic and logics that extend it, cut elimination can be understood as a process that is completely independent from logical information: only the shape of the proof, determined by its structural information (creation, duplication and erasing of atoms) matters. Logical information, such as the connectives in formulae, *do not matter*. This hints at a deeper nature of normalisation than what we thought so far. It seems that normalisation is a primitive, simple phenomenon that manifests itself in more or less complicated ways that depend more on the choice of representation for proofs rather than their true mathematical nature.

4 Pragmatic Properties

I will concentrate here on a crucial aspect of proofs, namely their size. This is interesting in proof complexity, because proof size is intimately connected to the problem of NP vs coNP. It is also interesting for the automated deduction community, because the size of proofs affects the size of the proof search space, and so it has a direct effect on the time it takes to find proofs.

Quantification in deep inference is not different from quantification in the Gentzen theory, or, at least, nothing significantly different has been discovered so far. Therefore we can limit the discussion to the propositional case. The situation can be described in a few words: in [8] we proved that deep inference has an exponential speed-up over Gentzen on analytic proof systems. In particular, one can consider Statman tautologies [28], which only have exponential-size proofs in the cut-free sequent calculus, and show that they have polynomial proofs in cut-free deep inference.

Obviously, at first sight it might seem that the subformula property does not hold in deep inference, and so that the notion of cut free-ness is weaker than in Gentzen. However, the issue is subtle and it turns out that the differences with Gentzen are surprisingly small. As Anupam Das proved in [10], only a very limited amount of deep inference is sufficient to completely capture the exponential speed-up. More precisely, any cut-free deep-inference system that can access at most depth 2 in formulae can polynomially simulate proof systems of unbounded depth, such as the system presented in this tutorial. In other words, the same depth visibility of hypersequents is sufficient to obtain small proofs. This means that for the same impact that hypersequents have on the branching factor in the proof search space, we can obtain much smaller proofs than in Gentzen systems, thanks to the better proof representation in deep inference. I will show an example here, by reasoning on the first three Statman tautologies (see [8, 28] for formal definitions):

$$S_1 = (a \wedge b) \vee \bar{a} \vee \bar{b} \quad ,$$

$$S_2 = (c \wedge d) \vee \left(\left[\bar{c} \vee \bar{d} \right] \wedge a \wedge \left[\bar{c} \vee \bar{d} \right] \wedge b \right) \vee \bar{a} \vee \bar{b} \quad ,$$

$$S_3 = (e \wedge f) \vee \left(\left[\bar{e} \vee \bar{f} \right] \wedge c \wedge \left[\bar{e} \vee \bar{f} \right] \wedge d \right) \vee$$
$$\left(\left[\bar{e} \vee \bar{f} \right] \wedge \left[\bar{c} \vee \bar{d} \right] \wedge a \wedge \left[\bar{e} \vee \bar{f} \right] \wedge \left[\bar{c} \vee \bar{d} \right] \wedge b \right) \vee \bar{a} \vee \bar{b} \quad .$$

It is well known, and the reader will have no difficulty in seeing it, that the size of cut-free sequent proofs of S_n grows exponentially with n. The structural reason is that the external connectives in formulae force repeated duplication of the context. Let us see what happens if we could just access connectives immediately below the external ones.

For S_1 we have a trivial cut-free proof in SKS:

For S_2 we can obtain:

Here we see how the external atoms c and d are 'brought inside' the tautology and two proofs similar to those for S_1 are performed inside a conjunction inside the external disjunction.

Finally, in Figure 1 we can see a proof of S_3, where the above principle is repeated and clearly gives rise to a sequence of proofs for S_n that grows polynomially over n instead of exponentially.

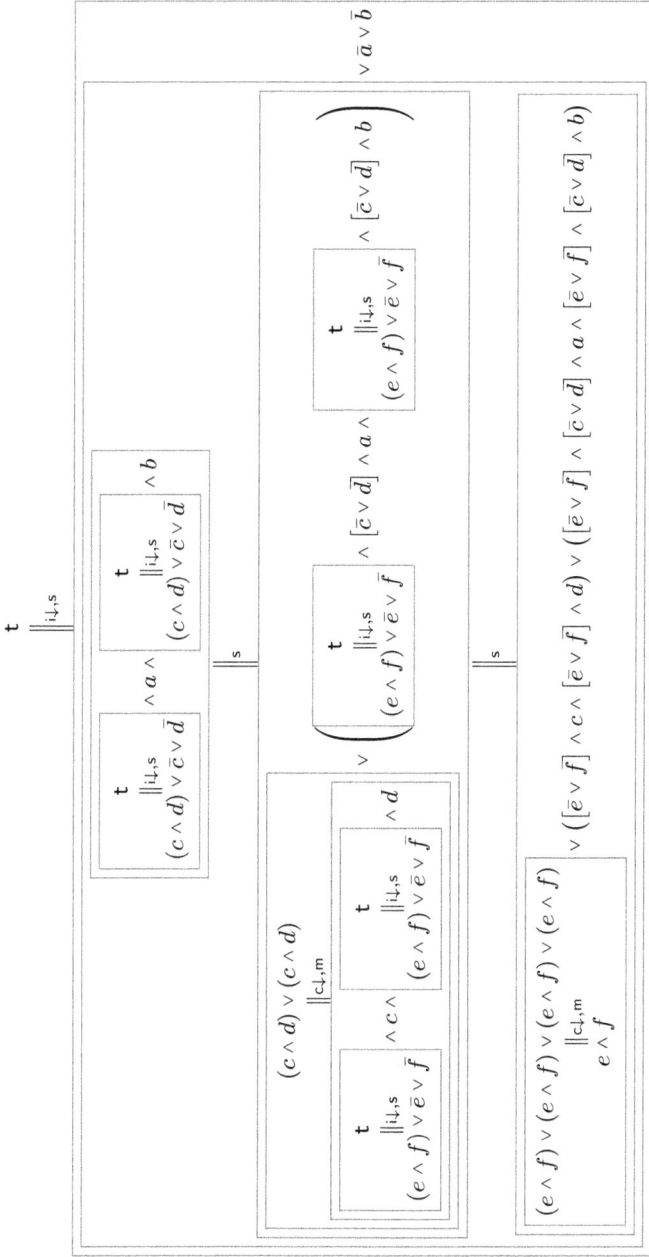

Fig. 1. Cut-free SKS proof of Statman tautology S_3.

5 Trends and Open Problems

The future of deep inference tends towards proof complexity, combinatorics and the study of proofs via algebraic topology. One of the most important open problems that deep inference intends to solve is that of the *identity of proofs* (sometimes called *Hilbert's 24th problem* [31]); this is related to the equally open problem of the *identity of algorithms* [1].

Acknowledgments. This work has been supported by the EPSRC Project *Efficient and Natural Proof Systems* (EP/K018868/1).

References

1. Andreas Blass, Nachum Dershowitz, and Yuri Gurevich. When are two algorithms the same? Technical Report MSR-TR-2008-20, Microsoft Research, 2008.
2. Richard F. Blute, Alessio Guglielmi, Ivan T. Ivanov, Prakash Panangaden, and Lutz Straßburger. A logical basis for quantum evolution and entanglement. In Claudia Casadio, Bob Coecke, Michael Moortgat, and Philip Scott, editors, *Categories and Types in Logic, Language, and Physics*, volume 8222 of *Lecture Notes in Computer Science*, pages 90–107. Springer-Verlag, 2014.
3. Kai Brünnler. *Deep Inference and Symmetry in Classical Proofs*. Logos Verlag, Berlin, 2004.
4. Kai Brünnler. Cut elimination inside a deep inference system for classical predicate logic. *Studia Logica*, 82(1):51–71, 2006.
5. Kai Brünnler. Nested sequents. Habilitation Thesis, 2010.
6. Kai Brünnler and Alwen Fernanto Tiu. A local system for classical logic. In R. Nieuwenhuis and Andrei Voronkov, editors, *Logic for Programming, Artificial Intelligence, and Reasoning (LPAR)*, volume 2250 of *Lecture Notes in Computer Science*, pages 347–361. Springer-Verlag, 2001.
7. Paola Bruscoli. A purely logical account of sequentiality in proof search. In Peter J. Stuckey, editor, *Logic Programming, 18th International Conference (ICLP)*, volume 2401 of *Lecture Notes in Computer Science*, pages 302–316. Springer-Verlag, 2002.
8. Paola Bruscoli and Alessio Guglielmi. On the proof complexity of deep inference. *ACM Transactions on Computational Logic*, 10(2):14:1–34, 2009.
9. Paola Bruscoli, Alessio Guglielmi, Tom Gundersen, and Michel Parigot. A quasipolynomial cut-elimination procedure in deep inference via atomic flows and threshold formulae. In Edmund M. Clarke and Andrei Voronkov, editors, *Logic for Programming, Artificial Intelligence, and Reasoning (LPAR-16)*, volume 6355 of *Lecture Notes in Computer Science*, pages 136–153. Springer-Verlag, 2010.
10. Anupam Das. On the proof complexity of cut-free bounded deep inference. In Kai Brünnler and George Metcalfe, editors, *Tableaux 2011*, volume 6793 of *Lecture Notes in Artificial Intelligence*, pages 134–148. Springer-Verlag, 2011.
11. Gerhard Gentzen. Investigations into logical deduction. In M.E. Szabo, editor, *The Collected Papers of Gerhard Gentzen*, pages 68–131. North-Holland, Amsterdam, 1969.
12. Jean-Yves Girard. Linear logic. *Theoretical Computer Science*, 50:1–102, 1987.
13. Alessio Guglielmi. Subatomic logic, 2002.

14. Alessio Guglielmi. Some news on subatomic logic, 2005.
15. Alessio Guglielmi. A system of interaction and structure. *ACM Transactions on Computational Logic*, 8(1):1:1–64, 2007.
16. Alessio Guglielmi and Tom Gundersen. Normalisation control in deep inference via atomic flows. *Logical Methods in Computer Science*, 4(1):9:1–36, 2008.
17. Alessio Guglielmi, Tom Gundersen, and Michel Parigot. A proof calculus which reduces syntactic bureaucracy. In Christopher Lynch, editor, *21st International Conference on Rewriting Techniques and Applications (RTA)*, volume 6 of *Leibniz International Proceedings in Informatics (LIPIcs)*, pages 135–150. Schloss Dagstuhl–Leibniz-Zentrum für Informatik, 2010.
18. Alessio Guglielmi, Tom Gundersen, and Lutz Straßburger. Breaking paths in atomic flows for classical logic. In Jean-Pierre Jouannaud, editor, *25th Annual IEEE Symposium on Logic in Computer Science (LICS)*, pages 284–293. IEEE, 2010.
19. Alessio Guglielmi and Lutz Straßburger. Non-commutativity and MELL in the calculus of structures. In L. Fribourg, editor, *Computer Science Logic (CSL)*, volume 2142 of *Lecture Notes in Computer Science*, pages 54–68. Springer-Verlag, 2001.
20. Tom Gundersen. *A General View of Normalisation Through Atomic Flows*. PhD thesis, University of Bath, 2009.
21. Tom Gundersen, Willem Heijltjes, and Michel Parigot. Atomic lambda calculus: A typed lambda-calculus with explicit sharing. In Orna Kupferman, editor, *28th Annual IEEE Symposium on Logic in Computer Science (LICS)*, pages 311–320. IEEE, 2013.
22. Tom Gundersen, Willem Heijltjes, and Michel Parigot. A proof of strong normalisation of the typed atomic lambda-calculus. In Ken McMillan, Aart Middeldorp, and Andrei Voronkov, editors, *Logic for Programming, Artificial Intelligence, and Reasoning (LPAR-19)*, volume 8312 of *Lecture Notes in Computer Science*, pages 340–354. Springer-Verlag, 2013.
23. Ozan Kahramanoğulları. Towards planning as concurrency. In M.H. Hamza, editor, *Artificial Intelligence and Applications (AIA)*, pages 197–202. ACTA Press, 2005.
24. Ozan Kahramanoğulları. On linear logic planning and concurrency. *Information and Computation*, 207(11):1229–1258, 2009.
25. Ozan Kahramanoğulları. Interaction and depth against nondeterminism in deep inference proof search. *Logical Methods in Computer Science*, 10(2):5:1–49, 2014.
26. Luca Roversi. Linear lambda calculus and deep inference. In Luke Ong, editor, *Typed Lambda Calculi and Applications*, volume 6690 of *Lecture Notes in Computer Science*, pages 184–197. Springer-Verlag, 2011.
27. Luca Roversi. A deep inference system with a self-dual binder which is complete for linear lambda calculus. *Journal of Logic and Computation*, 2014. To appear.
28. Richard Statman. Bounds for proof-search and speed-up in the predicate calculus. *Annals of Mathematical Logic*, 15:225–287, 1978.
29. Lutz Straßburger. *Linear Logic and Noncommutativity in the Calculus of Structures*. PhD thesis, Technische Universität Dresden, 2003.
30. Lutz Straßburger. From deep inference to proof nets via cut elimination. *Journal of Logic and Computation*, 21(4):589–624, 2011.
31. Rüdiger Thiele. Hilbert's twenty-fourth problem. *American Mathematical Monthly*, 110:1–24, 2003.
32. Alwen Tiu. A system of interaction and structure II: The need for deep inference. *Logical Methods in Computer Science*, 2(2):4:1–24, 2006.

On proof mining by cut-elimination

Alexander Leitsch

Vienna University of Technology
leitsch@logic.at

Abstract. We present cut-elimination as a method of proof mining, in the sense that hidden mathematical information can be extracted by eliminating lemmas from proofs. We present reductive methods for cut-elimination and the method `ceres` (cut-elimination by resolution). A comparison of `ceres` with reductive methods is given and it is shown that the asymptotic behavior of `ceres` is superior to that of reductive methods (nonelementary speed-up). It is illustrated, how `ceres` can be extended and applied in practice for analyzing mathematical proofs. Finally we give an application of `ceres` to a well-known proof of the infinitude of primes by Fürstenberg; this proof uses topological lemmas based on arithmetic progressions. These topological lemmas of the proof are eliminated by `ceres` and Euclid's construction of primes is extracted. We also touch the problem of cut-elimination by resolution on induction proofs and discuss the limits of the method.

1 Introduction

What is a mathematical proof? Just a *verification* of a statement or a key to *understand* a theorem? One and the same theorem may have several, possibly very different mathematical proofs, and each of them contains a specific form of *mathematical information*.

Mathematics in general is based on the structuring of reasoning by intermediate statements, the *lemmas*: this strongly increases the efficiency of mathematical thinking as the mathematician is not forced to have a proof of a lemma in mind when he makes use of it. He even might not know any proof of the lemma, but simply trusts other mathematicians concerning its truth.

The drawback of the use of lemmas is, however, that only their truth but not their proofs are reflected in the derivations of their end-statements. One of the most important insights in mathematical logic is Gentzen's Hauptsatz [17]. It states that lemmas (cuts) can be algorithmically eliminated from given first-order derivations. The result is a streamlined (though potentially quite long and incomprehensible) lemma-free proof combining all subproofs of the original derivation: the cut-free derivation. Gentzen's ground-breaking result has been motivated by Hilbert's distinction of *ideal* and *real* objects in mathematics, where the lemmas are supposed to encode properties of ideal objects. Hence, Gentzen's result validated Hilbert's idea that any proof about real objects can be proven without statements involving ideal objects. In fact, proofs making use of ideal objects can be transformed to proofs without them (by Gentzen's cut-elimination method).

The removal of cuts corresponds to the elimination of intermediate statements (lemmas) from proofs resulting in a proof which is *analytic* in the sense that all statements in the proof are subformulas of the result. Therefore, the proof of a combinatorial statement (possibly using theories outside the theory of the statement itself) is converted into a purely combinatorial proof.

While Gentzen's cut-elimination theorem found its immediate applications in abstract proof theory (in particular in proving consistency results), the technique of cut-elimination turned out to be fruitful in the analysis of "real" mathematical proofs. A famous application of cut-elimination to mathematical proofs is Girard's analysis of a topological proof of van der Waerden's theorem (given a partition of $\mathbb{N} = C_1 \cup \cdots \cup C_k$, one of the sets C_i contains arbitrarily long arithmetic progressions) [18]. In a formal sense Girard's analysis of van der Waerden's theorem is the application of cut-elimination to the topological proof of Fürstenberg/Weiss with the "perspective" of obtaining van der Waerden's (combinatorial) proof. Naturally, an application of a complex proof transformation like cut-elimination by humans requires a goal-oriented strategy.

The development of the method `ceres` [8] (cut-elimination by resolution) was inspired by the idea to fully automate cut-elimination on real mathematical proofs, with the aim of obtaining new interesting elementary proofs. A fully automated treatment proved successful for mathematical proofs of moderate complexity; we mention the tape proof [3] and the lattice proof [21] described in more detail in Section 4. On the other hand, more complex mathematical proofs required an interactive use of `ceres`; this way we successfully analyzed Fürstenberg's proof of the infinitude of primes (see [5] and [1]) and obtained Euclid's argument of prime construction. This proof, though much simpler than the proof of the Fürstenberg/Weiss proof of van der Waerden's theorem, is sufficiently "complex" in the sense that it proves a number theoretic result by a topological argument. By the use of `ceres` the topological proof was transformed into a number theoretic one, namely to that of Euclid. Though the analysis by `ceres` could not be fully automated, even its interactive use proved to be superior to the reductive cut-elimination method (based on Gentzen's proof) due to additional structural information given by the characteristic clause set (to be defined in Section 3).

`ceres` [8, 10] is a cut-elimination method that is based on resolution. The method roughly works as follows: The structure of the proof containing cuts is encoded in an unsatisfiable set of clauses $\mathcal{C}$ (the *characteristic clause set*). A resolution refutation γ of $\mathcal{C}$, which is obtained using a first-order theorem prover, serves as a skeleton for an atomic cut normal form ψ, a new proof which contains at most atomic cuts. γ is transformed to ψ by replacing the leaves of γ by so-called *proof projections*, which are essentially cut-free parts of the original proof. This method of cut-elimination has been implemented in the system `ceres`[1]. The system is capable of dealing with formal proofs in an extended version of Gentzen's

---

[1] available at `http://www.logic.at/ceres/`

sequent calculus **LK**, among them also very large ones (the largest proof processed so far has 10651 nodes).

Cut-elimination is not the only tool to *mine* proofs. An alternative method, the extraction of functionals, is based on Gödel's dialectica interpretation [19] and allows the construction of programs from proofs (see [12] and [13] for applications to mathematical proofs). Not only the result of the functional extraction method is different, also its range of applicability. Its advantage is the handling of the induction rule, which poses serious problems to cut-elimination; its disadvantage is the restriction to proofs of Π_2-statements (statements of the form $\forall x.\exists y.A(x, y)$ for a quantifier-free formula A), while cut-elimination can be applied to arbitrary statements. Both methods have in common that they can reveal hidden structures in proofs and provide new mathematical information by proof-transformation.

2 A Proof System for Cut-Elimination

As a basis for our investigations we use the sequent calculus **LK²** (defined by Gerhard Gentzen [17]). In our version of **LK** we do not use an exchange rule as our sequents are based on the multi-set structure. There are several extensions of **LK** which are useful for analyzing mathematical proofs; we will mention some of them in Section 4.

Definition 1. *Let $\mathcal{A}$ and $\mathcal{B}$ be two multi-sets of formulas and $\vdash$ be a symbol not belonging to the logical language. Then $\mathcal{A} \vdash \mathcal{B}$ is called a sequent.*

Definition 2. *Let $S\colon A_1, \ldots, A_n \vdash B_1, \ldots, B_m$ be a sequent and $\mathcal{M}$ be an interpretation over the signature of $\{A_1, \ldots, A_n, B_1, \ldots B_m\}$. Then S is valid in $\mathcal{M}$ if the formula $(A_1 \wedge \cdots \wedge A_n) \to (B_1 \vee \cdots \vee B_m)$ is valid in $\mathcal{M}$. S is called valid if S is valid in all interpretations.*

Definition 3 (LK). *As axioms of the calculus we take the sequents $A \vdash A$ for atomic formulas A.*

There are two groups of rules, the logical and the structural ones. All rules with the exception of the cut rule have left and right versions; left versions are denoted by $\xi : l$, right versions by $\xi : r$. A and B denote formulas, $\Gamma, \Delta, \Pi, \Lambda$ multi-sets of formulas

The logical rules:

- $\wedge$-*introduction:*

$$\frac{A, \Gamma \vdash \Delta}{A \wedge B, \Gamma \vdash \Delta} \wedge\colon l_1 \qquad \frac{B, \Gamma \vdash \Delta}{A \wedge B, \Gamma \vdash \Delta} \wedge\colon l_2 \qquad \frac{\Gamma \vdash A \quad \Gamma \vdash B}{\Gamma \vdash \Delta, A \wedge B} \wedge\colon r$$

- $\vee$-*introduction:*

$$\frac{A, \Gamma \vdash \Delta \quad B, \Gamma \vdash \Delta}{A \vee B, \Gamma \vdash \Delta} \vee\colon l \qquad \frac{\Gamma \vdash \Delta, A}{\Gamma \vdash \Delta, A \vee B} \vee\colon r1 \qquad \frac{\Gamma \vdash \Delta, B}{\Gamma \vdash \Delta, A \vee B} \vee\colon r2$$

---

² **LK** is an acronym for Logik-Kalkül

- →-*introduction:*

$$\frac{\Gamma \vdash \Delta, A \quad B, \Pi \vdash \Lambda}{A \to B, \Gamma, \Pi \vdash \Delta, \Lambda} \to : l \qquad \frac{A, \Gamma \vdash \Delta, B}{\Gamma \vdash \Delta, A \to B} \to : r$$

- ¬-*introduction:*

$$\frac{\Gamma \vdash \Delta, A}{\neg A, \Gamma \vdash \Delta} \neg : l \qquad \frac{A, \Gamma \vdash \Delta}{\Gamma \vdash \Delta, \neg A} \neg : r$$

- ∀-*introduction:* [3] [4]

$$\frac{A(x/t), \Gamma \vdash \Delta}{(\forall x) A(x), \Gamma \vdash \Delta} \forall : l \qquad \frac{\Gamma \vdash \Delta, A(x/y)}{\Gamma \vdash \Delta, (\forall x) A(x)} \forall : r$$

- *The logical rules for* ∃-*introduction (the variable conditions for* ∃ : *l are these for* ∀ : *r, and similarly for* ∃ : *r and* ∀ : *l):*

$$\frac{A(x/y), \Gamma \vdash \Delta}{(\exists x) A(x), \Gamma \vdash \Delta} \exists : l \qquad \frac{\Gamma \vdash \Delta, A(x/t)}{\Gamma \vdash \Delta, (\exists x) A(x)} \exists : r$$

Concerning the semantics of ∃ : *l, note that the sequents* $(\exists x) A(x), \Pi \vdash \Lambda$ *and* $A(x/y), \Pi \vdash \Lambda$ *are equivalent because they are both valid; in fact,* **LK** *is sound and derives only valid sequents given valid axiom sequents of the form* $A \vdash A$.

The structural rules:

- weakening*:*

$$\frac{\Gamma \vdash \Delta}{\Gamma \vdash \Delta, A} w : r \qquad \frac{\Gamma \vdash \Delta}{A, \Gamma \vdash \Delta} w : l$$

- contraction:

$$\frac{A, A, \Gamma \vdash \Delta}{A, \Gamma \vdash \Delta} c : l \qquad \frac{\Gamma \vdash \Delta, A, A}{\Gamma \vdash \Delta, A} c : r$$

- *The* cut *rule:*

$$\frac{\Gamma \vdash \Delta, A \quad A, \Pi \vdash \Lambda}{\Gamma, \Pi \vdash \Delta, \Lambda} cut(A)$$

An **LK**-proof is a tree, where the nodes are labeled with sequents, the edges with **LK**-rules applied to derive sequents. The leaves of the tree are labeled by axioms, and the root by the end-sequent.

LK is particularly suited for proof analysis, as the cut-rule (representing the use of lemmas) can be constructively eliminated from proofs; the resulting proof is analytic, i.e. the whole material of the proof consists of subformulas of the end-sequent. In typical Hilbert-type calculi the only rules are modus ponens and the generalization rule. There is no way to eliminate modus ponens from a typical Hilbert type calculus, and thus the elimination of lemmas cannot be described in such a framework.

---

[3] t is an arbitrary term containing only free variables.

[4] y is a free variable which may not occur in Γ, Δ. y is called an eigenvariable.

Example 1. We give two proofs of the same sequent, one with cut, the other without it.

Let $\varphi =$

$$
\cfrac{
  \cfrac{
    \cfrac{
      \cfrac{P(a) \vdash P(a)}{P(a) \vdash P(a) \vee Q(a)} \vee : r_1
    }{P(a) \vdash \exists y(P(y) \vee Q(y))} \exists : r
    \qquad
    \cfrac{
      \cfrac{Q(b) \vdash Q(b)}{Q(b) \vdash P(b) \vee Q(b)} \vee : r_2
    }{Q(b) \vdash \exists y(P(y) \vee Q(y))} \exists : r
  }{P(a) \vee Q(b) \vdash \exists y(P(y) \vee Q(y))} \vee : l
  \qquad
  \cfrac{(\chi)}{\exists y(P(y) \vee Q(y)), \forall x. \neg P(x) \vdash \exists z. Q(z)}
}{P(a) \vee Q(b), \forall x. \neg P(x) \vdash \exists z. Q(z)} \; cut
$$

for $\chi =$

$$
\cfrac{
  \cfrac{
    \cfrac{
      \cfrac{
        \cfrac{
          \cfrac{P(\alpha) \vdash P(\alpha)}{P(\alpha), \neg P(\alpha) \vdash} \neg : l
        }{P(\alpha), \neg P(\alpha) \vdash Q(\alpha)} w : r
        \qquad
        \cfrac{
          \cfrac{Q(\alpha) \vdash Q(\alpha)}{Q(\alpha), \neg P(\alpha) \vdash Q(\alpha)} w : l
        }{}
      }{P(\alpha) \vee Q(\alpha), \neg P(\alpha) \vdash Q(\alpha)} \vee : l
    }{P(\alpha) \vee Q(\alpha), \neg P(\alpha) \vdash \exists z. Q(z)} \exists : r
  }{P(\alpha) \vee Q(\alpha), \forall x. \neg P(x) \vdash \exists z. Q(z)} \forall : l
}{\exists y(P(y) \vee Q(y)), \forall x. \neg P(x) \vdash \exists z. Q(z)} \exists : l
$$

where α is an eigenvariable. When we search for a witness for the z in $\exists z. Q(z)$ and trace the proof part χ via the ancestors of $\exists z. Q(z)$ we see that no direct answer can be obtained. In fact we can trace α until it "disappears" by the $\exists$: *l*-rule. The following cut-free proof, which can be obtained via Gentzen-type cut-elimination, provides more information about z:

$\psi =$

$$
\cfrac{
  \cfrac{
    \cfrac{
      \cfrac{
        \cfrac{
          \cfrac{P(a) \vdash P(a)}{P(a), \neg P(a) \vdash} \neg : l
        }{P(a), \neg P(a) \vdash Q(b)} w : r
        \qquad
        \cfrac{Q(b) \vdash Q(b)}{Q(b), \neg P(a) \vdash Q(b)} w : l
      }{P(a) \vee Q(b), \neg P(a) \vdash Q(b)} \vee : l
    }{P(a) \vee Q(b), \neg P(a) \vdash \exists z. Q(z)} \exists : r
  }{P(a) \vee Q(b), \forall x. \neg P(x) \vdash \exists z. Q(z)} \forall : l
}{}
$$

Here we see that z was replaced by b in $\exists$: r (reading the proof backwards); moreover, ψ contains a so-called *Herbrand sequent*

$$
S_H \colon P(a) \vee Q(b), \neg P(a) \vdash Q(b)
$$

which is valid and can be obtained by instantiation of the quantified formulas from the end-sequent. So we get an *explicit* information about the "right" z in the cut-free proof. Note that, in general, not a single witness is obtained but rather a set of witnesses (Herbrand "disjunction"). Gentzen [17] has given a proof transformation on cut-free proofs to obtain Herbrand sequents (which he called mid-sequents, because the upper part of the obtained proof consists only of structural and propositional, the lower only of quantifier- and structural rules). We describe this theorem in more detail in Section 3.

3 Proof-Theoretical Aspects of Cut-Elimination

The main and basic theorem of proof theory is the so-called *Hauptsatz*:

Theorem 1 (Gentzen 1934). *Let φ be an **LK**-proof of a sequent S. Then there exists an **LK**-proof ψ of S (effectively constructible from φ) without application of the cut-rule.*

Gentzen's proof is based on a proof transformation method which will be described in more detail below. A cut-free proof is a *normal form* under this transformations. We will illustrate the benefits of this normal form below.

Proof transformations to normal form (cut-free proofs in **LK**, normal proofs in natural deduction) essentially change the nature of the proof in making implicit information explicit (see the simple Example 1). In case of cut-free **LK**-proofs of prenex end-sequents we obtain a *Herbrand sequent* describing all instantiations of quantifiers in the proof and reducing the problem to a propositional one. This abstraction from propositional reasoning allows mathematical interpretations of complex cut-free proofs obtained via cut-elimination (see [21] and [34]). For simplicity we define Herbrand sequents for prenex sequents only. Instead of working with a sequent S we can consider the skolemized form $sk(S)$, a form where the so-called strong quantifiers are eliminated via the introduction of terms in a new signature. This transformation is standard in automated deduction and crucial to the transformation into clause form. Skolemization can also be applied to whole proofs. In fact, every proof of φ of S can be transformed in to a proof φ' of $sk(S)$ by a merely quadratic proof transformation (the skolemization of proofs, see [11]). On the other hand, every cut-free proof of $sk(S)$ can be transformed into a cut-free proof of S; this transformation is polynomial if S is prenex and exponential in general [6]. For simplicity we assume that the sequent S is in prenex form. Then $sk(S)$ is of the form $A_1, \ldots, A_n \vdash B_1, \ldots, B_m$ where the A_i are of the form $\forall x_1, \ldots, \forall x_k.E$ (a Π_1-formula), and the B_j of the form $\exists y_1, \ldots, \exists y_l.F$ (a Σ_1-formula), where E and F are quantifier-free. The specific form of these sequents motivates the following definition:

Definition 4. *A sequent $S\colon A_1, \ldots, A_n \vdash B_1, \ldots, B_m$, where the A_i are Π_1- and the B_j Σ_1-formulas is called a Σ_1-sequent.*

The essence of Herbrand's theorem consists in the replacement of quantified formulas by instances of these formula, resulting in a quantifier-free formula which is validity-equivalent. We formulate this theorem in form of prenex sequents.

Definition 5. *Let A be a Σ_1- or a Π_1-formula of the form $Qx_1, \ldots, Qx_n.E$ and $t_1, \ldots, t_n$ be terms. Then $E\{x_1/t_1, \ldots, x_n/t_n\}$ is called an instantiation of A.*

Definition 6 (Herbrand sequent). *Let $S\colon A_1, \ldots, A_n \vdash B_1, \ldots, B_m$ be a provable Σ_1-sequent. For any A_i (B_j) let $\mathcal{A}_i$ ($\mathcal{B}_j$) be sequences of instantiations of A_i (B_j). Then $S'\colon \mathcal{A}_1, \ldots, \mathcal{A}_n \vdash \mathcal{B}_1, \ldots, \mathcal{B}_m$ is called a Herbrand sequent of S if S' is propositionally valid.*

178

Herbrand sequents can be constructed from proofs. We first eliminate all cuts down to atomic ones (full cut-elimination is not required) and then construct a sequent consisting only of instances of formulas of the end-sequent.

A cut is called *atomic* if the cut formula in the rule is an atomic formula.

Theorem 2 (mid-sequent theorem). *Let φ be an **LK**-proof of a Σ_1-sequent S with at most atomic cuts (if φ contains cuts then they are atomic). Then φ can be transformed into a proof φ' with the same number of logical inferences and atomic cuts and with the following property. φ' contains the derivation of a sequent S' (the mid-sequent), s.t. all propositional inferences and atomic cuts in φ are above S', and below S' there are only unary structural rules and quantifier-rules from φ.*

Proof. In [17] a step-wise proof transformation is given transforming a proof of a prenex sequent into a proof containing a mid-sequent. If we are only interested in the mid-sequent itself, which by our definition is a Herbrand sequent, it suffices to read off the instances from the quantifier-introduction rules and collect them in a sequent. This procedure can be performed in linear time. □

Corollary 1. *Let φ be a proof of a Σ_1-sequent S with at most atomic cuts and let S' be a mid-sequent as defined in Theorem 2. Then S' is a Herbrand sequent of S.*

Proof. Let φ' be the proof obtained from φ by the transformation of Theorem 2. Then the subproof ψ of φ' deriving the midsequent S' contains only propositional and structural rules. By the soundness of **LK** S' is propositionally valid. □

Example 2. Let ψ be the proof from Example 1:

$$
\cfrac{
  \cfrac{
    \cfrac{
      \cfrac{
        \cfrac{P(a) \vdash P(a)}{P(a), \neg P(a) \vdash} \ \neg\!:l
      }{P(a), \neg P(a) \vdash Q(b)} \ w\!:r
      \qquad
      \cfrac{
        \cfrac{Q(b) \vdash Q(b)}{Q(b), \neg P(a) \vdash Q(b)} \ w\!:l
      }{}
    }{P(a) \vee Q(b), \neg P(a) \vdash Q(b)} \ \vee\!:l
  }{P(a) \vee Q(b), \neg P(a) \vdash \exists z.Q(z)} \ \exists\!:r
}{P(a) \vee Q(b), \forall x.\neg P(x) \vdash \exists z.Q(z)} \ \forall\!:l
$$

This proof is already in midsequent form and does not need any transformation. Its midsequent is

$$S'\!: P(a) \vee Q(b), \neg P(a) \vdash Q(b).$$

S' is a Herbrand sequent of $P(a) \vee Q(b), \forall x.\neg P(x) \vdash \exists z.Q(z)$.

The concept of Herbrand sequent can be generalized to non-prenex sequents (see [7]); efficient algorithms for the computation of these more general sequents are given in [34]. Herbrand sequents can also be represented as expansion trees [26]. The construction of Herbrand sequents works not only for cut-free proofs but also for proofs containing only cuts with quantifier-free formulas.

This is one of the reasons why quantifier-free cuts are frequently called "inessential" [32]. So the essence of cut-elimination lies in the elimination of *quantified* cuts. In the method `ceres` defined below, we transform arbitrary **LK**-proofs into **LK**-proofs with only atomic cuts; these cuts are inessential and we may speak about cut-elimination, even if inessential cuts are still present.

As cut-elimination is of high (in the worst-case nonelementary complexity) the specific choice of algorithms is crucial. The proof reduction method of Gentzen (defined by reductions in the corresponding proof rewrite system, see [11]) turns out to be very redundant and expensive. The radically different method `ceres` (cut-elimination by resolution) has been developed in [8] and [10]. `ceres` is a cut-elimination method that is based on resolution. The method roughly works as follows: The structure of the proof containing cuts is mapped to a clause term which evaluates to an unsatisfiable set of clauses C (the *characteristic clause set*). A resolution refutation of C, which is obtained using a first-order theorem prover, serves as a skeleton for the new proof which contains only atomic cuts. In a final step also these atomic cuts can be eliminated, provided the (atomic) axioms are valid sequents (or, at least, are closed under cut); but this step is of minor mathematical interest only. In the system CERES this method of cut-elimination has been implemented. The system is capable of dealing with formal proofs in **LK**, among them also very large ones (of course, also the CERES-method cannot do better on the worst-case proof sequences of Orevkov [28] and Statman [31]). CERES was also used in the analysis of Fürstenberg's proof of the infinitude of primes to be described in Section 5.

3.1 Gentzen's method

Gentzen's proof of cut-elimination in **LK** is based on a reduction relation which selects an uppermost cut and reduces its complexity. There are two possibilities:

1. the cut formulas on both sides are introduced by rules immediately over the cut in the proof. Then the cut is simplified to one or two cuts of lower formula complexity (or the cut is deleted at all). This is called a *grade* reduction.
2. One or both of the cut formulas are not introduced immediately above the cut. Then the cut is shifted upwards and a rule permutation is performed. This reduction is called a *rank* reduction. Rank reduction rules serve the purpose to enforce a situation in which a grade reduction rule can be carried out.

From the rank and grade reductions in Gentzen's proof a set of proof rewrite rules $\mathcal{R}$ can be extracted. Further refinements of $\mathcal{R}$, e.g. restricting the rewriting to uppermost cuts in the proof (Gentzen reduction $>_G$) guarantee that the rewriting relation is terminating. A terminating sequence of proof reduction yields a so-called Gentzen normal form (which is not unique as $>_G$ is not confluent). For the complete list of $\mathcal{R}$ we refer to [11]; here we list just two of them to illustrate the nature of the method:

– a rank-reduction rule which shifts the cut rule over an $\vee: l$-rule. The proof

$$
\cfrac{\cfrac{(\varphi_1) \qquad (\varphi_2)}{\cfrac{B, \Gamma \vdash \Delta, A \quad C, \Gamma \vdash \Delta, A}{B \vee C, \Gamma \vdash \Delta, A}} \vee: l \qquad \cfrac{(\psi)}{A, \Pi \vdash \Lambda}}{B \vee C, \Gamma, \Pi \vdash \Delta, \Lambda} \; cut
$$

reduces to

$$
\cfrac{\cfrac{(\varphi_1) \qquad (\psi)}{\cfrac{B, \Gamma \vdash \Delta, A \quad A, \Pi \vdash \Lambda}{B, \Gamma, \Pi \vdash \Delta, \Lambda}} \; cut \qquad \cfrac{(\varphi_2) \qquad (\psi)}{\cfrac{C, \Gamma \vdash \Delta, A \quad A, \Pi \vdash \Lambda}{C, \Gamma, \Pi \vdash \Delta, \Lambda}} \; cut}{B \vee C, \Gamma, \Pi \vdash \Delta, \Lambda} \; \vee: l
$$

– A grade reduction rule which reduces the logical complexity of the cut formula. The proof

$$
\cfrac{\cfrac{(\varphi_1)}{\cfrac{\Gamma \vdash \Delta, A(x/t)}{\Gamma \vdash \Delta, \exists x.A(x)}} \exists: r \qquad \cfrac{(\varphi_2(y))}{\cfrac{A(x/y), \Pi \vdash \Lambda}{\exists x.A(x), \Pi \vdash \Lambda}} \exists: l}{\Gamma, \Pi \vdash \Delta, \Lambda} \; cut
$$

reduces to

$$
\cfrac{(\varphi_1) \qquad (\varphi_2(t))}{\cfrac{\Gamma \vdash \Delta, A(x/t) \quad A(x/t), \Pi \vdash \Lambda}{\Gamma, \Pi \vdash \Delta, \Lambda}} \; cut
$$

Here, y is an eigenvariable which does not occur in Π, Λ, but occurs free in the proof $\varphi_2(y)$. $\varphi_2(t)$ is defined by replacing y in the whole proof $\varphi_2(y)$ by t. To ensure soundness of this proof substitution t may not contain variables bound in φ_2; this property can be guaranteed by distinguishing free and bound variables [17].

Example 3. We consider the proof φ in Example 1 (where some subproofs are abbreviated by $\varphi_1, \varphi_2, \varphi_3$):

$$
\cfrac{\cfrac{\cfrac{(\varphi_1)}{\cfrac{P(a) \vdash P(a) \vee Q(a)}{P(a) \vdash \exists y(P(y) \vee Q(y))} \exists: r \qquad \cfrac{(\varphi_2)}{Q(b) \vdash \exists y(P(y) \vee Q(y))}}{P(a) \vee Q(b) \vdash \exists y(P(y) \vee Q(y))} \vee: l \qquad \cfrac{(\varphi_3(\alpha))}{\cfrac{P(\alpha) \vee Q(\alpha), \forall x. \neg P(x) \vdash \exists z.Q(z)}{\exists y(P(y) \vee Q(y)), \forall x. \neg P(x) \vdash \exists z.Q(z)} \exists: l}}{P(a) \vee Q(b), \forall x. \neg P(x) \vdash \exists z.Q(z)} \; cut
$$

We see that the cut formula is immediately introduced in the right side of the proof, but not in the left one. Therefore we apply the rank reduction rule for $\vee: l$ defined above and obtain the proof $\varphi' =$

$$
\cfrac{(\eta_1) \qquad (\eta_2)}{\cfrac{P(a), \forall x. \neg P(x) \vdash \exists z.Q(z) \quad Q(b), \forall x. \neg P(x) \vdash \exists z.Q(z)}{P(a) \vee Q(b), \forall x. \neg P(x) \vdash \exists z.Q(z)}} \; \vee: l
$$

where $\eta_1 =$

$$\cfrac{\cfrac{(\varphi_1)}{\cfrac{P(a) \vdash P(a) \vee Q(a)}{P(a) \vdash \exists y(P(y) \vee Q(y))} \ \exists\colon r} \qquad \cfrac{(\varphi_3(\alpha))}{\cfrac{P(\alpha) \vee Q(\alpha), \forall x.\neg P(x) \vdash \exists z.Q(z)}{\exists y(P(y) \vee Q(y)), \forall x.\neg P(x) \vdash \exists z.Q(z)} \ \exists\colon l}}{P(a), \forall x.\neg P(x) \vdash \exists z.Q(z)} \ cut$$

and $\eta_2 =$

$$\cfrac{\cfrac{(\varphi_2)}{Q(b) \vdash \exists y(P(y) \vee Q(y))} \qquad \cfrac{\cfrac{(\varphi_3(\alpha))}{P(\alpha) \vee Q(\alpha), \forall x.\neg P(x) \vdash \exists z.Q(z)}}{\exists y(P(y) \vee Q(y)), \forall x.\neg P(x) \vdash \exists z.Q(z)} \ \exists\colon l}{Q(b), \forall x.\neg P(x) \vdash \exists z.Q(z)} \ cut$$

Now we locate the leftmost uppermost cut in φ'; the corresponding subproof ψ is

$$\cfrac{\cfrac{(\varphi_1)}{\cfrac{P(a) \vdash P(a) \vee Q(a)}{P(a) \vdash \exists y(P(y) \vee Q(y))} \ \exists\colon r} \qquad \cfrac{(\varphi_3(\alpha))}{\cfrac{P(\alpha) \vee Q(\alpha), \forall x.\neg P(x) \vdash \exists z.Q(z)}{\exists y(P(y) \vee Q(y)), \forall x.\neg P(x) \vdash \exists z.Q(z)} \ \exists\colon l}}{P(a), \forall x.\neg P(x) \vdash \exists z.Q(z)} \ cut$$

In the next step we obtain the proof φ'' by replacing ψ by ψ', which is obtained via the grade reduction rule for $\exists$-cuts defined above. ψ' is

$$\cfrac{\cfrac{(\varphi_1)}{P(a) \vdash P(a) \vee Q(a)} \qquad \cfrac{(\varphi_3(a))}{P(a) \vee Q(a), \forall x.\neg P(x) \vdash \exists z.Q(z)}}{P(a), \forall x.\neg P(x) \vdash \exists z.Q(z)} \ cut$$

So, in one part of the proof φ' we have broken down a cut with $\exists y(P(y) \vee Q(y))$ to a cut with $P(a) \vee Q(a)$, which is of lower logical complexity. We can do a similar thing for the remaining cut in φ'' with $\exists y(P(y) \vee Q(y))$ in φ'' (with a new cut formula $P(b) \vee Q(b)$). There are several steps more before all cuts are eliminated.

3.2 The Method CERES

The method ceres is based on a totally different approach: We analyze the proof φ first and extract a structure from the binary rules in the proof, the characteristic clause set $\mathrm{CL}(\varphi)$. In the second step we compute the so-called proof projections to the clauses in $\mathrm{CL}(\varphi)$; these are cut-free proofs obtained by skipping rules inferring ancestors of the cut rule in φ. The third step consists in a resolution refutation of $\mathrm{CL}(\varphi)$; the last one in plugging the resolution refutation together with the projections. This, finally, yields a proof with only atomic cuts. Below we give a rather informal description of ceres (but we will provide a formal definition of the characteristic clause set, which is the most important structure within the ceres-method).

Example 4. Let us consider again the proof φ from Example 1:

$$
\cfrac{
  \cfrac{
    \cfrac{
      \cfrac{P(a) \vdash P(a)^{\star}}{P(a) \vdash P(a) \vee Q(a)^{\star}} \vee : r_1
    }{P(a) \vdash \exists y(P(y) \vee Q(y))^{\star}} \exists : r
    \quad
    \cfrac{
      \cfrac{Q(b) \vdash Q(b)^{\star}}{Q(b) \vdash P(b) \vee Q(b)^{\star}} \vee : r_2
    }{Q(b) \vdash \exists y(P(y) \vee Q(y))^{\star}} \exists : r
  }{P(a) \vee Q(b) \vdash \exists y(P(y) \vee Q(y))^{\star}} \vee : l
  \quad
  \cfrac{(\chi)}{\exists y(P(y) \vee Q(y))^{\star}, \forall x.\neg P(x) \vdash \exists z.Q(z)}
}{P(a) \vee Q(b), \forall x.\neg P(x) \vdash \exists z.Q(z)} cut
$$

for $\chi =$

$$
\cfrac{
  \cfrac{
    \cfrac{
      \cfrac{
        \cfrac{\dfrac{P(\alpha)^{\star} \vdash P(\alpha)}{P(\alpha)^{\star}, \neg P(\alpha) \vdash} \neg : l}{P(\alpha)^{\star}, \neg P(\alpha) \vdash Q(\alpha)} w : r
        \quad
        \cfrac{\dfrac{Q(\alpha)^{\star} \vdash Q(\alpha)}{Q(\alpha)^{\star}, \neg P(\alpha) \vdash Q(\alpha)} w : l}{}
      }{P(\alpha) \vee Q(\alpha)^{\star}, \neg P(\alpha) \vdash Q(\alpha)} \vee : l
    }{P(\alpha) \vee Q(\alpha)^{\star}, \neg P(\alpha) \vdash \exists z.Q(z)} \exists : r
  }{P(\alpha) \vee Q(\alpha)^{\star}, \forall x.\neg P(x) \vdash \exists z.Q(z)} \forall : l
}{\exists y(P(y) \vee Q(y))^{\star}, \forall x.\neg P(x) \vdash \exists z.Q(z)} \exists : l
$$

where all cut-ancestors were marked with $\star$. We trace the ancestors up to the axioms where we find $C_1 \colon \ \vdash P(a)$, $C_2 \colon \ \vdash Q(b)$ on the left-hand-side, and $C_3 \colon P(\alpha) \vdash, C_4 \colon Q(\alpha) \vdash$ on the right-hand-side of the proof. There is one binary inference $\vee : l$ on the left side, which goes into the end-sequent, by which we merge C_1, C_2 to $C_5 \colon \ \vdash P(a), Q(b)$. On the right side of the proof the binary inference $\vee : l$ operates on ancestors of the cut (and does go into the cut), and we union the clauses to the set $\{C_3 \colon P(\alpha) \vdash, C_4 \colon Q(\alpha) \vdash\}$. Finally the cut itself is binary rule "going into the cut", and we take the union of all clauses generated so far; the result is

$$
\mathcal{C} = \{\vdash P(a), Q(b), \quad P(\alpha) \vdash, \quad Q(\alpha) \vdash\}.
$$

Note that α is a variable. $\mathcal{C}$ is called the *characteristic clause set* of φ. $\mathcal{C}$ is unsatisfiable and has the following resolution refutation R:

$$
\cfrac{Q(\beta) \vdash \quad \cfrac{\vdash P(a), Q(b) \quad P(\alpha) \vdash}{\vdash Q(b)}}{\vdash}
$$

Note that, as common in resolution theorem proving, we may always rename the variables in a clause. In fact, clauses in first-order logic represent universally closed disjunctions and bound variables can always be renamed. By applying the substitution $\{\alpha/a, \beta/b\}$ we obtain a propositional refutation R' of $\{\vdash P(a), Q(b), \quad P(a) \vdash, \quad Q(b) \vdash\}$ of the form

$$
\cfrac{Q(b) \vdash \quad \cfrac{\vdash P(a), Q(b) \quad P(a) \vdash}{\vdash Q(b)}}{\vdash}
$$

This proof R' can be taken as a *skeleton* of a proof with only atomic cuts of the end sequent $S \colon P(a) \vee Q(b), \forall x.\neg P(x) \vdash \exists z.Q(z)$. In this skeleton we will fill in

183

the so-called proof projections; these are cut-free proofs of the end-sequent + an instance of a characteristic clause. The idea of a proof projection is to skip all inferences going into the cut; all inferences going into the end-sequent are performed. Skipping binary rules going into the cut is achieved by weakening. We consider the clause C_5: $\vdash P(a), Q(b)$ and the corresponding projection π_1 (built from the left part of the proof):

$$\cfrac{\cfrac{\cfrac{P(a) \vdash P(a)}{P(a) \vdash P(a), Q(b)} \; w{:}\,r \qquad \cfrac{Q(b) \vdash Q(b)}{Q(b) \vdash P(a), Q(b)} \; w{:}\,r}{P(a) \vee Q(b) \vdash P(a), Q(b)} \; \vee{:}\,l}{P(a) \vee Q(b), \forall x.\neg P(x) \vdash \exists z.Q(z), P(a), Q(b)} \; w^*$$

From the right part of the proof and the instances $P(a) \vdash$ and $Q(b) \vdash$ of the clauses C_3, C_4 we get (by instantiating α in φ by a and b) the projections $\pi_2 =$

$$\cfrac{\cfrac{\cfrac{\cfrac{P(a) \vdash P(a)}{P(a), \neg P(a) \vdash} \; \neg{:}\,l}{P(a), \forall x.\neg P(x) \vdash} \; \forall{:}\,l}{P(a), P(a) \vee Q(b), \forall x.\neg P(x) \vdash \exists z.Q(z)} \; w^*}{}$$

and $\pi_3 =$

$$\cfrac{\cfrac{\cfrac{\cfrac{\cfrac{Q(b) \vdash Q(b)}{Q(b), \neg P(a) \vdash Q(b)} \; w{:}\,l}{Q(b), \neg P(a) \vdash \exists z.Q(z)} \; \exists{:}\,r}{Q(b), \forall x.\neg P(x) \vdash \exists z.Q(z)} \; \forall{:}\,l}{Q(b), P(a) \vee Q(b), \forall x.\neg P(x) \vdash \exists z.Q(z)} \; w{:}\,l}{}$$

Let $S \circ C$ be the sequent S merged with the sequent C^5. Then the projections are cut-free proofs π_1 of $S \circ \vdash P(a), P(b)$, π_2 of $P(a) \vdash \circ S$ and π_3 of $Q(b) \vdash \circ S$. By replacing the clauses in R' by the proofs π_1, π_2, π_3 we get the proof:

$$\cfrac{\cfrac{(\pi_3)}{Q(b) \vdash \circ S} \qquad \cfrac{\cfrac{(\pi_1)}{\vdash P(a), Q(b) \circ S} \qquad \cfrac{(\pi_2)}{P(a) \vdash \circ S}}{\vdash Q(b) \circ S \circ S} \; cut}{\cfrac{S \circ S \circ S}{S} \; c^*} \; cut$$

where c^* denotes a sequence of contractions.

Note that propositional resolution and atomic cut are the same rules; indeed, in our formalism, propositional resolution is just a sub-calculus of **LK**.

The general definition of a characteristic clause set is the following:

Definition 7. *We define clause sets $\mathcal{C}_\nu$ for every node of a proof tree:*

---

[5] The merge $\Gamma_1 \vdash \Delta_1 \circ \Gamma_2 \vdash \Delta_2$ is defined as $\Gamma_1, \Gamma_2 \vdash \Delta_1, \Delta_2$

- *in the axioms S select the subsequents S' consisting of atoms which are ancestors of a cut, and construct $\{S'\}$.*
- *Assume that a clause set $\mathcal{C}_\nu$ is already constructed at premise node ν of a unary inference yielding the conclusion ν'. Then $\mathcal{C}_{\nu'} = \mathcal{C}_\nu$ (unary inferences do not change the clause set).*
- *Assume that $\mathcal{C}_{\nu_1}$ and $\mathcal{C}_{\nu_2}$ for the premises ν_1, ν_2 are already defined. We distinguish two cases:*
  - *The inferred formula in the consequent ν is an ancestor of a cut (or does not exist as the rule itself is a cut). Then $\mathcal{C}_\nu = \mathcal{C}_{\nu_1} \cup \mathcal{C}_{\nu_2}$.*
  - *The inferred formula in the consequent ν is an ancestor of the end-sequent. Then $\mathcal{C}_\nu = \mathcal{C}_{\nu_1} \times \mathcal{C}_{\nu_2}$, where $\mathcal{C} \times \mathcal{D} = \{C \circ D \mid C \in \mathcal{C}, D \in \mathcal{D}\}$.*

If ν_0 is the root of the proof φ then $\mathrm{CL}(\varphi) = \mathcal{C}_{\nu_0}$. $\mathrm{CL}(\varphi)$ is called the character-istic clause set *of φ*

For a formal definition of the proof projections we refer to [10] and [11]. As already illustrated in Example 4, a projection of a proof φ of S to a characteristic clause C is a cut-free proof of the sequent $S \circ C$ with less inferences than φ (where inferences going into cuts are dropped). Projections of the given proof φ together with a grounded resolution refutation of $\mathrm{CL}(\varphi)$ give the **ceres**-normal form. The whole procedure on a given (skolemized) **LK**-proof of a sequent S is as follows:

- construct $\mathrm{CL}(\varphi)$,
- compute the projections $\varphi[C]$ of φ to C for $C \in \mathrm{CL}(\varphi)$,
- construct a resolution refutation γ of $\mathrm{CL}(\varphi)$,
- compute a ground resolution refutation γ' from γ by applying a global unifier,
- replace the leaves C' of γ' by the projections $\varphi[C']$ (which is an instance of $\varphi[C]$); maintain the same cuts as in γ' and contract multiple occurrences of formulas in S. The result is a proof $\varphi(\gamma')$ of S with only atomic cuts.

The proof $\varphi(\gamma')$ is called a **ceres** *normal form of φ.*

ceres, in its original version, requires proof skolemization as a preprocessing; without skolemization of the proof, projections may violate eigenvariable conditions. However, the extension of **ceres** to higher-order logic (see [20]) yields (as a kind of side effect) a version of first-order **ceres** without skolemization, which is relatively complex. In fact, as proof projections on non-skolemized proofs may violate eigenvariable conditions, we obtain unsound proofs in a first step. How-ever, due to the special type of these violations, they can be repaired globally by a proof transformation; finally a sound cut-free proof in higher-order **LK** can be obtained.

A comparison of **ceres** and the Gentzen method (and, more general, of every cut-elimination method using the Gentzen proof-rewriting rules $\mathcal{R}$) shows that **ceres** is capable of producing much shorter proofs in the following exact asymp-totic sense, that of nonelementary improvement. Nonelementary improvement (to be formally defined below) is a natural measure in comparing cut-elimination methods as the complexity of cut-elimination itself is nonelementary.

Definition 8. *Let* $e : \mathbb{N}^2 \to \mathbb{N}$ *be the following function*

$$e(0, m) = m$$
$$e(n + 1, m) = 2^{e(n,m)}.$$

A function $f : \mathbb{N}^k \to \mathbb{N}^m$ *for* $k, m \geq 1$ *is called* elementary *if there exists an* $n \in \mathbb{N}$ *and a Turing machine* T *computing* f *s.t. the computing time of* T *on input* $(l_1, \ldots, l_k)$ *is less than or equal to* $e(n, |(l_1, \ldots, l_k)|)$ *where* $|\ |$ *denotes the maximum norm on* $\mathbb{N}^k$ *(see also [14]).*

The function $s : \mathbb{N} \to \mathbb{N}$ *is defined as* $s(n) = e(n, 1)$ *for* $n \in \mathbb{N}$.

A function which is not elementary is called nonelementary.

Remark 1. The notion of elementary function is robust under use of different models of Turing machines. In fact, it does not matter whether we consider machines with just one tape or several ones, or machines with unary or k-ary alphabets for $k > 1$.

Note that the functions s and e are nonelementary. In general, any function f which grows "too fast", i.e. for which there exists no number k s.t.

$$f(n) \leq e(k, n),$$

is nonelementary.

Every exponential function $f(x, y)$ of the form $p(x)^{q(y)}$ for polynomials p and q is elementary. It is easy to prove that there exists a Turing machine T computing f and number k s.t. the computing time of T on (x, y) is less than $e(k, |(x, y)|)$.

Definition 9. *Let* $\zeta : (x_n)_{n \in \mathbb{N}}$ *and* $\eta : (y_n)_{n \in \mathbb{N}}$ *two sequences of natural numbers. We say that* ζ *is elementary in* η *if there exists a number* k *s.t. for all* $n \in \mathbb{N}$: $x_n \leq e(k, y_n)$; *otherwise* ζ *is called* nonelementary in η.

For complexity analysis we use two measures:

- the symbolic complexity $\|\ \|$ (the number of symbol occurrences), and
- $l(\psi)$, the length of proof ψ (the number of inference nodes in the proof tree).

In [31] R. Statman proved that there exists a sequence of short proofs γ_n of sequents $S_n : \Gamma \vdash A_n$ s.t. the Herbrand complexity $\mathrm{HC}(S_n)$ of S_n (which is the minimal symbol complexity of a Herbrand sequent of a cut-free proof of S_n) is inherently nonelementary in $l(\gamma_n)$ (it is also nonelementary in $\|\gamma_n\|$); in fact, Statman did not explicitly address a specific formal calculus, leaving the formalization of the proof sequence to the reader. Independently V. Orevkov [28] proved the nonelementary complexity of cut-elimination for function-free predicate logic without equality. The proof sequences of Statman and Orevkov are different, but both encode the principle of iterated exponentiation best described by P. Pudlak in [29].

Theorem 3 (Statman, Orevkov). *There exists a sequence S_n of sequents with the following properties:*

- *There is a constant a s.t. for every n there exists an* **LK***-proof φ_n of S_n with $\|\varphi_n\| \leq 2^{a*n}$.*
- *For every n let $c(n) = \min\{\|\psi\| \mid \psi$ is a cut-free proof of $S_n\}$. Then $(c_n)_{n\in\mathbb{N}}$ is not elementary in $(\|\varphi_n\|)_{n\in\mathbb{N}}$.*

Proof. In [31] and [28]. Note that the Herbrand complexity $\mathrm{HC}(S_n)$ of S_n defines a lower bound on $c(n)$; on the other hand, $c(n)$ is at most exponential (and thus elementary) in $\mathrm{HC}(S_n)$). So it does not matter, whether we speak of the symbolic lengths of shortest cut-free proofs or about Herbrand complexity. $\square$

Below we give a definition which provides a basis for comparing reductive cut-elimination methods and `ceres`. Thereby, reductive cut-elimination is described as sequence of proofs θ obtained via a proof reduction relation $>_x$ based on $\mathcal{R}$, starting with a proof φ and ending in a proof φ' with at most atomic cuts. Such a sequence θ is called an $>_x$-cut-elimination sequence on φ.

Definition 10. *Let $>_x$ be a proof reduction relation based on $\mathcal{R}$. We say that* `ceres` *NE-improves $>_x$ if there exists a sequence of proofs $(\varphi_n)_{n\in\mathbb{N}}$ s.t.*

- *there exists a sequence of resolution refutations $(\gamma_n)_{n\in\mathbb{N}}$ of the sequence of the corresponding characteristic clause sets $(\mathrm{CL}(\varphi_n))_{n\in\mathbb{N}}$ such that $(\|\gamma_n\|)_{n\in\mathbb{N}}$ is elementary in $(\|\varphi_n\|)_{n\in\mathbb{N}}$.*
- *Let $g(n) = \min\{\|\theta\| \mid \theta$ is an $>_x -cut$-elimination sequence on $\varphi_n\}$, for $n \in \mathbb{N}$. Then $(g(n))_{n\in\mathbb{N}}$ is nonelementary in $(\|\varphi_n\|)_{n\in\mathbb{N}}$.*

Similarly we define that $>_x$ NE-improves `ceres` *if there exists a sequence of proofs $(\varphi_n)_{n\in\mathbb{N}}$ s.t.*

- *there exists a sequence of $>_x$-cut-elimination sequences $(\theta_n)_{n\in\mathbb{N}}$ on $(\varphi_n)_{n\in\mathbb{N}}$ s.t. $(\|\theta_n\|)_{n\in\mathbb{N}}$ is elementary in $(\|\varphi_n\|)_{n\in\mathbb{N}}$.*
- *For all n let $h(n) = \min\{\|\gamma\| \mid \gamma$ is a resolution refutation of $\mathrm{CL}(\varphi_n)\}$. Then $(h(n))_{n\in\mathbb{N}}$ is nonelementary in $(\|\varphi_n\|)_{n\in\mathbb{N}}$.*

Remark 2. Comparing the size of the resolution refutations in `ceres` with the total size of cut-elimination sequences is justified, as the resolution refutations of characteristic clause sets are the main source of complexity in `ceres`; in fact, the computation time of a sequence of `ceres` normal forms grows nonelementarily in the size of the input proofs iff this holds for the computation of the resolution refutations. So, for this asymptotic comparison, the computation of the characteristic clause sets and the projections do not matter. Also mathematically the core of the `ceres`-method is the resolution refutation of the characteristic clause set.

Theorem 4. `ceres` *NE-improves the Gentzen method of cut-elimination*

Proof. We give a modified version of the proof in [11]. Let $(\psi_n)_{n\in\mathbb{N}}$ be a sequence of proofs for $\psi_n =$

$$
\dfrac{
\dfrac{
\dfrac{A \vdash A}{A, \Delta_n \vdash A}\ w\colon l \quad
\dfrac{\genfrac{}{}{0pt}{}{(\gamma_n)}{\Delta_n \vdash D_n}}{A, \Delta_n \vdash D_n}\ w\colon l
}{A, \Delta_n \vdash A \wedge D_n}\ \wedge\colon r \quad
\dfrac{
\dfrac{A \vdash A \quad A \vdash A}{A, A \to A \vdash A}\ \to\colon l
}{A \wedge D_n, A \to A \vdash A}\ \wedge\colon l
}{A, \Delta_n, A \to A \vdash A}\ cut
$$

where γ_n is Statman's worst-case sequence admitting only nonelementary cut-elimination (no matter which method is applied); for details in the definition of γ_n see [11]. In the method of Gentzen we always select an uppermost cut. As all cuts in γ_n are above the cut with $A \wedge D_n$, Gentzen's method eliminates all the cuts in γ_n before eliminating the cut with formula $A \wedge D_n$; thus it constructs a cut-free proof of $\Delta_n \vdash D_n$, which is of nonelementary size in $\|\gamma_n\|$ and also in $\|\psi_n\|$.

Let us turn to `ceres` on ψ_n. The characteristic clause sets are

$$
\mathrm{CL}(\psi_n) = \{\vdash A;\ A \vdash\} \cup \mathrm{CL}(\gamma_n).
$$

Trivially every $\mathrm{CL}(\psi_n)$ has the resolution refutation $\rho =$

$$
\dfrac{\vdash A \quad A \vdash}{\vdash}
$$

which is of constant length and, by defining $\rho_n = \rho$ for all n, we get $\|\rho_n\| = 5$. Trivially $(\|\rho_n\|)_{n\in\mathbb{N}}$ is elementary in $(\|\psi_n\|)_{n\in\mathbb{N}}$. $\qquad\square$

A similar result also holds for the Tait-method, another method of cut-elimination based on $\mathcal{R}$ [11]. A nonelementary speed-up in the other direction is impossible – for every method based on $\mathcal{R}$.

Theorem 5. *No reductive method based on $\mathcal{R}$ NE-improves* `ceres`*; in particular the Gentzen method does not NE-improve* `ceres`*.*

Proof. In [10] and [11]. $\qquad\square$

The proof of this theorem is based on a result showing that reductive cut-elimination has only redundant effects on the characteristic clause set. Surprisingly, this redundancy is defined by subsumption, a common redundancy-elimination principle of automated deduction (see e.g. [23]):

Theorem 6. *Let φ be an **LK**-derivation and ψ be a normal form of φ under a cut reduction relation $>_{\mathcal{R}}$ based on $\mathcal{R}$. Then $\mathrm{CL}(\varphi) \leq_{ss} \mathrm{CL}(\psi)$.*

Proof. In [10] and [11]. $\qquad\square$

The theorems above show that, from a complexity theoretic point of view, `ceres` is superior to reductive methods of cut-elimination. It pays out that, prior to cut-elimination, we analyze the proof and make use of the extracted structure of characteristic clause set (which then is analyzed via a resolution refutation). In contrast, the reductive methods are just local (they focus on the upmost operators of cut-formulas) and cannot take into account the global structure of proofs. `ceres` is much less redundant than the reductive methods, but also "less" confluent. Note than also Gentzen's method is not confluent, and there is no unique normal form under the reduction rules. `ceres`, in turn, can produce much more different normal forms (and corresponding Herbrand sequents) than reductive methods.

`ceres` is basically a method for cut-elimination in *classical logic*. The generalization to finitely valued logics is unproblematic [9]; there exists also a `ceres`-method for Gödel logic [2] and for subclasses of intuitionistic logic [24]. An advantage of reductive methods is their flexibility concerning structural restrictions; note that the reductive Gentzen method is virtually the same for classical and for intuitionistic logic.

4 Cut-Elimination in Practice

Using cut-elimination in practice requires first to *formalize* a mathematical proof as an **LK**-proof. In the formalization it is crucial to avoid unnecessary cuts, as then the characteristic clause sets becomes too complex, which can make it unfeasible for the theorem prover to refute them. This step is generally delicate as there is nothing like a unique formal representation of an informal mathematical proof. We address this problem once more in the application chapter where we describe the analysis of Fürstenberg's proof on the infinitude of primes. Reductive methods fail in practice because of the sheer size of the cut-free proofs and the high number of reduction steps. The `ceres` method, which is asymptotically superior as described in Section 3, is also much better in practice. This can be immediately seen by investigating the characteristic clause set of a real problem. A typical characteristic clause set of an **LK**-proof with cuts contains numerous tautologies and subsumed clauses, which remain undetected by reductive methods. In `ceres`, however, which is based on resolution theorem proving, tautologies and subsumed clauses can be eliminated without loss of completeness. In fact, the proofs obtained by Gentzen's method correspond to very redundant resolution derivations using lots of tautologies and subsumed clauses. Though the proofs found by `ceres` are generally much smaller and more compact, the size of the `ceres` normal forms still remains a barrier: remember that our aim is to analyze proofs, the main goal being the *interpretation* of the obtained cut-free proof. For huge `ceres` normal forms such an interpretation cannot be found by just reading down the proof. Further compressions of information are required. The ideal concept for compression is the Herbrand sequent of a proof constructible from all proofs with only atomic cuts (see Corollary 1). Herbrand sequents abstract from propositional reasoning and represent the first-order content of a proof in a very compact way; the Herbrand sequent of a proof essentially describes the instantiations of the formulas needed to prove the theorem. After this post-processing proofs become much more readable and it is much easier to mine the very mathematical content of it. In some cases (see the application chapter) the refutation of the characteristic clause alone contains the main mathematical information, and it is not even necessary to produce an atomic cut normal form or even a cut-free proof.

Gentzen's **LK** is the original calculus for which cut-elimination was defined. In formalizing mathematical proofs it turns out that **LK** (and also natural deduction) are not sufficiently close to real mathematical inference.

First of all, the calculus **LK** lacks a specific handling of equality to implement equality reasoning equality axioms have to be added to the end-sequent. Due to the importance of equality this defect was already apparent to proof theorists; e.g. Takeuti [32] defined an extension of **LK** to a calculus $\mathbf{LK}_=$, adding atomic equality axioms to the standard axioms of the form $A \vdash A$. The advantage of $\mathbf{LK}_=$ over **LK** is that no new axioms have to be added to the end-sequent; on the other hand, in presence of the equality axioms, full cut-elimination is no longer possible, but merely reduction to *atomic cut*. As we are not interested in eliminating atomic cuts this causes no problems. But still $\mathbf{LK}_=$ uses the same rules as **LK**; in fact, in $\mathbf{LK}_=$, equality is *axiomatized*, i.e. additional atomic (non-tautological) sequents are admitted as axioms. On the other hand, in formalizing mathematical proofs, using equality as a *rule* is much more natural and concise (for a general discussion on rules versus axioms see [27]). For this reason we choose the most natural equality rule, which is strongly related to paramodulation in automated theorem proving. Our approach differs from this in [33], where a unary equality rule is used (which does not directly correspond to paramodulation). In the *equality rules* below we mark the auxiliary formulas by $+$ and the principal formula by $*$.

$$\frac{\Gamma_1 \vdash \Delta_1, s = t^+ \quad A[s]_\Lambda^+, \Gamma_2 \vdash \Delta_2}{A[t]_\Lambda^*, \Gamma_1, \Gamma_2 \vdash \Delta_1, \Delta_2} =: l1 \qquad \frac{\Gamma_1 \vdash \Delta_1, t = s^+ \quad A[s]_\Lambda^+, \Gamma_2 \vdash \Delta_2}{A[t]_\Lambda^*, \Gamma_1, \Gamma_2 \vdash \Delta_1, \Delta_2} =: l2$$

for inference on the left and

$$\frac{\Gamma_1 \vdash \Delta_1, s = t^+ \quad \Gamma_2 \vdash \Delta_2, A[s]_\Lambda^+}{\Gamma_1, \Gamma_2 \vdash \Delta_1, \Delta_2, A[t]_\Lambda^*} =: r1 \qquad \frac{\Gamma_1 \vdash \Delta_1, t = s^+ \quad \Gamma_2 \vdash \Delta_2, A[s]_\Lambda^+}{\Gamma_1, \Gamma_2 \vdash \Delta_1, \Delta_2, A[t]_\Lambda^*} =: r2$$

on the right, where Λ denotes a set of positions of subterms where replacement of s by t has to be performed. We call $s = t$ the *active equation* of the rules.

Furthermore, as the only axiomatic extension, we need the set of reflexivity axioms

$$\text{REF}: \ \vdash s = s$$

for all terms s.

Definition 11. *The calculus* **LKe** *is* **LK** *extended by the axioms* REF *and by the rules*

$$=: l1, \ =: l2, \ =: r1, \ and \ =: r2.$$

The calculus **LKe** contains additional rules and any cut-elimination method has to be adapted accordingly. For `ceres` this adaption consists in the extension of the clausal calculus from resolution to resolution + paramodulation. Note that **LKe**, without use of logical rules, coincides with resolution and paramodulation on clause logic, provided most general unification has already been carried out. The definitions of characteristic clause set and proof-projections remain the same, only we have to handle the equality rules as binary rules going into the end-sequent; for details see [11]. In contrast, the reduction rules of Gentzen cannot be adapted in an easy way. If we want to use reductive cut-elimination methods we must transform the whole proof prior to cut-elimination: all equality rules have to be shifted upwards in the proof s.t. they apply to atoms only. After this transformation the Gentzen method can be applied to the part of the proof below the equality rules to get a proof with only atomic cuts.

Also Herbrand sequent extraction can be generalized to **LKe**-proofs. We obtain a more general version of the midsequent theorem:

Theorem 7 (mid-sequent theorem for LKe). *Let φ be an* **LKe**-*proof of a Σ_1-sequent S with at most atomic cuts s.t. all equality rules in φ are only applied to atoms. Then φ can be transformed into a proof φ' with the same number of logical inferences, equality rules and atomic cuts and with the following property: φ' contains the derivation of a sequent S' (the mid-sequent), s.t. all propositional inferences, atomic cuts and equational rules in φ are above S', and below S' there are only unary structural rules and quantifier-rules from φ.*

Proof. The proof transformation is essentially the same as in Theorem 2 as (still) the formulas in the end-sequent are prenex and there are no equality rules applied to quantified formulas. □

Remark 3. Note that the sequents S' obtained in Theorem 7 are no longer valid in general, but just E-valid, i.e. valid in equality-interpretations (the predicate symbol = is interpreted as equality over a domain). In fact, the equality rules of **LKe** are only sound w.r.t. equality-interpretations.

A further generalization is useful in practice. Instead of using just axioms of the form $A \vdash A$ we may also allow equational axioms of the form $\vdash s = t$. Then the set of axioms is still consistent and its models are equational theories. The method `ceres` remains exactly the same as it can be applied to any proof with only atomic axioms. Moreover, for facilitating the specification of proofs, definition-rules can be added to the calculus; the **LK**-version using the equality rules defined above and the definition rules is called **LKDe** (for details see [4]). Characteristic clause sets and their refutations are not affected by the extension by definitions.

The equational `ceres`-method based on **LKDe** has been applied to analyze several (rather simple) mathematical proofs fully automatically. We mention the analysis of the *tape proof* [3] and of the *lattice proof* [21].

The tape proof φ expresses a very simple statement about an infinite tape, each cell containing either 0 or 1, namely that, on such a tape, there are two cells containing the same number. φ uses the lemma that either the tape contains infinitely many 0's (I_0) or it contains infinitely many 1's (I_1); φ also contains two cuts, one with I_0, the other with I_1. The application of ceres to φ, first with positive- and then with negative hyperresolution yielded two different mathematical arguments (both simple arguments without detour over the lemma).

The lattice proof is based on three equivalent definitions of a lattice. The first definition L_1 defined it as a semi-lattice with the property

$$\forall x \forall y (x \cap y = x \leftrightarrow x \cup y = y),$$

the second (L_2) as a semi-lattice fulfilling the absorption laws

$$\forall x \forall y (x \cap y) \cup x = x \quad \text{and} \quad \forall x \forall y (x \cup y) \cap x = x.$$

In the third definition L_3 it was defined as a partially ordered set with a greatest lower and least upper bound. From proofs φ_1 of $L_1 \vdash L_3$ (all L_1-lattices are L_3-lattices) and φ_2 of $L_3 \vdash L_2$ (all L_3-lattices are L_1-lattices) a proof φ of $L_1 \vdash L_2$ (all L_1-lattices are L_2-lattices) could be obtained via cut on L_3. Cut-elimination via ceres and Herbrand sequent extraction produced a simple direct proof that L_1-lattices are L_2-lattices, without detour over partially ordered sets.

These applications illustrate that ceres is a suitable tool for *mining* proofs, i.e. to extract "hidden" mathematical information from proofs.

5 An Application of Cut-Elimination

We apply ceres to Fürstenberg's proof of the existence of infinitely many primes. The arguments of this proof are of topological nature, which form the synthetic notions of this synthetic proof. The proof is synthetic in the sense that it contains the ideal objects of topology in proving a property of real objects, the natural numbers. A natural formalization of this argument in second-order arithmetic is constructed and then translated to many-sorted first-order logic. In order to avoid induction axioms, the proof is eventually formalized as a scheme representing an infinite sequence of ordinary first-order proofs, demonstrating the existence of more and more primes. We show that the analytic proof schema corresponding to Euclid's proof belongs to the solution space of the schema of topological proofs.

In 1955 the renowned mathematician H. Fürstenberg published a proof of the infinity of primes by topological means [16] (see also [1]): He proved the infinity of primes using a topology induced by arithmetic progressions over the integers.

We give a proof with a topology over the natural numbers in order to have a simpler formulation of the proof later on. We start with the definition of a topological space:

Definition 12 (Topological Space). *A topological space is a set X together with a collection T of subsets of X satisfying the following axioms:*

1. *The empty set and X are in T.*
2. *The union of any collection of sets in T is also in T.*
3. *The intersection of any pair of sets in T is also in T.*

The collection T is called a topology on X. The sets in T are the open sets, *and their complements in X are the* closed sets.

The arithmetic progressions can be used as a basis for a topology over the natural numbers. We will denote an arithmetic progression by

$$\nu(a,b) = \{a + bn \mid n \in \mathbb{N}\}$$

for $a \in \mathbb{N}$ and $b \in \mathbb{N} \setminus \{0\}$.

Proposition 1. *By defining a set $A \subseteq \mathbb{N}$ as open, when A is either empty or for each $x \in A$ there exists an $a \in \mathbb{N} \setminus \{0\}$ such that $\nu(x, a) \subseteq A$, one obtains a topology over $\mathbb{N}$.*

Proof. We check definition 12:

1. The empty set and $\mathbb{N}$ are open. Trivial.
2. The union of a collection of open sets is also open. Trivial.
3. The intersection of two open sets is also open.
   Let A and B two open sets. If $x \in A \cap B$, then there exist $a, b > 0$ such that $\nu(x, a) \subseteq A$ and $\nu(x, b) \subseteq B$ holds. Let c be the least common multiple of a and b, then $\nu(x, c) \subseteq \nu(x, a)$ and $\nu(x, c) \subseteq \nu(x, b)$, and hence $\nu(x, c) \subseteq A \cap B$.
   $\square$

A nice property of this topology is that every arithmetic progression starting at 0 is not only open but closed as well. Indeed this property holds for every progression $\nu(a, b)$ where $a < b$, but this is not needed for the theorem.

Lemma 1. *Every arithmetic progression starting at 0 is closed.*

Proof. Let $A = \nu(0, b)$ be an arithmetic progression. Then the complement of A is a union of arithmetic progressions:

$$\bar{A} = \bigcup_{i=1}^{b-1} \nu(i, b).$$

The sets $\nu(i, b)$ are open, and the union of any collection of open sets is open; therefore $\bar{A}$ is open, hence A is closed. $\square$

Theorem 8. *There are infinitely many primes.*

Proof. Denote with P the set of all primes and assume P is finite. Let $X = \bigcup \{\nu(0, p) \mid p \in P\}$. By Lemma 1 every $\nu(0, p)$ for $p \in P$ is closed, so X is a finite union of closed sets and therefore closed as well. As every number different from 1 has a prime divisor we get $\bar{X} = \{1\}$. Being a complement of a closed set, $\bar{X}$ is open. But $\{1\}$ is neither empty nor does it contain an arithmetic progression, and so $\{1\}$ is not open. Contradiction! We conclude that P must be infinite. $\square$

The automated processing of Fürstenberg's proof requires a nontrivial logical preprocessing by humans. The first important step consists in the right choice of the logical language. As Fürstenberg's proof contains a topology defined over natural numbers and topological lemmas (with quantification over set-variables), an adequate candidate is *second-order arithmetic*. The formalization in [5] started with a formalization of the proof in second-order arithmetic. In a second step this specification was translated into a scheme of sorted first-order definitions and proofs. For the details we refer to [5], but we present the main steps and formal definitions here. We start with the formalization in second-order arithmetic:

(a) $m \in \nu(k, l) \equiv \exists n(m = k + n * l)$.
(b) $\mathrm{DIV}(l, k) \equiv \exists m.l * m = k$.
(c) $\mathrm{PRIME}(k) \equiv 1 < k \wedge \forall l(\mathrm{DIV}(l, k) \rightarrow (l = 1 \vee l = k))$.
(d) $X \subseteq Y \equiv \forall n(n \in X \rightarrow n \in Y)$, and $X = Y \equiv X \subseteq Y \wedge Y \subseteq X$.
(e) $n \in \overline{X} \equiv n \notin X$.
(f) A function $p \colon \mathbb{N} \rightarrow \mathbb{N}$ which enumerates primes is one that fulfills the property:
$$\forall i \forall k(p(i) = k \rightarrow \mathrm{PRIME}(k)).$$

For the definition of p the comprehension principle is needed; for information about function definitions in second-order arithmetic see [30].
(g) $n \in \mathrm{S}[l] \equiv \exists m(m \leq l \wedge n \in \nu(0, p(m)))$.
S[l] describes the set of all elements n which occur in some $\nu(0, k)$, where k is one of the first $l + 1$ primes enumerated by p. In mathematical notation we get
$$\mathrm{S}[l] = \bigcup_{m=0}^{l} \nu(0, p(m)).$$

(h) $\mathrm{F}[l] \equiv \forall k(\mathrm{PRIME}(k) \leftrightarrow \exists m(m \leq l \wedge k = p(m)))$.
F[l] is a formula which asserts that there are only $l + 1$ primes, namely $\{p(0), \ldots, p(l)\}$.
(i) $\mathrm{O}(X) \equiv \forall m(m \in X \rightarrow \exists l \, \nu(m, l + 1) \subseteq X)$.
(j) $\mathrm{C}(X) \equiv \mathrm{O}(\overline{X})$.
(k) $\infty(X) \equiv \forall k \exists l \, k + l + 1 \in X$.

Let (*) be the assumption that all primes occur in the set $M \colon \{p(0), \ldots, p(l)\}$. The first lemma in Fürstenberg's proof states that, under the assumption (*), every natural number different from 1 occurs in some $\nu(0, m)$ for $m \in M$. The corresponding formula is

$$\text{(I)} \ \forall l(\mathrm{F}[l] \rightarrow \mathrm{S}[l] = \overline{\{1\}}).$$

The second lemma states that, under the assumption (*), the set S[l] is closed. The formula expressing this lemma is

$$\text{(II)} \ \forall l(\mathrm{F}[l] \rightarrow \mathrm{C}(\mathrm{S}[l])).$$

Proofs of (I) and (II) can easily combined to a proof of

$$\text{(III)} \ \forall l(\text{F}[l] \to \text{C}(\overline{\{1\}})).$$

The proofs of (II) and (III) in second order arithmetic require induction. By (j) it is straightforward to prove

$$\text{(IV)} \ \forall l(\text{F}[l] \to \text{O}(\{1\})).$$

Another main lemma of the proof states that nonempty open sets are infinite:

$$\text{(V)} \ \forall X(\text{O}(X) \wedge X \neq \emptyset \to \infty(X)).$$

While (I), (II), (III) and (IV) can be directly translated to first order logic (via the definitions), (V) is genuinely second order. Using (V) we show that $\infty(\{1\})$ holds giving a contradiction to $\neg\infty(\{1\})$, which is easily derivable in second order arithmetic.

To formulate Fürstenberg's proof in **LKDe** it is necessary to schematize it in order to avoid induction. In particular, induction is needed to prove the lemmas (II) and (III) above. The tool `hlk` [22] allows to define an infinite sequence of **LKDe**-proofs by specifying a proof scheme. The k-th proof can then be generated automatically from the scheme for any k.

The k-th proof shows that there cannot be $\leq k+1$ prime numbers.

To compile the second-order formulation to first order we work in a two-sorted logic containing sorts for 1. the natural numbers (denoted by $k, l, m, n, \ldots$ as before) and 2. sets of natural numbers (denoted by $x, y, \ldots$). Addition $(+)$, multiplication $(*)$ and the less-than relation $(<)$ in the natural numbers are axiomatized. The background theory is purely universal and thus can be expressed as a set of clauses **AX**; It contains 34 clauses, among them associativity, commutativity and distributivity laws plus some derived laws like e.g. the cancellation law $(k + l = m + l \vdash k = m)$. For the full list of axioms see the documentation on the web[6]. All of these axiom clauses are valid axiom sequents for the **LKDe**-proof.

Some of the definitions (a) to (k) given above can be taken over without change. This holds for (a), (b) and (c). For the others we get:

(d') $x \subseteq y \equiv \forall n(n \in x \to n \in y)$, and $x = y \equiv x \subseteq y \wedge y \subseteq x$. Here we only replaced the set variables by variables of the sort "set of natural numbers".
(e') $n \in \overline{x} \equiv n \notin x$.
(f') Instead of p we introduce a finite set $\text{P}[k]$ defined by

$$\text{P}[k] \equiv \{p_0\} \cup \cdots \cup \{p_k\}.$$

where the p_i are constant symbols denoting primes. Note that the k appearing in the definition is a metavariable, not an object variable as l in the definition of $\text{F}[l]$ and $\text{S}[l]$.

---

[6] `http://www.logic.at/ceres/examples/prime.php`

(g') $S[k] \equiv \nu(0, p_0) \cup \cdots \cup \nu(0, p_k)$. Note that, in place of the object variable l in the definition (g), we have the metavariable k of the scheme.

(h') $F[k] \equiv \forall m(\text{PRIME}(m) \leftrightarrow m \in P[k])$.

(i') $O(x) \equiv \forall m(m \in x \rightarrow \exists l \; \nu(m, l+1) \subseteq x)$.

(j') $C(x) \equiv O(\overline{x})$.

(k') $\infty(x) \equiv \forall k \exists l \; k + l + 1 \in x$.

In order to avoid induction we also introduce three axioms (which can be proven in Peano arithmetic): (1) Every number greater than 0 has a predecessor, (2) every number is in a remainder class modulo l and (3) every number has a prime divisor. These axioms will be carried down to the antecedent of the end sequent of the **LKDe**-proof.

$$(1) \; \text{PRE} \equiv \forall k(0 < k \rightarrow \exists m \; k = m + 1)$$
$$(2) \; \text{REM} \equiv \forall l(0 < l \rightarrow \forall m \exists k(k < l \wedge m \in \nu(k, l)))$$
$$(3) \; \text{PRIME-DIV} \equiv \forall m(m \neq 1 \rightarrow \exists l(\text{PRIME}(l) \wedge \text{DIV}(l, m)))$$

We now formulate a proof $\varphi_1(k)$ which proves the translation of (IV) above:

$\varphi_1(k) :=$

$$
\begin{array}{c}
\psi_1(k) \qquad\qquad\qquad\qquad \psi_2(k) \\
\vdots \qquad\qquad\qquad\qquad\qquad \vdots \\
\cfrac{
\cfrac{F[k], \text{PRIME-DIV} \vdash S[k] = \overline{\{1\}} \quad F[k], \text{PRE}, \text{REM} \vdash C(S[k])}{F[k], \Gamma \vdash C(\overline{\{1\}})} \quad
\begin{array}{c} \vdots \\ C(\overline{\{1\}}) \vdash O(\{1\}) \end{array}
}{F[k], \Gamma \vdash O(\{1\})} \; cut
\end{array}
$$

with $=: r$ on the upper inference.

The proof $\psi_1(k)$ shows that if there are $\leq k+1$ primes, then by the prime divisor axiom PRIME-DIV, the complement of all multiples of these primes is $\{1\}$, and the proof $\psi_2(k)$ demonstrates (under the assumption of $\leq k+1$ primes and the remainder axiom REM) that the set of these multiples is closed. With the help of these lemmas we can show that the set $\{1\}$ is open — if there are $\leq k+1$ primes.

The proof φ_2 (which does not depend on k) shows that every (non-empty) open set is infinite. This lemma yields that, under the assumption of the set of primes being finite, the set $\{1\}$ must be either empty or infinite; of course neither is the case, which is easily shown. Hence we get our end-sequent, stating that there cannot be $\leq k+1$ primes:

$\varphi(k) :=$

$$
\begin{array}{c}
\varphi_2 \\
\varphi_1(k) \qquad\qquad\qquad \vdots \\
\vdots \qquad\qquad \vdash \forall x((O(x) \wedge x \neq \emptyset) \rightarrow \infty(x)) \quad \vdots \\
\cfrac{\vdash \{1\} \neq \emptyset \quad \cfrac{F[k], \Gamma \vdash O(\{1\}) \quad \cfrac{O(\{1\}), \{1\} \neq \emptyset \vdash \infty(\{1\})}{}}{\cfrac{\{1\} \neq \emptyset, F[k], \Gamma \vdash \infty(\{1\})}{F[k], \Gamma \vdash \infty(\{1\})}} \quad \cfrac{\vdots}{\infty(\{1\}) \vdash}}{\underbrace{\cfrac{F[k], \Gamma \vdash}{\text{PRIME-DIV}, \text{PRE}, \text{REM} \vdash \neg F[k]}}_{\Gamma}}
\end{array}
$$

with cut inferences and $\neg : r$.

The proof-schema φ_k above was then subjected to skolemization and the characteristic clause sets $\mathrm{CL}(\varphi_k)$ were computed for a large interval $[0, k]$ (for $k > 10$). The clause sets were surprisingly simple and could be strongly reduced in size by subsumption and tautology-deletion. The next (human-based) step consisted in the generalization of the clause pattern. The resulting sequence of clause sets (after redundancy-elimination) was

$$\mathrm{CL}_r := \mathcal{C}_r \cup \mathrm{AX}$$

where

$$\mathcal{C}_r := A \cup \bigcup_{i=0}^{r} B_i \cup \{C_r\}$$

for

$$C_r := \ \vdash m_0 = 1, s_1(m_0) = p_0, \ldots, s_1(m_0) = p_r,$$

$B_i :=$

$$0 < p_i \vdash p_i = s_7(p_i) + 1$$
$$0 < p_i \vdash t_0 = s_5(p_i, t_0) + (s_6(p_i, t_0) * p_i)$$
$$0 < p_i, s_5(p_i, t_0) = 0 \vdash t_0 = 0 + (s_6(p_i, t_0) * p_i)$$
$$0 < p_i \vdash s_5(p_i, t_0) < p_i$$
$$t_0 = p_i, m_0 * n_0 = t_0 \vdash m_0 = 1, m_0 = t_0$$
$$t_0 = p_i \vdash 1 < t_0$$
$$t_0 = p_i, 1 = n_0 * t_0 \vdash$$

and $A :=$

$$\vdash m_0 = 1, s_1(m_0) * s_4(m_0) = m_0$$

$$\vdash m_0 + (((k * (l_0 + (1 + 1))) + (l_0 * (m_0 + 1))) + 1) =$$
$$k + ((k + (m_0 + 1)) * (l_0 + 1))$$

$$m_0 = k_0 + (r_0 * ((t_0 + 1) * (t_1 + 1)))$$
$$\vdash m_0 = k_0 + ((r_0 * (t_0 + 1)) * (t_1 + 1))$$

$$m_0 = k_0 + (r_0 * ((t_0 + 1) * (t_1 + 1)))$$
$$\vdash m_0 = k_0 + ((r_0 * (t_1 + 1)) * (t_0 + 1))$$

$$\vdash (((t_0 + 1) * t_1) + t_0) + 1 = (t_0 + 1) * (t_1 + 1)$$

For this structurally simple sequence of characteristic clause sets a schema of resolution refutations was defined. In this refutation schema the crucial (schematic) clause

$$E_r : 1 < t_r \vdash$$

for $t_r = p_0 * \ldots * p_r + 1$ was derivable.

By several steps of paramodulations, E_r was transformed into E'_r: $1 < (s_r + 1) + 1 \vdash$ for some term s_r. From axiom clauses one could derive the clause G: $\vdash 1 < (w + 1) + 1$ (w being a variable). G and E'_r finally resolve to $\vdash$, the empty clause. The term t_r obtained in the derivation by resolution and paramodulation reflects exactly the construction in Euclid's proof of the infinitude of primes. We see that the elimination of topological arguments (performed by resolution and paramodulation in ceres) reveals a hidden property of Fürstenberg's proof, in the sense that it yields a *construction method for primes*, while the original proof does not. It is a delicate question, whether Fürstenberg's and Euclid's proof share some *common core* which was eventually revealed by ceres or not; perhaps, Fürstenberg found his proof by playing around with Euclid's proof and transforming it in an ingenious way. But it may also be that, in some sense, Fürstenberg's proof was constructed completely independently of Euclid's argument (and the latter is not "contained" in Fürstenberg's proof); in this case ceres could potentially find *new* arguments originally not present in a proof (clearly, in this special case, the argument is quite old and well-known). Finding an answer to this question by purely mathematical means seems to be impossible; there is a deep philosophical problem behind, beyond the scope of this paper (and beyond the expertise of the author).

6   Open Problems

As most mathematical proofs about numbers or discrete structures use inductive arguments, it is desirable to apply cut-elimination methods also to these proofs. However, cut-elimination in presence of an induction rule is either impossible [32] or does not produce a proof with the subformula property (see e.g. [25] where a cut-elimination method is defined for a logic with induction). But note that a universal sequent S: $\Gamma \vdash \forall x.A(x)$, where x is supposed to range over the natural numbers, can be replaced by a sequence S_n: $A(\bar{n})_{n \in \mathbb{N}}$ for numerals $\bar{n}$. If, instead of proving S by a single proof φ (using induction), we consider a proof sequence φ_n of S_n (like in the analysis of the Fürstenberg proof), we can use cut-elimination methods like ceres on the sequence. But that makes sense only if we succeed to describe the resolution refutations on the sequence of characteristic clause sets *uniformly* and (at least) to obtain a uniform description of the sequence of corresponding Herbrand sequents. A general method of this type capable of handling Peano arithmetic is intrinsically very complex (in fact cut-elimination of this type proves the consistency of Peano arithmetic) and far outside of any means of automation. A partial solution of this problem for simple kind of inductions can be found in [15]. The elimination of lemmas in inductive proofs (resulting in some type of analytic proofs) remains one of the major challenges of proof analysis.

7 Conclusion

We have presented a method of analyzing proofs via cut-elimination by resolution. As the core of the method consists of a theorem proving problem (the resolution refutation of a characteristic clause set) the real efficiency of the method is closely tied to that of automated theorem provers. Not only the problem of *finding* a proof of a theorem, but also the problem of cut-elimination on proofs, i.e. *finding proofs of a specific form* from existing proofs of a theorem, can be a hard challenging problem. In theorem proving the main focus is on the production of *some proof* of a theorem (in a frequently unreadable form); however, less emphasis is laid on post-processing of the proof output and its interpretations by humans. Proofs like that of Fürstenberg cannot be obtained by automated theorem provers (due to the weak lemma structure of resolution and paramodulation in clause logic). We claim that the investigation of the relation between complex abstract proofs (using complex lemmas) and elementary proofs of a mathematical theorem may lead to deep insights into a mathematical theory, far beyond the knowledge that the theorem simply holds.

Acknowledgements

Thanks to the reviewers the original version of this paper could be substantially improved. One of the reviewers directed me to additional literature on the topic and pointed to some foundational issues which were not clearly addressed. In a "write-only" society this service cannot be overestimated.

References

1. M. Aigner, G. M. Ziegler. Proofs from THE BOOK. Springer, 1998.
2. M. Baaz, A. Ciabattoni, C.G. Fermüller: Cut Elimination for First Order Gödel Logic by Hyperclause Resolution. Proc. of *LPAR'2008*. LNCS 5330, pp. 451–466, 2008.
3. M. Baaz, S. Hetzl, A. Leitsch, C. Richter, H. Spohr: Cut-Elimination: Experiments with CERES. *LPAR' 2004*, LNCS 3452, pp. 481–495, 2004.
4. M. Baaz, S. Hetzl, A. Leitsch, C. Richter, H. Spohr: Proof Transformation by CERES. *MKM 2006*, LNCS 4108, pp. 82–93, 2006.
5. M. Baaz, S. Hetzl, A. Leitsch, C. Richter, H. Spohr: CERES: An Analysis of Fürstenberg's Proof of the Infinity of Primes. *Theoretical Computer Science*, 403 (2–3), pp. 160-175, 2008.
6. M. Baaz, S. Hetzl, D. Weller: On the complexity of proof skolemization *Journal of Symbolic Logic*, 77(2), pp. 669-686, 2012.
7. M. Baaz, A. Leitsch: On skolemization and proof complexity. *Fundamenta Informaticae*, 20/4, pp. 353-379, 1994.
8. M. Baaz, A. Leitsch: Cut-Elimination and Redundancy-Elimination by Resolution. *Journal of Symbolic Computation*, 29, pp. 149-176, 2000.
9. M. Baaz, A. Leitsch: CERES in Many-Valued Logics. *Proceedings of LPAR'2004*, LNAI 3452, pp. 1–20, 2004.
10. M. Baaz, A. Leitsch: Towards a Clausal Analysis of Cut-Elimination. *Journal of Symbolic Computation*, 41, pp. 381–410, 2006.

11. M. Baaz, A. Leitsch: Methods of Cut-Elimination. *Trends in Logic 34*. Springer, 2011.
12. U. Berger, W. Buchholz, H. Schwichtenberg: Refined Program Extraction from Classical Proofs. *Annals of Pure and Applied Logic*, 114(1-3), pp. 3–25, 2002.
13. U. Berger, S. Berghofer, P. Letouzey, H. Schwichtenberg: Program Extraction from Normalization Proofs. *Studia Logica*, 82(1), pp. 25–49, 2006.
14. W.S. Brianerd, L.H. Landweber: Theory of Computation. John Wiley & Sons, 1974.
15. C. Dunchev, A. Leitsch, M. Rukhaia, D. Weller: CERES for first-order schemata. CoRR abs/1303.4257 (2013).
16. H. Fürstenberg: On the infinitude of primes. *American Mathematical Monthly* 62, p. 353, 1955.
17. G. Gentzen: Untersuchungen über das logische Schließen. *Mathematische Zeitschrift* 39, pp. 405–431, 1934–1935.
18. J.Y. Girard: Proof Theory and Logical Complexity. in *Studies in Proof Theory*, Bibliopolis, Napoli, 1987.
19. K. Gödel: Über eine bisher noch nicht benützte Erweiterung des finiten Standpunktes. *Dialectica* 12, pp. 280–287, 1958.
20. S. Hetzl, A. Leitsch, D. Weller: CERES in Higher-order Logic *Annals of Pure and Applied Logic* 162(12), pp. 1001–1034, 2011
21. S. Hetzl, A. Leitsch, D. Weller, B. Woltzenlogel Paleo: Herbrand Sequent Extraction. AISC/Calculemus/MKM 2008, LNAI 5144, pp. 462-477, 2008.
22. S. Hetzl, A. Leitsch, D. Weller, B. Woltzenlogel Paleo: Proof Analysis with HLK, CERES and ProofTool: Current Status and Future Directions. Proceedings of the CICM Workshop on Empirically Successful Automated Reasoning in Mathematics, CEUR Workshop Proceedings Vol-378 (2008), ISSN 1613-0073, 2008.
23. A. Leitsch: The Resolution Calculus. *EATCS Texts in Theoretical Computer Science*, Springer, Berlin, 1997.
24. A. Leitsch, G. Reis, B. Woltzenlogel Paleo: Towards CERes in intuitionistic logic. CSL 2012, pp. 485-499, 2012.
25. R. MacDowell, D. Miller: Cut-Elimination for a Logic with Definitions and Induction. *Theoretical Computer Science* 232, pp. 91-119, 2000
26. D. Miller: A Compact Representation of Proofs. *Studia Logica* 46/4, pp. 347–370, 1987.
27. S. Negri, J. von Plato: Structural Proof Theory. Cambridge University Press, 2001.
28. V. P. Orevkov: Lower Bounds for Increasing Complexity of Derivations after Cut Elimination. *J. Soviet Mathematics*, pp. 2337–2350, 1982.
29. P. Pudlak: The lengths of proofs. In: *Handbook of Proof Theory*, S.R. Buss (ed), Elsevier 1998.
30. S.G. Simpson: Subsystems of Second Order Arithmetic. Springer, 1999.
31. R. Statman: Lower bounds on Herbrand's theorem. *Proc. of the Amer. Math. Soc.* 75, pp. 104–107, 1979.
32. G. Takeuti: Proof Theory. North-Holland, Amsterdam, 2nd edition, 1987.
33. A. Degtyarev and A. Voronkov: Equality Reasoning in Sequent-Based Calculi. *Handbook of Automated Reasoning* vol. I, ed. by A. Robinson and A. Voronkov, chapter 10, pp. 611-706, Elsevier Science, 2001.
34. B. Woltzenlogel Paleo: Herbrand Sequent Extraction. VDM Verlag Dr.Müller e.K. (February 7, 2008), ISBN-10: 3836461528.

Definition of a Mathematical Language Together with its Proof System in Event-B

Jean-Raymond Abrial

Marseille
jrabrial@neuf.fr

1   Introduction

Our application domain with the B Method and Event-B is the modeling and development of industrial (embedded) systems. We have been working in this area for more than 20 years. More is described in [1] where the evolution from Z to B and Event-B is described with full details.The starting application of this approach was the driverless metro system for the metro line 14 in Paris: the B Method was used in the development of the safety critical part of the corresponding controller. In the B Method and in Event-B, the *prover part* is *absolutely central* together with the notion of *refinement*. In fact, in the beginning in the nineties, the Atelier B tool was developed together with the line 14 system itself. Later, Event-B [2] and the corresponding Rodin Platform [3] was developed with some heavy fundings from the European Commission.

In order to explain how all this emerged technically, this paper first contains an important definition of the *Mathematical Language* we used in Event-B. This is done in section 2. The reason for this important section is that we want to explain how we can interface our platform [3] with many different provers. The idea is quite simple: all the provers we use are first order predicate calculus with equality provers. The idea is then to translate set theoretic statements into predicate calculus statement as explained in sub-section 2.6.

Section 2 is made of six sub-sections. The first one contains a preliminary definition of sequents, inference rules, and proofs. Then we have the presentation of our Mathematical Language. It is defined as follows: the Propositional Language (section 2.3), the Predicate Language (section 2.4), the Equality Language (section 2.5), and the Set-theoretic Language (section 2.6). Each of these languages will be presented as an *extension* of the previous one.

In section 3, we develop some of the technologies used in the Rodin Platform [3] prover: the connection to some external automatic provers in section 3.1, the idea of reasoners and tactics in section 3.2, the notion of tactic profiles in section 3.3, the interactive prover in section 3.4, and finally the "Theory" plug-in in section 3.5.

Various sections (4 to 8) give then more information and comments on our approach with the proving system of Event-B. "Some Results" in section 4, "Using Proofs" in section 5, "Comparison of Proofs" in section 6, and finally "Trends and Open Problems" in section 7. We conclude in section 8.

2 A Mathematical Language Formal Construction

This section is essentially a reprint of chapter 9 of [2]

2.1 Sequent Calculus

Definitions In this section, we give some definitions which will be helpful to present the Sequent Calculus.

(1) A *sequent* is a generic name for "something we want to prove". For the moment, this is just an informally defined notion, which we shall refine later in section 2.1. The important thing to note at this point is that we can associate a *proof* with a sequent. For the moment, we do not know what a proof is however. It will only be defined at the end of this section.

(2) An *inference rule* is a device used to construct proofs of sequents. It is made of two parts: the *antecedent* part and the *consequent* part. The antecedent denotes a finite set of sequents while the consequent denotes a single sequent. An inference rule, named say **R1**, with antecedent A and consequent C is usually written as follows:

$$\frac{\mathsf{A}}{C} \quad \textbf{R1}$$

It is to be read:

Inference Rule **R1** yields a proof of sequent C as soon as we have proofs of each sequent of A.

The antecedent A might be empty. In this case, the inference rule, named say **R2**, is written as follows:

$$\frac{}{C} \quad \textbf{R2}$$

It is to be read:

Inference Rule **R2** yields a proof of sequent C.

(3) A *theory* is a set of inference rules.

(4) The *proof of a sequent* within a theory is simply a finite tree with certain constraints. The nodes of such a tree have two components: a sequent s and a rule r of the theory. Here are the constraints for each node of the form (s, r): the consequent of the rule r is s, and the children of this node are nodes whose sequents are exactly all the sequents of the antecedent of rule r. As a consequence, the leaves of the tree contain rules with no antecedent. Moreover, the root node of the tree contains the sequent to be proved. As an example, let be given the following theory involving sequents $S1$ to $S7$ and rules **R1** to **R7**:

$$\frac{}{S2}\ \textbf{R1} \qquad \frac{S7}{S4}\ \textbf{R2} \qquad \frac{S2\quad S3\quad S4}{S1}\ \textbf{R3} \qquad \frac{}{S5}\ \textbf{R4}$$

$$\frac{S5\quad S6}{S3}\ \textbf{R5} \qquad \frac{}{S6}\ \textbf{R6} \qquad \frac{}{S7}\ \textbf{R7}$$

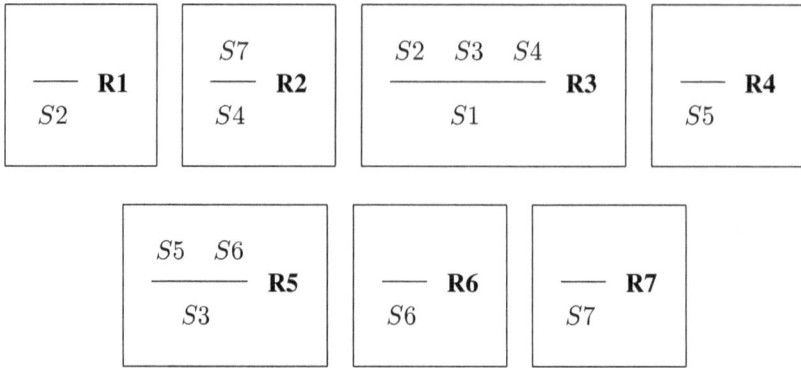

In figure 1 you can see a proof of sequent $S1$:

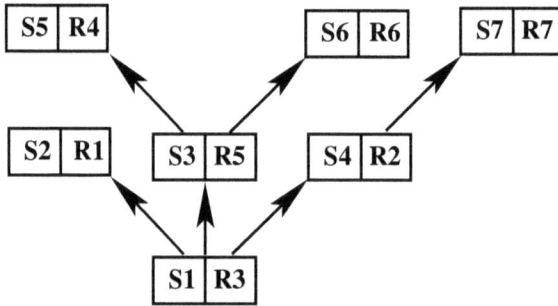

Fig. 1. A Proof

As can be seen, the root of the tree contains sequent $S1$, which is the one we want to prove. And it is easy to check that each node, say node $(S3, \textbf{R5})$, is indeed such that the consequent of its rule is the sequent of the node. More precisely, $S3$ in this case, is the consequent of rule $\textbf{R5}$. Moreover, we can check that the sequents of the child nodes of node $(S3, \textbf{R5})$, namely, $S5$ and $S6$, are exactly the sequents forming the antecedents of rule $\textbf{R5}$.

This tree can be interpreted as follows: In order to prove $S1$, we prove $S2$, $S3$, and $S4$, according to rule $\textbf{R3}$. In order to prove $S2$ we prove nothing more, according to rule $\textbf{R1}$. In order to prove $S3$ we prove $S5$ and $S6$, according to $\textbf{R5}$. And so on. This tree can be represented horizontally: this is indicated in figure 2. We shall now adopt this representation.

Sequents for a Mathematical Language We now refine our notion of sequent in order to define the way we shall make proofs with our Mathematical Language. Such a language contains constructs called *Predicates*. For the moment, this is all what we know about our Mathematical Language. Within this framework, a sequent S, as defined

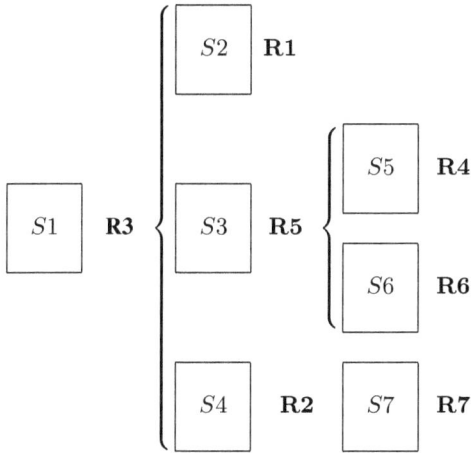

Fig. 2. Another Representation of the Proof Tree

in the previous section, now becomes a more complex object. It is made of two parts: the *hypotheses* part and the *goal* part. The hypothesis part denotes a finite set of predicates while the goal part denotes a single predicate. A sequent with hypotheses H and goal G is written as follows:

$$H \vdash G$$

This sequent is to be read as follows:

> Goal G holds under the set of hypotheses H

This is the sort of sequents we want to prove. It is also the sort of sequents we shall have in the theories associated with our Mathematical Language. Note that the set of hypotheses of a sequent might be empty and that the order and repetition of hypotheses in the set H is meaningless.

Initial Theory We now have enough elements at our disposal to define the first rules of our proving theory. Note again that we still do not know what a predicate is. We just know that predicates are constructs we shall be able to define within our future Mathematical Language. We start with three basic rules which we first state informally and then define more rigorously. They are called **HYP**, **MON**, and **CUT**. Here are their definitions:

- **HYP**: If the goal P of a sequent belongs to the set of hypotheses of this sequent, then it is proved.

$$\frac{}{\text{H, } P \ \vdash \ P} \ \textbf{HYP}$$

– **MON**: In order to prove a sequent, it is sufficient to prove another sequent with the same goal but with less hypotheses.

$$\frac{\text{H } \vdash \ Q}{\text{H, } P \ \vdash \ Q} \ \textbf{MON}$$

– **CUT**: If you succeed in proving a predicate P under a set of hypotheses H, then P can be added to the set of hypotheses H for proving a goal Q.

$$\frac{\text{H } \vdash \ P \qquad \text{H, } P \ \vdash \ Q}{\text{H } \vdash \ Q} \ \textbf{CUT}$$

2.2 Rule Schema

Note that in the rules defined in the previous section, the letter H, P and Q are, so-called, *meta-variables*. The letter H is a meta-variable standing for a finite set of predicates, whereas the letter P and Q are meta-variables standing for predicates. Clearly then each of the previous "rules" stands for more than just one rule: it is better to call them *rule schemas*. This will always be the case in what follows.

2.3 The Propositional Language

In this section we present a first simple version of our Mathematical Language, it is called the Propositional Language. It will be later refined to more complete versions: Predicate Language (section 2.4), Equality Language (section 2.5), Set-theoretic Language (section 2.6).

Syntax Our first version is built around five constructs called *falsity, negation, conjunction, disjunction,* and *implication*. Given two predicates P and Q, we can construct their conjunction $P \wedge Q$, their disjunction $P \vee Q$, and their implication $P \Rightarrow Q$. And given a predicate P, we can construct its negation $\neg P$. This can be formalized by means of the following syntax:

$$
\begin{array}{ll}
predicate & ::= \bot \\
& \neg\,predicate \\
& predicate \,\wedge\, predicate \\
& predicate \,\vee\, predicate \\
& predicate \,\Rightarrow\, predicate
\end{array}
$$

This syntax is clearly ambiguous, but we do not care about it at this stage. Only note that conjunction and disjunction operators have stronger syntactic priorities than the implication operator. Moreover, conjunction and disjunction have the same syntactic priorities so that parentheses will always be necessary when several such distinct operators are following each other. Also note that this syntax does not contain any "base" predicate (except $\bot$): such predicates will come later in sections 2.5 and 2.6.

Enlarging the Initial Theory The initial theory of section 2.1 is enlarged with the following inference rules:

$$
\frac{}{\mathsf{H}, \bot \;\vdash\; P} \;\textbf{FALSE\_L}
\qquad
\frac{\mathsf{H} \;\vdash\; P \qquad \mathsf{H} \;\vdash\; \neg P}{\mathsf{H} \;\vdash\; \bot} \;\textbf{FALSE\_R}
$$

$$
\frac{\mathsf{H}, \neg Q \;\vdash\; P}{\mathsf{H}, \neg P \;\vdash\; Q} \;\textbf{NOT\_L}
\qquad
\frac{\mathsf{H}, P \;\vdash\; \bot}{\mathsf{H} \;\vdash\; \neg P} \;\textbf{NOT\_R}
$$

$$
\frac{\mathsf{H}, P, Q \;\vdash\; R}{\mathsf{H}, P \wedge Q \;\vdash\; R} \;\textbf{AND\_L}
\qquad
\frac{\mathsf{H} \;\vdash\; P \qquad \mathsf{H} \;\vdash\; Q}{\mathsf{H} \;\vdash\; P \wedge Q} \;\textbf{AND\_R}
$$

$$
\frac{\mathsf{H}, P \;\vdash\; R \qquad \mathsf{H}, Q \;\vdash\; R}{\mathsf{H}, P \vee Q \;\vdash\; R} \;\textbf{OR\_L}
\qquad
\frac{\mathsf{H}, \neg P \;\vdash\; Q}{\mathsf{H} \;\vdash\; P \vee Q} \;\textbf{OR\_R}
$$

$$
\frac{\mathsf{H}, P, Q \;\vdash\; R}{\mathsf{H}, P, P \Rightarrow Q \;\vdash\; R} \;\textbf{IMP\_L}
\qquad
\frac{\mathsf{H}, P \;\vdash\; Q}{\mathsf{H} \;\vdash\; P \Rightarrow Q} \;\textbf{IMP\_R}
$$

As can be seen, each kind of predicates, namely falsity, negation, conjunction, disjunction, and implication, is given two rules: a left rule, labelled with _**L**, and a right rule, labelled with _**R**. This corresponds to the predicate appearing either in the hypothesis part (left) or in the goal part (right) of the consequent of the rule.

It is important to notice that we do not "define" the various propositional calculus operators with any kind of "truth table". We rather say how sequents involving such operators can be proved.

Derived Rules Besides the previous rules the following *derived* rule (among many others) is quite useful. It says that for proving a goal P it is sufficient to prove it first under hypothesis Q and then under hypothesis $\neg Q$.

$$\frac{H, Q \;\vdash\; P \qquad H, \neg Q \;\vdash\; P}{H \;\vdash\; P} \quad \textbf{CASE}$$

For proving a derived rule, we assume its antecedents (if any) and prove its consequent. With this in mind, here is the proof of derived rule **CASE**:

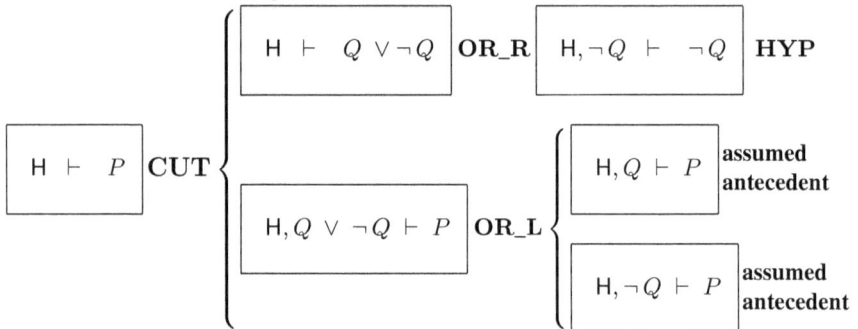

Methodology The method we are going to use to build our Mathematical Language must start to be clearer: it will be very systematic. It is made of two steps: first we augment our syntax. Then either the extension corresponds to a simple facility. In that case, we give simply the definition of the new construct in terms of previous ones. Or the new construct is not related to any previous constructs. In that case, we augment our current theory.

Extending the Propositional Language The Propositional Language is now extended by adding one more construct called *equivalence*. Given two predicates P and Q, we can construct their equivalence $P \Leftrightarrow Q$. We also add one predicate: $\top$. As a consequence, our syntax is now the following:

$$
\begin{array}{rcl}
predicate & ::= & \bot \\
& & \top \\
& & \neg\, predicate \\
& & predicate \,\wedge\, predicate \\
& & predicate \,\vee\, predicate \\
& & predicate \,\Rightarrow\, predicate \\
& & predicate \,\Leftrightarrow\, predicate
\end{array}
$$

Note that implication and equivalence operators have the same syntactic priorities so that parentheses will be necessary when several such distinct operators are following each other. Such extensions are defined in terms of previous ones by mere rewriting rules:

Predicate	Rewritten
$\top$	$\neg\ \bot$
$P \Leftrightarrow Q$	$(P \Rightarrow Q) \wedge (Q \Rightarrow P)$

The following derived rules can be proved easily:

$$
\frac{\mathsf{H} \vdash P}{\mathsf{H}, \top \vdash P}\ \textbf{TRUE\_L}
\qquad\qquad
\frac{}{\mathsf{H} \vdash \top}\ \textbf{TRUE\_R}
$$

Note that rule **TRUE_L** can be proved using rule **MON** but the reverse rule (exchanging antecedent and consequent), which holds as well, cannot. We leave it as an exercise to the reader to prove these rules.

2.4 The Predicate Language

Syntax In this section, we introduce the Predicate Language. The syntax is extended with a number of new kinds of predicates and also with the introduction of two new syntactic categories called *expression* and *variable*. A *variable* is a simple identifier. Given a non-empty list of variables x made of pairwise distinct identifiers and a predicate P, the construct $\forall x \cdot P$ is called a *universally quantified predicate*. Likewise, given a non-empty list of variables x made of pairwise distinct identifiers and a predicate P, the construct $\exists x \cdot P$ is called an *existentially quantified predicate*. An *expression* is either a variable or else a *paired expression* $E \mapsto F$, where E and F are two expressions. Here is this new syntax:

208

$$
\begin{array}{rcl}
predicate & ::= & \bot \\
& & \top \\
& & \neg\, predicate \\
& & predicate \;\wedge\; predicate \\
& & predicate \;\vee\; predicate \\
& & predicate \;\Rightarrow\; predicate \\
& & predicate \;\Leftrightarrow\; predicate \\
& & \forall var\_list \cdot predicate \\
& & \exists var\_list \cdot predicate \\
\\
expression & ::= & variable \\
& & expression \mapsto expression \\
\\
var\_list & ::= & variable \\
& & variable,\ var\_list
\end{array}
$$

This syntax is also ambiguous. Note however that the scope of the universal or existential quantifiers extends to the right as much as they can, the limitation being expressed either by the end of the formula or by means of enclosing parentheses.

Predicates and Expressions It might be useful at this point to clarify the difference between a predicate and an expression. A predicate P is a piece of formal text which can be *proved* when embedded within a sequent as in:

$$
\mathrm{H} \vdash P
$$

A predicate does not denote anything. This is not the case of an expression which always denotes an *object*. An expression cannot be "proved". Hence predicates and expressions are incompatible. Note that for the moment the possible expressions we can define are quite limited. This will be considerably extended in the Set-theoretic Language defined in Section 2.6.

Inference Rules for Universally Quantified Predicates The universally and existentially quantified predicates require introducing corresponding rules of inference. As for propositional calculus, in both cases we need two rules: one for quantified assumptions (left rule) and one for a quantified goal (right rule). Here are these rules for universally quantified predicates:

$$\frac{\text{H}, \ \forall x \cdot P, \ [x := E]P \ \vdash \ Q}{\text{H}, \ \forall x \cdot P \ \vdash \ Q} \qquad \textbf{ALL\_L}$$

$$\frac{\text{H} \ \vdash \ P}{\text{H} \ \vdash \ \forall x \cdot P} \qquad \begin{array}{l}\textbf{ALL\_R}\\ (x \text{ not free in H})\end{array}$$

The first rule (ALL_L) allows us to add another assumption when we have a universally quantified one. This new assumption is obtained by instantiating the quantified variable x by any expression E in the predicate P: this is denoted by $[x := E]P$. The second rule (ALL_R) allows us to remove the "$\forall$" quantifier appearing in the goal. This can be done however only if the quantified variable (here x) *does not appear free* in the the set of assumptions H: this requirement is called a *side condition*. In the sequel we shall write x nfin P to mean that variable x is not free in predicate P. The same notation is used with an expression E. We omit in this presentation to develop the syntactic rules allowing us to compute non-freeness as well as substitutions. We have similar rules for existentially quantified predicates:

$$\frac{\text{H}, \ P \ \vdash \ Q}{\text{H}, \ \exists x \cdot P \ \vdash \ Q} \qquad \begin{array}{l}\textbf{XST\_L}\\ (x \text{ not free in H and } Q)\end{array}$$

$$\frac{\text{H} \ \vdash \ [x := E]P}{\text{H} \ \vdash \ \exists x \cdot P} \qquad \textbf{XST\_R}$$

As an example, we prove now the following sequent:

$$\forall x \cdot (\exists y \cdot P_{x,y}) \Rightarrow Q_x \ \vdash \ \forall x \cdot (\forall y \cdot P_{x,y} \Rightarrow Q_x)$$

where $P_{x,y}$ stands for a predicate containing variables x and y only as free variables, and Q_x stands for a predicate containing variable x only as a free variable.

$\forall x \cdot (\exists y \cdot P_{x,y}) \Rightarrow Q_x$ $\vdash$ $\forall x \cdot (\forall y \cdot P_{x,y} \Rightarrow Q_x)$	ALL_R ALL_R IMP_R

$$\begin{array}{l}\forall x \cdot (\exists y \cdot P_{x,y}) \Rightarrow Q_x\\ P_{x,y}\\ \vdash\\ Q_x\end{array} \qquad \textbf{CUT} \ldots$$

$$\begin{array}{l} \forall x \cdot (\exists y \cdot P_{x,y}) \;\Rightarrow\; Q_x \\ P_{x,y} \\ \vdash \\ \exists y \cdot P_{x,y} \end{array}$$

XST_R

$$\begin{array}{l} \forall x \cdot (\exists y \cdot P_{x,y}) \;\Rightarrow\; Q_x \\ P_{x,y} \\ \vdash \\ P_{x,y} \end{array}$$

HYP

…

$$\begin{array}{l} \forall x \cdot (\exists y \cdot P_{x,y}) \;\Rightarrow\; Q_x \\ P_{x,y} \\ \exists y \cdot P_{x,y} \\ \vdash \\ Q_x \end{array}$$

ALL_L
IMP_L

$$\begin{array}{l} \forall x \cdot (\exists y \cdot P_{x,y}) \;\Rightarrow\; Q_x \\ Q_x \\ P_{x,y} \\ \exists y \cdot P_{x,y} \\ \vdash \\ Q_x \end{array}$$

HYP

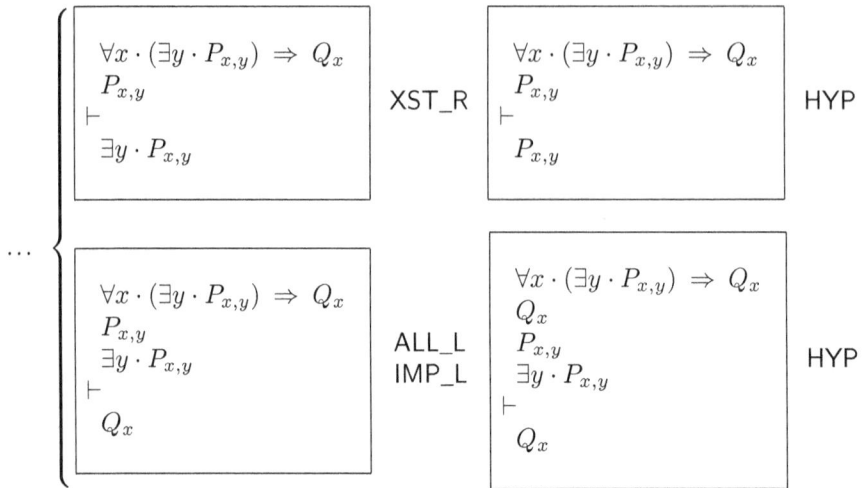

The proof of the following sequent is left to the reader:

$$\forall x \cdot (\forall y \cdot P_{x,y} \;\Rightarrow\; Q_x) \quad \vdash \quad \forall x \cdot (\exists y \cdot P_{x,y}) \;\Rightarrow\; Q_x$$

2.5 Introducing Equality

The Predicate Language is once again extended by adding a new predicate, the *equality predicate*. Given two expressions E and F, we define their equality by means of the construct $E = F$. Here is the extension of our syntax:

$$\begin{array}{lcl} predicate & ::= & \bot \\ & & \top \\ & & \neg\, predicate \\ & & predicate \;\wedge\; predicate \\ & & predicate \;\vee\; predicate \\ & & predicate \;\Rightarrow\; predicate \\ & & predicate \;\Leftrightarrow\; predicate \\ & & \forall var\_list \cdot predicate \\ & & \exists var\_list \cdot predicate \\ & & expression = expression \\ \\ expression & ::= & variable \\ & & expression \mapsto expression \end{array}$$

Notice the syntax for pairs of *expression* is: *expression* $\mapsto$ *expression*. The common mathematical notation for it is usually $(expression, expression)$. We did not retain this notation because it is ambiguous. For example it is not clear whether $\{(1,2)\}$ means the set made of 1 and 2 or the singleton set made of the pair of 1 and 2.

Note that we shall use the operator $\neq$ in the sequel to mean, as is usual, the negation of equality. The inference rules for equality are the following:

$$
\frac{[x := F]\mathrm{H},\ E = F\ \vdash\ [x := F]P}{[x := E]\mathrm{H},\ E = F\ \vdash\ [x := E]P}\qquad \textbf{EQ\_LR}
$$

$$
\frac{[x := E]\mathrm{H},\ E = F\ \vdash\ [x := E]P}{[x := F]\mathrm{H},\ E = F\ \vdash\ [x := F]P}\qquad \textbf{EQ\_RL}
$$

It allows us to *apply* an equality assumption in the remaining assumptions and in the goal. This can be made by using the equality from left to right or from right to left. Subsequent rules correspond to the reflexivity of equality and to the equality of pairs. They are both defined by some rewriting rules as follows:

Operator	Predicate	Rewritten
Equality	$E = E$	$\top$
Equality of pairs	$E \mapsto F = G \mapsto H$	$E = G \wedge F = H$

The following rewriting rules, within which x is supposed to be not free in E, are easy to prove. They are called the *one point rules*:

Predicate	Rewritten
$\forall x \cdot x = E \Rightarrow P$	$[x := E]P$
$\exists x \cdot x = E \wedge P$	$[x := E]P$

The notation $[x := E]P$ stands for P where occurrences of the free variable x have been replaced by E.

2.6 The Set-theoretic Language

Our next language, the Set-theoretic Language, is now presented as an extension to the previous Predicate Language.

Syntax In this extension, we introduce some special kind of expressions called *sets*. Note that not all expressions are set: for instance a pair is not a set. However, in the coming syntax we shall not make any distinction between expressions which are sets and expressions which are not.

We introduce another predicate the *membership predicate*. Given an expression E and a set S, the construct $E \in S$ is a membership predicate which says that expression E is a *member* of set S.

We also introduce the basic set constructs. Given two sets S and T, the construct $S \times T$ is a set called the *Cartesian product* of S and T. Given a set S, the construct $\mathbb{P}(S)$ is a set called the *power set of S*. Finally, given a list of variables x with pairwise distinct identifiers, a predicate P, and an expression E, the construct $\{x \cdot P \mid E\}$ is called a *set defined in comprehension*. Here is our new syntax:

$$
\begin{array}{lll}
predicate & ::= & \ldots \\
& & expression \ \in \ expression \\
\\
expression & ::= & variable \\
& & expression \mapsto expression \\
& & expression \times expression \\
& & \mathbb{P}(expression) \\
& & \{\, var\_list \cdot predicate \mid expression \,\}
\end{array}
$$

Note that we shall use the operator $\notin$ in the sequel to mean, as is usual, the negation of set membership.

Axioms of Set Theory The axioms of the set-theoretic Language are given under the form of equivalences to various set memberships. They are all defined in terms of rewriting rules. Note that the last of these rules defines equality for sets. It is called the *Extensionality Axiom*.

Operator	Predicate	Rewritten	Side Cond.
Cartesian product	$E \mapsto F \in S \times T$	$E \in S \ \wedge \ F \in T$	
Power set	$E \in \mathbb{P}(S)$	$\forall x \cdot x \in E \ \Rightarrow \ x \in S$	$x \ \underline{\text{nfin}} \ E$ $x \ \underline{\text{nfin}} \ S$

Operator	Predicate	Rewritten	Side Cond.
Comprehension	$E \in \{\, x \cdot P \mid F \,\}$	$\exists x \cdot P \;\wedge\; E = F$	x <u>nfin</u> E
Equality	$S = T$	$S \in \mathbb{P}(T) \;\wedge\; T \in \mathbb{P}(S)$	

The notation $\{\, x \cdot P \mid F \,\}$ although given a firm definition in the above axiom, can be a bit misleading. It stands for the set of "objects" of the form F for all x such that P holds. For example, the set:

$$\{x \, . \, x \in \mathbb{N} \mid 2x + 3\}$$

stands for all "objects" (natural numbers in this case) of the form $2x + 3$ for $x \in \mathbb{N}$, that is $\{3, 5, 7, 9, 11, \ldots\}$. Indeed 11 belongs to this set because:

$$\exists x \cdot x \in \mathbb{N} \;\wedge\; 11 = 2x + 3$$

As a special case, set comprehension can sometimes be written $\{\, F \mid P \,\}$, which can be read as follows: "the set of objects with shape F when P holds". However, as you can see, the list of variables x has now disappeared. In fact, these variables are then *implicitly determined* as being all the free variables in F. When we want that x represent only *some*, but not all, of these free variables we cannot use this shorthand.

A more special case is one where the expression F is exactly a single variable x, that is $\{\, x \cdot P \mid x \,\}$. As a shorthand, this can be written $\{\, x \mid P \,\}$, which is very common in informally written mathematics. And then $E \in \{\, x \mid P \,\}$ becomes $[x := E]P$ according to the second "one point rule" of section 2.5.

From now on, all extensions of the Set-theoretic Language will take the form of "simple facilities", as explained in section 2.3 (under the header **Methodology**). And most of them are extensions of the *set* syntactic category. As a consequence, the new syntax will be presented differentially. The new constructs will be presented under the form of rewriting rules. And since most of the new construct are sets, the rewriting rules will transform some set membership predicates into simpler ones.

Elementary Set Operators In this section, we introduce the classical set operators: inclusion, union, intersection, difference, extension, and the empty set.

$$
\begin{array}{lll}
predicate & ::= & \ldots \\
& & expression \subseteq expression \\
\\
expression & ::= & \ldots \\
& & expression \cup expression \\
& & expression \cap expression \\
& & expression \setminus expression \\
& & \{expression\_list\} \\
& & \varnothing \\
\\
expression\_list & ::= & expression \\
& & expression,\ expression\_list
\end{array}
$$

Notice that the expressions in an *expression_list* are not necessarily distinct.

Operator	Predicate	Rewritten
Inclusion	$S \subseteq T$	$S \in \mathbb{P}(T)$
Union	$E \in S \cup T$	$E \in S \ \lor \ E \in T$
Intersection	$E \in S \cap T$	$E \in S \ \land \ E \in T$
Difference	$E \in S \setminus T$	$E \in S \ \land \ \neg(E \in T)$
Set extension	$E \in \{a, \ldots, b\}$	$E = a \ \lor \ \ldots \ \lor \ E = b$
Empty set	$E \in \varnothing$	$\bot$

Generalization of Elementary Set Operators The next series of operators consists in generalizing union and intersection to sets of sets. This takes the forms either of an operator acting on a set or of a quantifier.

```
. . .

expression ::= . . .
                union(expression)
                ⋃ var_list · predicate | expression
                inter(expression)
                ⋂ var_list · predicate | expression
```

Operator	Predicate	Rewritten	Side Cnd.
Generalized union	$E \in \text{union}(S)$	$\exists s \cdot s \in S \wedge E \in s$	s nfin S s nfin E
Quantified union	$E \in \bigcup x \cdot P \mid T$	$\exists x \cdot P \wedge E \in T$	x nfin E
Generalized intersection	$E \in \text{inter}(S)$	$\forall s \cdot s \in S \Rightarrow E \in s$	s nfin S s nfin E
Quantified intersection	$E \in \bigcap x \cdot P \mid T$	$\forall x \cdot P \Rightarrow E \in T$	x nfin E

The last two rewriting rules require that the set $\text{inter}(S)$ and $\bigcap x \cdot P \mid T$ be *well defined*. This is presented in the following table:

Set construction	Well-definedness condition
$\text{inter}(S)$	$S \neq \varnothing$
$\bigcap x \cdot P \mid T$	$\exists x \cdot P$

Binary Relation Operators We now define a first series of binary relation operators: the set of binary relations built on two sets, the domain and range of a binary relation, and then various sets of binary relations.

216

```
. . .

expression ::= . . .
                expression ↔ expression
                dom(expression)
                ran(expression)
                expression ⇹ expression
                expression ↠ expression
                expression ⤖ expression
```

Operator	Predicate	Rewritten	Side Cnd.
Binary relations	$r \in S \leftrightarrow T$	$r \subseteq S \times T$	
Domain	$E \in \mathrm{dom}\,(r)$	$\exists y \cdot E \mapsto y \in r$	$y\ \underline{\mathsf{nfin}}\ E$ $y\ \underline{\mathsf{nfin}}\ r$
Range	$F \in \mathrm{ran}\,(r)$	$\exists x \cdot x \mapsto F \in r$	$x\ \underline{\mathsf{nfin}}\ F$ $x\ \underline{\mathsf{nfin}}\ r$
Total relations	$r \in S \leftrightarrow T$	$r \in S \leftrightarrow T \wedge \mathrm{dom}\,(r) = S$	
Surjective relations	$r \in S \leftrightarrow\!\!\!\to T$	$r \in S \leftrightarrow T \wedge \mathrm{ran}\,(r) = T$	

The next series of binary relation operators define the converse of a relation, various relation restrictions and the image of a set under a relation.

```
expression ::= . . .
                expression⁻¹
                expression ◁ expression
                expression ▷ expression
                expression ◁ expression
                expression ▷ expression
                expression[expression]
```

Operator	Predicate	Rewritten	Side Cnd.
Converse	$E \mapsto F \in r^{-1}$	$F \mapsto E \in r$	
Domain restriction	$E \mapsto F \in S \lhd r$	$E \in S \ \wedge \ E \mapsto F \in r$	
Range restriction Domain subtraction	$E \mapsto F \in r \rhd T$ $E \mapsto F \in S \lhd\!\!\!- r$	$E \mapsto F \in r \ \wedge \ F \in T$ $\neg E \in S \ \wedge \ E \mapsto F \in r$	
Range subtraction	$E \mapsto F \in r \rhd\!\!\!- T$	$E \mapsto F \in r \ \wedge \ \neg F \in T$	
Relational Image	$F \in r[U]$	$\exists x \cdot x \in U \ \wedge \ x \mapsto F \in r$	$x \ \underline{\mathsf{nfin}} \ F$ $x \ \underline{\mathsf{nfin}} \ r$ $x \ \underline{\mathsf{nfin}} \ U$

Let us illustrate the relational image. Given a binary relation r from a set S to a set T, the image of a subset U of S under the relation r is a subset of T. The image of U under r is denoted by $r[U]$. Here is its definition:

$$r[U] \ = \ \{\, y \mid \exists x \cdot x \in U \ \wedge \ x \mapsto y \ \in \ r \,\}$$

This is illustrated in figure 3. As can be seen on this figure, the image of the set $\{a, b\}$ under relation r is the set $\{m, n, p\}$.

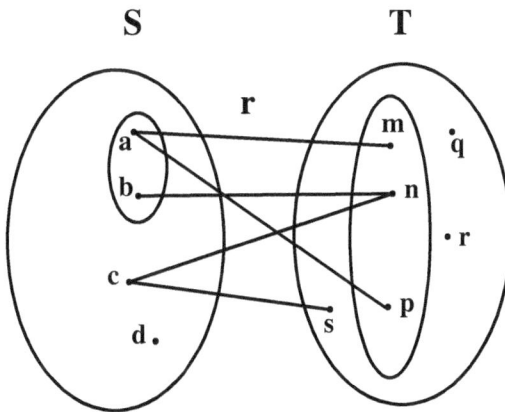

Fig. 3. Image of a Set under a Relation

218

Our next series of operators defines the composition of two binary relations, the over-riding of a relation by another one, and the direct and parallel products of two relations.

$$
\begin{aligned}
expression ::= \ & \dots \\
& expression \, ; expression \\
& expression \circ expression \\
& expression \lhd\!\!\!\!- \, expression \\
& expression \otimes expression \\
& expression \parallel expression
\end{aligned}
$$

Operator	Predicate	Rewritten	Side Cnd.
Forward composition	$E \mapsto F \in f \,; g$	$\exists x \cdot E \mapsto x \in f \;\wedge\; x \mapsto F \in g$	x <u>nfin</u> E x <u>nfin</u> F x <u>nfin</u> f x <u>nfin</u> g
Backward composition	$E \mapsto F \in g \circ f$	$E \mapsto F \in f \,; g$	

Given a relation f from S to T and a relation g from T to U, the forward relational composition of f and g is a relation from S to U. It is denoted by the construct $f \,; g$. Sometimes it is denoted the other way around as $g \circ f$, in which case is is said to be the backward composition.

Operator	Predicate	Rewritten
Overriding	$E \mapsto F \in f \lhd\!\!\!\!- \, g$	$E \mapsto F \in (\text{dom}\,(g) \lhd f) \, \cup \, g$
Direct product	$E \mapsto (F \mapsto G) \in f \otimes g$	$E \mapsto F \in f \;\wedge\; E \mapsto G \in g$
Parallel product	$(E \mapsto F) \mapsto (G \mapsto H) \in f \parallel g$	$E \mapsto G \in f \;\wedge\; F \mapsto H \in g$

The overriding operator is applicable in general to a relation f from, say, a set S to a set T, and a relation g also from S to T. When f is a function and g is the singleton function $\{x \mapsto E\}$, then $f \lhd\!\!\!\!- \, \{x \mapsto E\}$ replaces in f the pair $x \mapsto f(x)$ by the pair $x \mapsto E$. Notice that in the case where x is not in the domain of f, then $f \lhd\!\!\!\!- \, \{x \mapsto E\}$ simply adds the pair $x \mapsto E$ to the function f. In this case, it is thus equal to $f \cup \{x \mapsto E\}$.

Functional Operators In this section we define various function operators: the sets of all partial and total functions, partial and total injections, partial and total surjections, and bijections. We also introduce the two projection functions as well as the identity function.

$$
\begin{aligned}
expression ::= \;&\ldots \\
&\mathrm{id} \\
&expression \twoheadrightarrow expression \\
&expression \rightarrow expression \\
&expression \rightarrowtail\mkern-14mu\rightarrow expression \\
&expression \rightarrowtail expression \\
&expression \twoheadrightarrow\mkern-18mu\cdot\; expression \\
&expression \rightarrow\mkern-16mu\rightarrow expression \\
&expression \rightarrowtail\mkern-14mu\twoheadrightarrow expression \\
&\mathrm{prj}_1 \\
&\mathrm{prj}_2
\end{aligned}
$$

Operator	Predicate	Rewritten
Identity	$E \mapsto F \in \mathrm{id}$	$E = F$
Partial functions	$f \in S \twoheadrightarrow T$	$f \in S \leftrightarrow T \;\wedge\; (f^{-1}\,;f) \subseteq \mathrm{id}$
Total functions	$f \in S \rightarrow T$	$f \in S \twoheadrightarrow T \;\wedge\; S = \mathrm{dom}\,(f)$
Partial injections	$f \in S \rightarrowtail\mkern-14mu\rightarrow T$	$f \in S \twoheadrightarrow T \;\wedge\; f^{-1} \in T \twoheadrightarrow S$
Total injections	$f \in S \rightarrowtail T$	$f \in S \rightarrow T \;\wedge\; f^{-1} \in T \twoheadrightarrow S$
Partial surjections	$f \in S \twoheadrightarrow\mkern-18mu\cdot\; T$	$f \in S \twoheadrightarrow T \;\wedge\; T = \mathrm{ran}\,(f)$
Total surjections	$f \in S \rightarrow\mkern-16mu\rightarrow T$	$f \in S \rightarrow T \;\wedge\; T = \mathrm{ran}\,(f)$

Operator	Predicate	Rewritten
Bijections	$f \in S \rightarrowtail\!\!\!\rightarrow T$	$f \in S \rightarrowtail T \;\land\; f \in S \twoheadrightarrow T$
First projection	$(E \mapsto F) \mapsto G \in \mathrm{prj}_1$	$G = E$
Second projection	$(E \mapsto F) \mapsto G \in \mathrm{prj}_2$	$G = F$

Lambda Abstraction and Function Invocation We now define *lambda abstraction*, which is a way to construct functions, and also function invocation, which is a way to call functions. But first we have to define the notion of *pattern of variables*. A pattern of variables is either an identifier or a pair made of two patterns of variables. Moreover, all variables composing the pattern must be distinct. For example, here are three patterns of variables:

$$abc$$

$$abc \mapsto def$$

$$abc \mapsto (def \mapsto ghi)$$

Given a pattern of variables x, a predicate P, and an expression E, the construct $\lambda x \cdot P \,|\, E$ is a lambda abstraction, which is a function. Given a function f and an expression E, the construct $f(E)$ is an expression denoting a function invocation. Here is our new syntax:

$$
\begin{aligned}
expression \;::=\;& \dots \\
& expression(expression) \\
& \lambda\, pattern \cdot predicate \,|\, expression \\[4pt]
pattern \quad ::=\;& variable \\
& pattern \mapsto pattern
\end{aligned}
$$

In the following table, l stands for the list of variables in the pattern L.

221

Operator	Predicate	Rewritten
Lambda abstraction	$F \in \lambda L \cdot P \mid E$	$F \in \{l \cdot P \mid L \mapsto E\}$
Function invocation	$F = f(E)$	$E \mapsto F \in f$

The function invocation construct $f(E)$ requires a well-definedness condition, which is the following:

Expression	Well-definedness condition
$f(E)$	$f^{-1} ; f \subseteq \mathrm{id} \ \land \ E \in \mathrm{dom}(f)$

3 Prover Technologies Used in the Rodin Platform

3.1 Connecting to Various External Provers

On the Rodin Platform [3], we use various "external" provers. As explained in the introduction, the connection to these external provers is made by means of a translation of the set theoretic statement (defining the predicate to be proved) into a predicate calculus statement. As a very simple example, suppose we have to prove a statement like $S \subseteq T$ where S and T stand for some set expressions. It is translated as follows: $\forall x \cdot x \in S' \ \Rightarrow \ x \in T'$ where S' and T' stand for the translations of S and T respectively.

The external provers we have on the Rodin Platform are the, so called, Predicate Prover (an internally developed predicate calculus prover) and some SMT provers (Alt-Ergo, CVC3, VeriT, Z3) [7]. There is even a plug-in for translating Event-B sequents to an embedding in HOL which allows to perform proofs with the Isabelle proof assistant.

3.2 Reasoners and Tactics

The text of this section is a copy from [6].

Like the rest of the Rodin platform, the prover has been designed for openness. The main code of the prover just maintains a proof tree in Sequent Calculus and does not contain any reasoning capability. It is extensible through reasoners and tactics. A reasoner is a piece of code that, given an input sequent, either fails or succeeds. In case of success, the reasoner produces a proof rule which is applied to the current proof tree node.

Reasoners could be applied interactively. However this would be very tedious. Reasoner application can thus be automated by using tactics that take a more global view of the proof tree and organise the running of reasoners. Tactics can also backtrack the proof tree, that is undo some reasoner applications in case the prover entered a dead-end.

The core platform contains a small set of reasoners written in Java that either implement the basic proving rules (HYP, CUT, FALSE_L, etc.), or perform some simple clean-up on sequents such as normalisation or unit propagation (generalised modus-ponens). These reasoners allow to discharge the most simple proof obligations. As already explained, they are complemented by reasoners that link the Rodin platform to external provers, such as those of Atelier B (ML and PP) and SMT solvers (Alt-Ergo, CVC3, VeriT, Z3, etc.) [7].

3.3 Tactic Profile

As explained in the previous section, the Rodin Platform is provided with some elementary tactics including calls to external provers. Such tactics can be put together in, so-called, *tactic profiles*. The user can define several such profiles and attach them interactively (in the Rodin Platform preference framework) to the prover. These has the effect of automatising proofs.

3.4 Interactive Proofs

When the automatic treatment of the prover fails, the user can perform an interactive proof. The idea is to give the possibility to the user to call some elementary tactics explicitly. In practice, this is done by clicking on some emphasised symbols either on the goal part or the hypothesis part of the sequent to be proved.

For example, if the following sequent $H \vdash S \subseteq T$ is to be proved, then by clicking on the emphasised symbol $\subseteq$, the sequent to be proved is transformed to $H \vdash \forall x \cdot x \in S \Rightarrow x \in T$. The user can then click on the emphasised symbol $\forall$: this has the effect of transforming the present sequent to $H \vdash x \in S \Rightarrow x \in T$. Finally, by clicking on the emphasised symbol $\Rightarrow$, we obtain the following sequent $H, x \in S \vdash x \in T$, and so on. As can be seen, activating interactively the sequent to be proved, has the effect of decomposing gradually this sequent into smaller ones.

3.5 Theory Plug-in

The text of this section is a copy from the paper [6]. This plug-in of the Rodin Platform, defined in [4], allows one to extend the basic mathematical operators of Event-B. These operators are polymorphic. They can be defined explicitly in terms of existing ones. As examples of these extensions, we can define the concept of well-foundedness, that of fixpoint, that of relational closure, and so on.

It is also possible to give some axiomatic definitions only. An interesting outcome of this last feature allows one to define the set of Real numbers axiomatically. Moreover, the user of this plug-in can add some corresponding theorems, and inference or rewriting rules able to extend the provers. It is also possible to define new (possibly recursive) types. This very important plug-in has been developed by Issam Maamria and Asieh Salehi in Southampton University.

4 Some Results

The most common proof generating deduction tools are the Predicate Prover (developed internally) and also the SMT provers we mention earlier (Alt-Ergo, CVC3, VeriT, and Z3). We are very happy with these provers. In the future we will probably try other proof systems although it is not decided yet which ones. In section 3.4 we gave a small examples of an interactve proof. Here is another example, that of the proof of the following sequent:

$$r \in S \leftrightarrow T \vdash A1 \subseteq A2 \Rightarrow r[A1] \subseteq r[A2]$$

where $r[A1]$ or $r[a2]$ stand for the images of the set $A1$ or $A2$ under the relation r. The automatic proof goes as follows. Each line contains first an indication of the inference rule (or sometimes the rewriting rule) that is applied, then the goal of the sequent is shown after the ":" symbol (the hypotheses of the sequent are not shown):

$\Rightarrow$ in goal : $A1 \subseteq A2 \Rightarrow r[A1] \subseteq r[A2]$
  remove $\subseteq$ in goal : $r[A1] \subseteq r[A2]$
    remove $\in$ in goal : $\forall x \cdot x \in r[A1] \Rightarrow x \in r[A2]$
      $\forall$ goal (frees x) : $\forall x \cdot (\exists x0 \cdot x0 \in A1 \wedge x0 \mapsto x \in r) \Rightarrow \ldots$
        $\Rightarrow$ in goal : $(\exists x0 \cdot x0 \in A1 \wedge x0 \mapsto x \in r) \Rightarrow (\exists x0 \cdot x0 \in A2 \wedge x0 \mapsto x \in r)$
          $\exists$ hyp $(\exists x0 \cdot x0 \in A1 \wedge x0 \mapsto x \in r)$: $\exists x0 \cdot x0 \in A2 \wedge x0 \mapsto x \in r$
            $\exists$ goal (inst $x0$) : $\exists x0 \cdot x0 \in A2 \wedge x0 \mapsto x \in r$
              $\wedge$ goal : $x0 \in A2 \wedge x0 \mapsto x \in r$
                $\forall$ hyp p (inst $x0$) : $x0 \in A2$
                  hyp : $x0 \in A1$
                  hyp : $x0 \in A2$
                  hyp : $x0 \mapsto x \in r$

Notice that the goals are optional. In this case, we would obtain the following proof:

$\Rightarrow$ in goal
  remove $\subseteq$ in goal
    remove $\in$ in goal
      $\forall$ goal (frees x)
        $\Rightarrow$ in goal
          $\exists$ hyp $(\exists x0 \cdot x0 \in A1 \wedge x0 \mapsto x \in r)$
            $\exists$ goal (inst $x0$)
              $\wedge$ goal
                $\forall$ hyp p (inst $x0$)
                  hyp
                  hyp
                  hyp

As can be seen, the proof indentation reflects the tree structure of the proof. In an interactive proof, the user can easily navigate within this tree. By pointing to a node in the tree, the user can also get more details about the usage of the inference rule used in that node. It can also hide some parts of the tree or do some copy/paste of some sub-tree when a similar sub-proof is needed somewhere. The user can also "review" a node in the tree without providing a proof for it yet. It is very convenient when applying the cut rule in order to define a local lemma (to be proven later). We are satisfied with the generated proofs: they are detailed enough. However these proofs are not intended to be read, just used at some specific moment in the proof process. The simple proof we have just shown is clearly already a bit difficult to read. It is very frequent to have far bigger proofs, which are thus totally unreadable. In the shown proof, the entire sequent is not shown, it could have been of course, but it'll make the proof even more difficult to read.

At the moment, we are very satisfied with the system at hand. In fact, the introduction of SMT solvers into our proving system was a big step forward. Notice that these provers can be mentioned explicitly in a tactic profile, resulting in many proofs, that were not done automatically before this introduction, now becoming automatic proofs.

Concerning model-checking, it is already present in the Rodin Platform. It is called PROB [5]. It has been developed in the University of Duesseldorf in Germany. We see it has a very useful complement to our proof system. It is particularly interesting when the user finds some difficulties in doing a proof. Using the model-checker can easily help finding some counter-examples showing that the user tries to prove something that cannot be proven.

5   Using Proofs

Proofs are used to discharge proof obligations generated by the proof obligation generator, a tool in the Rodin Platform. This tool generates many different proof obligations, among which the more important are invariant proof obligations and refinement proof obligations. All this is defined in great details in [2].

Witnesses are directly defined in our models, thus avoiding many existential proofs. The only important properties we extract from a proof is whether it has been done automatically or interactively, or if it is not discharged or has some reviewed nodes still in it. Out of this, we determine some important statistics calculated on all the proofs of an entire development. As an example, the B development for the Paris metro line 14 generated 27,800 proofs. The system developed for the Charles de Gaule airport generated 43,600 proofs. More recently the development of an embedded system formalising an aircraft landing system with Event-B generated 2328 proofs on the Rodin Platform. All these proofs were discharged automatically.

The construction of code out of the last refinement of a model is independent of the proofs. We are filtering the last refinement of a model in order to determine whether it can be translated into code. The system for the line 14 metro system generated 86,000 lines of ADA. For the CDG shuttle, 158,000 lines of ADA were generated.

The proofs are not explicitly attached to the generated code but all proofs of a development can be consulted directly on the Rodin Platform, together with the mentioned statistics.

6 Comparison of Proofs

The question of comparing proofs is not relevant in our domain. The quality of a proof (essentially its length and the number of explicit quantified variable instantiations) is an important qualitative factor. This quality is a good indication of the overall quality of the corresponding formal model. The proportion of automatic proofs is also an important factor of the overall quality of the model.

7 Trends and Open Problems

Our current trend is to experiment with the "Theory" plug-in that has been developed recently (see section 3.5). Our idea is to incorporate more and more the usage of this plug-in in our future developments. In particular, this plug-in allows us to incorporate an axiomatisation of the Real Numbers. This is very important for extending the usage of Event-B to model hybrid systems.

Within the next ten years, our vision is to develop more systems using this approach, thus reducing the need for programming and replace it by the need for proving.

8 Conclusions

In this paper, we describe the proving approach we developed over the years in order to model computerised systems using a formal method (Event-B) based on refinement.

References

1. J-R. Abrial. *From Z to B and then Event-B: Assigning Proofs to Meaningful Programs.* In E.B. Johnsen and L. Petre, editors, IFM, volume 7940 of Lecture Notes in Computer Science, pages 1Ð15. Springer, 2013.
2. J.R. Abrial. *Modeling in Event-B: System and Software Engineering.* Cambridge University Press 2010.
3. http://www.event-b.org *Rodin Platform*
4. M. Butler and I. Maamria. *Practical theory extension in Event-B.* In Zhiming Liu, Jim Woodcock, and Huibiao Zhu, editors, Theories of Programming and Formal Methods, volume 8051 of Lecture Notes in Computer Science, pages 67 to 81. Springer Berlin Heidelberg, 2013.
5. M. Leuschel and M. Butler. *ProB: An Automated Analysis Toolset for the B Method.* International Journal on Software Tools for Technology Transfer 2008.
6. L. Voisin and J.R. Abrial *The Rodin Platform Has Turned Ten* ABZ 2014
7. D. Deharbe, P. Fontaine, Y. Guyot, and L. Voisin. *SMT solvers for Rodin.* In Proceedings of the Third International Conference on Abstract State Machines, Alloy, B, VDM, and Z, ABZ'12, pages 194 to 207, Berlin, Heidelberg, 2012. Springer-Verlag.

Computer-aided cryptography: some tools and applications

Gilles Barthe[1], François Dupressoir[1], Benjamin Grégoire[2], Benedikt Schmidt[1], and Pierre-Yves Strub[1]

[1] IMDEA Software Institute, Madrid, Spain
[2] INRIA Sophia-Antipolis Méditerranée, France

1 Introduction

The goal of modern cryptography is to design efficient constructions that simultaneously achieve some desired functionality and provable security against resource-bounded adversaries. Over the years, the realm of cryptography has expanded from basic functionalities such as encryption, authentication and key agreement, to elaborate functionalities such as zero-knowledge protocols, secure multi-party computation, and more recently verifiable computation. In many cases, these elaborate functionalities can only be achieved through cryptographic systems, in which several elementary constructions interact. As a consequence of the evolution towards more complex functionalities, cryptographic proofs have become significantly more involved, and more difficult to check. Several cryptographers have therefore advocated the use of tool-supported frameworks for building and verifying proofs; the most vivid recommendation for using computer support is elaborated in a farseeing article in which Halevi (2005) describes a potential approach for realizing this vision.

Besides increasing confidence in cryptographic proofs, tool-supported frameworks have the potential to address another prominent difficulty with provable security: because cryptographic proofs are very complex, it is common practice to reason about algorithmic descriptions of the cryptographic constructions, rather than about implementations. As a consequence, implementations of well-known and provably secure constructions are vulnerable to attacks, and regularly fail to provide their intended security guarantees. This uncomfortable gap between provable security and cryptographic engineering may be due to i. the mismatch between the powerful but abstract adversary models considered in proofs and "real-world" adversaries that may glean information about the secret data not only from the input and output to computations, but also from a host of side-channels (timing, power consumption or electromagnetic radiations...), and may even be able to interfere with the computation itself; ii. the fact that the object on which the security proof is performed is not the object that is implemented in practice, either due to a developer's (possibly malicious) mistake or even to an unjustified refinement when turning abstract algorithms into standard documents and recommendations. These points are the focus of the recent "real world" security approach to provable security, developed, most notably by Degabriele, Paterson, and Watson (2011). However, we believe that tool support

227

is essential for accomodating the additional complexity introduced by dealing with implementation-level descriptions of cryptographic constructions and complex adversary models. In particular, even though security proofs themselves are difficult to automate, some automation is useful to help deal with the low-level implementation details.

2 Tools

Since 2005, we have been actively working on developing foundations and proving tool support for building and verifying the security of cryptographic constructions. To date, we have constructed several tools, ranging from general frameworks that can be applied to many classes of constructions to specialized frameworks which target a single class of constructions. We review both kinds of tools below, and provide for each of them a brief account of the rationale behind their design and of their applications so far. We also discuss the status of proofs in these tools.

2.1 CertiCrypt

CertiCrypt (Barthe et al., 2009) is a machine-checked framework built on top of the Coq proof assistant (The Coq development team, 2004). It supports the game-based code-based approach to cryptography (Shoup, 2004; Bellare and Rogaway, 2004), in which security notions and assumptions are formalized as probabilistic programs, also called games, and proofs are organized as sequences or trees of games. The proof is then performed by bounding the total distance (defined as the upper bound on the probability that a bounded adversary can induce distinguishable input-output behaviours) between the initial game, which expresses the security of the construction under study, and some subset of the leaf games, that represent computational hardness assumptions. From the perspective of formalization, the advantages of the game-based code-based approach are two-fold. First, it offers a rigorous formalism based on programming languages, a field with a long history of formal verification and extensive tool support. Second, organizing proofs as sequences of games is essential to tame their complexity, and opens the possibility to identify high-level principles whose application could potentially be automated. Indeed, CertiCrypt supports the code-centric view adopted in the game-based code-based approach by providing a deep embedding of an extensible probabilistic imperative language. It provides a denotational semantics of the language, based on the ALEA[3] library by Audebaud and Paulin-Mohring (2009), as well as an instrumented semantics that is used for modelling the computational complexity of programs and for defining the class of probabilistic polynomial-time program. In addition, CertiCrypt supports common forms of reasoning in cryptographic proofs through a rich set of verification methods for probabilistic programs, including a probabilistic relational Hoare

---

[3] https://www.lri.fr/~paulin/ALEA/

logic (pRHL), certified program transformations, and techniques widely used in cryptographic proofs such as eager/lazy sampling and failure events. Verification methods and logical proof rules are implemented in Coq, and proven correct with respect to the program semantics.

The main rationale behind the development of CertiCrypt was to provide strong trust guarantees on cryptographic proofs, rigorously formalizing game-based code-based assumptions and security notions and increasing trust in the proofs relating them. CertiCrypt was developed between 2005 and 2011, and was used to prove the security of several prominent cryptographic constructions, including the Full Domain Hash signature, the OAEP padding scheme, the Boneh-Frankling identity-based encryption scheme, zero-knowledge protocols, and hash functions into elliptic curves. An extension of CertiCrypt was used to reason about differential privacy, a notion that formalizes strong privacy guarantees in the context of privacy-preserving data mining. However, even such small examples reached the practical limits of the tool, due to the lack of automation, and additional burdens due to the formalization of arithmetic in $\mathbb{R}$.

2.2 EasyCrypt Prototype

The early EasyCrypt prototype (Barthe et al., 2011) is a tool-assisted framework for reasoning about the security of cryptographic constructions. As reported by Barthe et al. (2011), two key goals of the design of EasyCrypt were to improve automation (when compared to CertiCrypt) and to reuse existing program verification technology, in particular SMT solvers, leveraging their recent and future improvements. Thus, the main components of the initial prototype were a verification condition generator for CertiCrypt's probabilistic relational Hoare logic (pRHL) and a back-end interface to multiple theorem provers and SMT solvers, via the Why3 platform (Bobot et al., 2013).

These components enabled the semi-automated verification of pRHL judgments by interactively generating for each judgment a set of verification conditions that were sent to SMT solvers. Moreover, the initial prototype implemented a rudimentary algorithm for inferring loop invariants and adversary specifications, and allowed the user to provide specifications for loops and adversaries to palliate the incompleteness of the inference algorithms. In order to reduce the Trusted Computing Base and increase trust in EasyCrypt proofs, the original implementation of the verification condition generator produced an independently verifiable CertiCrypt proof of the validity of pRHL judgments, assuming the validity of the formulae discharged by the SMT solvers, hoping to later rely on proof-producing SMT solvers to obtain completely certified proofs in CertiCrypt.

The development of EasyCrypt was initiated in 2009, and the initial prototype was used to prove the security of several constructions, including the Cramer-Shoup encryption scheme, the Merkle-Damgård iterative hash function design, and of the ZAEP encryption scheme. As was done for CertiCrypt, an extension of the EasyCrypt prototype was developed to reason about differential privacy and its variant against computationally bounded adversaries, and was used to verify a smart-metering protocol. Even though this early prototype greatly simplified the writing of fully formal security proofs for primitives, it was still ill-suited to dealing with the complex layered cryptographic systems in which cryptographers are interested. In particular, without any abstraction mechanism, parallel or successive reductions could not be proved in isolation and combined in abstract ways, which led to important modularity and scalability issues.

2.3 EasyCrypt 1.0

Starting from 2012, a complete reimplementation of EasyCrypt (https://www.easycrypt.info) was therefore initiated, with the goal to overcome these limitations. In addition to dealing with the scalability issues mentioned above, the goals of the reimplementation were three-fold: first, consolidate the prototype into a robust platform that can be maintained and extended with reasonable effort; second, provide a versatile platform that supports automated proofs but also allows users to perform complex interactive proofs that interleave program verification and formalization of mathematics, which are intimately intertwined when formalizing cryptographic proofs; third, develop and implement the necessary foundations required to apply standard cryptographic reasoning principles that were not supported by the EasyCrypt prototype. To achieve these goals, the current version of EasyCrypt implements a probabilistic Hoare logic pHL for bounding the probability of post-conditions, and embeds both pRHL and pHL into an ambient logic that can for instance be used to perform hybrid arguments involving equivalences on parameterized programs coupled with inductive arguments on the parameters. In addition, it implements a module system and a theory mechanism that support compositional proofs through quantification over programs (as modules) and over types and values (through theories); using the module system and the new logics, we have been able to formalize cryptographic proofs that were out of reach of the initial prototype, including proofs of security for secure function evaluation, verifiable computation and authenticated key exchange protocols.

Example: security of a stateful random generator. As an example of an EasyCrypt proof, we now discuss a proof of security for a simple stateful random generator based on a pseudo-random function (PRF). We start by defining the assumption and security notion, and give a high-level overview of the proof. This simple proof does not fully exercise the module system. A more complex, albeit slightly more contrived, example can be found in the EasyCrypt tutorial (Barthe et al., 2014a).

We consider a set of *seeds* $\mathcal{S}$ and a set of *outputs* $\mathcal{O}$, equipped with unspecified but proper distributions $d_{\mathcal{S}}$ and $d_{\mathcal{O}}$. We denote sampling a variable x in $d_{\mathcal{S}}$ (resp. $d_{\mathcal{O}}$) with $x \xleftarrow{\$} \mathcal{S}$ (resp. $x \xleftarrow{\$} \mathcal{O}$). We assume a family of functions F from $\mathbb{N}$ to $\mathcal{O}$ indexed by $\mathcal{S}$. We construct a stateful random generator by using F in counter mode. The EasyCrypt code for these declarations and definitions is shown in Listing 1.1. The SRG construction is defined as a *module*, which defines a memory space containing global variables (here a variable s of type

$\mathcal{S}$ and a variable c in $\mathbb{N}$) and two procedures: i. a procedure init that, when called, simply samples the seed s in $d_{\mathcal{S}}$ and initializes the counter c to 0; and ii. a procedure next that, when queried, computes its output by applying F_s to c before incrementing c.

```
type S, O.
op F: S → N → O.

module SRG = {
  var s:S
  var c:N

  proc init(): unit = {
    s ⟵$ S;
    c = 0;
  }

  proc next(): O = {
    var r = F s c;
    c = c + 1;
    return r;
  }
}.
```

Listing 1.1. A Stateful Random Generator

Our objective is to show that SRG is a secure pseudo-random generator (PRG), under the assumption that F is a secure pseudo-random function (PRF). We first express these two notions formally, starting with the assumption on F.

Secure PRF. We say that F is a secure PRF if it is *computationally indistinguishable*, when used with a seed s sampled in $d_{\mathcal{S}}$, from the lazily sampled random function displayed in Listing 1.2.[4] We use the type (α, β) *map* of *finite maps* from α to β, using map0 to denote the empty map, dom m to denote the set of elements where m is defined, and standard notations for map updates and reads.

```
module RF = {
  var m:(N,O) map

  proc init(): unit = { m = map0; }

  proc f(x:N): O = {
    if (x ∉ dom m) m[x] ⟵$ O;
    return m[x];
  }
}.
```

Listing 1.2. Random Function

---
[4] This is a generalization of the standard cryptographic notion.

Computational indistinguishability is defined using a security experiment, parameterized by a construction and a distinguisher. *Module types* specifying the set of procedures expected to be implemented by a module are used to define parameterized modules. Module types themselves can be parameterized, allowing us to define, for example, the type of PRF distinguishers as modules that must implement a boolean procedure that may make oracle queries to a procedure f that take an argument in $\mathbb{N}$ and return some output in $\mathcal{O}$. We wrap the function family F into a module whose init procedure samples the seed that is used as index to F to answer f queries.

```
module PRFr = {
  var s:S
  proc init(): unit = { s $← S; }
  proc f(x:ℕ): 𝒪 = { return F s x; }
}.

module type PRF = {
  proc init(): unit
  proc f(x:ℕ): 𝒪
}.

module type Distinguisher^{PRF} (P:PRF) = {
  proc distinguish(): bool { P.f }
}.

module IND^{PRF} (P:PRF,D:Distinguisher^{PRF}) = {
  proc main(): bool = {
    P.init();
    return D(P).distinguish();
  }
}.
```

Listing 1.3. Pseudo-Random Functions

We define the PRF advantage of a PRF distinguisher D against F as follows, where $\mathrm{IND}_M^{\mathrm{PRF}}(\cdot)$ stands for $\mathrm{IND}^{\mathrm{PRF}}(M, \cdot)$, and $\Pr[G : E]$ is the probability of event E in the distribution on final states (using res to denote the special state location holding the procedure's return value) produced when executing G in some fixed initial memory.

$$\mathrm{Adv}_F^{\mathrm{PRF}}(D) = \Pr\left[\mathrm{IND}_{\mathrm{PRFr}}^{\mathrm{PRF}}(D) : \mathsf{res}\right] - \Pr\left[\mathrm{IND}_{\mathrm{RF}}^{\mathrm{PRF}}(D) : \mathsf{res}\right]$$

Intuitively, F is a secure PRF whenever, for all "efficient" PRF distinguisher D, $\mathrm{Adv}_F^{\mathrm{PRF}}(D)$ is "small". We do not formalize what it means for an algorithm to be efficient or for an advantage to be small, but simply relate the advantages of various adversaries against various constructions. In practice, further work is needed to argue for security; in particular, the complexity of reductions often needs to be analyzed.

Secure PRG. We say that SRG is a secure PRG if it is computationally indistinguishable from the true random generator that samples its successive outputs in $d_{\mathcal{O}}$ (Listing 1.4).

```
module RG = {
  proc init(): unit = { }
  proc next(): O = {
    var r ⇐ O;
    return r;
  }
}.
```

Listing 1.4. Random Generator

We use similar module types and modules to those used to define PRF security and say that SRG is a secure PRG if, for all "efficient" PRG distinguisher D, the following quantity is "small".

$$\mathsf{Adv}_{\mathsf{SRG}}^{\mathsf{PRG}}(\mathsf{D}) = \Pr\left[\mathsf{IND}_{\mathsf{SRG}}^{\mathsf{PRG}}(\mathsf{D}) : \mathsf{res}\right] - \Pr\left[\mathsf{IND}_{\mathsf{RG}}^{\mathsf{PRG}}(\mathsf{D}) : \mathsf{res}\right]$$

```
module type PRG = {
  proc init(): unit
  proc next(): O
}.

module type Distinguisher^PRG (G:PRG) = {
  proc distinguish(): bool { G.next }
}.

module IND^PRG (G:PRG,D:Distinguisher^PRG) = {
  proc main(): bool = {
    G.init();
    return D(G).distinguish();
  }
}.
```

Listing 1.5. Pseudo-Random Generators

Security of SRG. We prove the security of the SRG construction by constructing, from any PRG distinguisher D, a PRF distinguisher D' whose advantage bounds that of D. This involves simulating PRG oracles using only the PRF oracles and public information, so the PRG distinguisher can be run, and using its result to break the PRF's security. In this case, the reduction is a simple reinterpretation of the SRG construction as part of an adversary against the underlying PRF and the security proof is a simple proof of equivalence. The adversary is defined as follows, first defining a module PRGp that simulates the SRG algorithm with only oracle-access to the PRF and then using the PRG security experiment as distinguisher. Note that the initialization of the PRF module's state is, crucially, left to the PRF experiment, allowing us to prove that, given a PRG distinguisher D, the module $\mathsf{D}_{\mathsf{D}}^{\mathsf{PRF}}$ is a valid PRF distinguisher.

```
module D^PRF (D:Distinguisher^PRG,P:PRF) = {
  module PRGp = {
    proc init() = { SRG.c = 0; }
    proc next() = {
      var r = P.f(SRG.c);
      SRG.c = SRG.c + 1;
      return r;
    }
  }

  proc distinguish = IND^PRG_PRGp(D).main
}.
```

Listing 1.6. Reduction

Indeed, D_D^{PRF} implements a boolean procedure distinguish that may make oracle queries only to the f procedure of its parameter P. This observation allows us to consider the modules $IND_P^{PRF}(D_D^{PRF})$ (for $P <: PRF$, and in particular for $P \in \{PRFr, RF\}$), since they are well-typed, and we can prove the following equalities.

$$\Pr\left[IND_{SRG}^{PRG}(D) : res\right] = \Pr\left[IND_{PRFr}^{PRF}(D_D^{PRF}) : res\right] \tag{1}$$

$$\Pr\left[IND_{RG}^{PRG}(D) : res\right] = \Pr\left[IND_{RF}^{PRF}(D_D^{PRF}) : res\right] \tag{2}$$

Equality (1) is an easy program equivalence: by inlining D_D^{PRF}.distinguish in the PRF security experiment, we see that the initialization code is the same in both programs. The PRG adversary is called on the left with the next oracle from module SRG, whereas it is called on the right using the next oracle from the PRGp simulation. We use EasyCrypt's adversary rule to reduce the equivalence of these two adversary calls to the observational equivalence of the oracles they query (with respect to some observation on the states). In this case, we use the fact that the value of SRG.s on the left (SRG.s{1}) is equal to the value of PRFr.k on the right (PRGr.k{2}), and that the counters are equal. This can be proved easily by inlining P.f.

The invariant used to prove Equality (2) is more involved. Indeed, the program on the left always returns freshly sampled randomness whereas the program on the right only returns freshly sampled randomness on fresh queries to the PRF. However, it is easy to see that the very structure of the D^{PRF} construction imposes that all queries made to the PRF oracle are indeed fresh. We therefore use the following invariant, easily discharged by inlining and the SMT solvers, which allows us to prove that the results of each query to the next oracle are equally distributed.

$$\forall x, x \in dom\ PRFi.m\{2\} \Leftrightarrow 0 \leq x \leq SRG.c\{1\}$$

We then conclude the proof by rewriting in the advantage definitions, establishing the following equality for all PRG distinguisher D.

$$Adv_{SRG}^{PRG}(D) = Adv_F^{PRF}(D_D^{PRF})$$

Comparison with the EasyCrypt prototype. EasyCrypt 1.0, learning from the prototype's shortcomings, focused on a proof engine that relies on constructing proof trees that include some SMT nodes. This move from semi-automated to "mostly-interactive" did not only provide a graceful fallback mechanism when automation fails and better prospects for the construction of independently-verifiable proof certificates (for example, through the use of proof-producing SMT solvers), but also enabled us to embed program logic judgments in pRHL, pHL and standard Hoare logic into the ambient logic, permitting explicit quantification over stateful programs. In addition, the implementation of a theory mechanism allows us to develop libraries of *data structures* and *security notions* that can be instantiated at will by the end user, instead of having to re-formalize them in the context of each particular proof. For example, the security notions for PRF and PRG security described in the example above can be made into abstract theories, and instantiated with concrete types and distributions as needed. The SRG construction itself is described on abstract sets $\mathcal{S}$ and $\mathcal{O}$ and an abstract function family F that can be instantiated, for example, with bitstrings of appropriate lengths and the AES block cipher. This additional abstraction mechanism allows us to think modularly about proofs of implementations, but also about proofs of complex constructions: a proof obtained in an abstract setting can be fully refined (obtaining a proof for some implementation code), or simply instantiated with the types, operators and distributions used in another abstract construction (obtaining a generic proof of composition).

In addition, highly modular proofs allow us to identify widely-used high-level principles in cryptographic proofs. In turn, developing specialized libraries for these high-level principles allows us to support their application with minimal use of EasyCrypt's interactive core, simply proving straightforward program equivalences when refactoring cryptographic constructions to enable the application of the desired high-level principle.

2.4 ZooCrypt

The ZooCrypt framework (Barthe et al., 2013) provides tools for automatically analyzing and synthesizing padding-based encryption schemes. The class of padding-based encryption schemes consists of public-key encryption schemes built from one-way trapdoor permutations and random oracles. In practice, these primitives are often instantiated with the RSA function and hash functions.

Even though these building blocks are relatively simple and well understood, it is surprisingly difficult to find constructions that are simple, minimize ciphertext expansion and support tight reductions to the security of the employed one-way function. For example, Bellare and Rogaway (1994) proved security against chosen-ciphertext attacks (IND-CCA) for the OAEP scheme shown in Listing 1.7 under the one-way assumption.

$r \xleftarrow{\$} \{0,1\}^k$; $s = H(r) \oplus (m \mid 0^l)$; $t = H(s) \oplus r$; **return** $f(s|t)$

Listing 1.7. OAEP encryption of message m

Later on, Shoup (2001) proved that it is impossible to reduce the security of OAEP to one-wayness and the proof must therefore be flawed. To regain confidence in the widely used OAEP scheme, Shoup (2001) and Fujisaki et al. (2001) developed new proofs for OAEP under stronger assumptions. Additionally, many schemes have been proposed that improve on various aspects of OAEP, for example by providing security under the weaker one-wayness assumption.

The goal of the ZooCrypt framework is to demonstrate that fully automated game-based proofs and computer-aided design are feasible in the domain of padding-based encryption schemes. ZooCrypt consists of two components: an analyzer that can decide efficiently whether an instance construction is secure, and a synthesizer that implements a smart generation algorithm for candidate instances. The analyzer combines efficient search procedures to prove the security of an instance using a custom proof system, and attack finding procedures based on symbolic models of cryptography. The custom proof system consists of a small number of high level proof rules that formalize the game hops used in such proofs. Using ZooCrypt, we have built a database that contains more than one million padding-based encryption schemes. To build this database, our tool has not only found many new schemes, it has also rediscovered most schemes from the literature (including proofs).

2.5 Generic Group Analyzer

The GenericGroupAnalyzer (Barthe et al., 2014b) is a tool to analyze cryptographic assumptions in generic group models. Barring a breakthrough in complexity theory, the hardness of cryptographic assumptions, such as the discrete logarithm problem in certain cyclic groups, cannot be proved in general models of computation. To sidestep this problem, a commonly used approach is to prove lower bounds on the runtime of *generic algorithms* that do not exploit the concrete representation of group elements. This approach was initiated by Nechaev (1994) and Shoup (1997) and a proof in the generic group model can be considered as a minimal requirement for a newly proposed cryptographic assumption.

The GenericGroupAnalyzer supports three types of problems: i. non-parametric problems where the group setting is fixed and the adversary obtains a fixed set of group elements; ii. parametric problems where the group setting and the size of the adversary input is parameterized; iii. interactive problems where the adversary can adaptively query oracles to obtain new group elements.

In the non-parametric mode, our tool takes a group setting and a specification of the right and left adversary input and returns either an upper bound on the winning probability or an algebraic attack on the assumption. The assumption shown in Listing 1.8 formalizes the decisional Diffie-Hellman problem in a bilinear group of Type II. A bilinear group of Type II consists of three prime-order groups $\mathbb{G}_1, \mathbb{G}_2, \mathbb{G}_T$ with generators g_1, g_2, g_T, an efficiently computable isomorphism $\psi : \mathbb{G}_2 \to \mathbb{G}_1$, and a bilinear map $e : \mathbb{G}_1 \times \mathbb{G}_2 \to \mathbb{G}_T$. To solve the decisional Diffie-Hellman problem, the adversary must distinguish the triple $(g_2^x, g_2^y, g_2^{x*y}) \in \mathbb{G}_2^3$ from the triple $(g_2^x, g_2^y, g_2^z) \in \mathbb{G}_2^3$ for randomly sampled

$x, y, z \in \mathbb{F}_p$ given blackbox access to the bilinear map e and the isomorphism ψ. For this input, our tool returns the distinguishing test $e(\psi(w_1), w_2) = e(g_1, w_3)$, where w_i denotes the i-th element of the triple.

```
iso G1 → G2. map G1 * G2 → GT.

input [ x, y ] in G2.

input_left [ x*y ] in G2.
input_right [ z ] in G2.
```

Listing 1.8. Decisional Diffie-Hellman problem in bilinear group of Type II

Listing 1.9 shows the Diffie-Hellman exponent problem, where the adversary is given $g_1^y, g_1, g_1^x, \ldots, g_1^{x^{n-1}}, g_1^{x^{n+1}}, \ldots, g_1^{x^{2n}} \in \mathbb{G}_1$ for random $x, y \in \mathbb{F}_p$ in a symmetric bilinear group. To win, the adversary must distinguish $g_2^{y*x^n}$ from a random element in the target group $\mathbb{G}_2$. Our tool confirms that the winning probability of the adversary is negligible.

```
setting symmetric. levels 2. problem_type decisional.

input [ y, forall i in [0, n − 1]: x^i, forall j in [n + 1, 2*n]: x^j ] in G1.

challenge y*x^n in G2.
```

Listing 1.9. Parametric n-Diffie-Hellman-exponent-problem

Listing 1.10 shows the interactive LRSW problem introduced by Lysyanskaya et al. (2000). Here, the adversary gets $g_1^x, g_1^y \in \mathbb{G}_1$ and can additionally query the oracle O with an element m_O to obtain the triple $(A, A^y, A^{x+m_O*x*y}) \in \mathbb{G}_1^3$ for a randomly sampled $A \in \mathbb{G}_1$. To win, the adversary must compute a tuple $(A', V, W) \in (\mathbb{G}_1 \setminus \{1\}) \times \mathbb{G}_1 \times \mathbb{G}_1$ that satisifies the same relation for an $m \in \mathbb{F}_p^*$ of his choice that has not been queried to O. Our tool confirms that the winning probability of the adversary is negligible for a polynomial number of queries.

```
input [x, y] in G1.

oracle O(mo:Fp) = sample a; return [a, y*a, a*(x + mo*x*y)] in G1.

win (A':G1, V:G1, W:G1, m:Fp) =
    V = A'*y ∧ W = A'*(x + m*x*y) ∧ A' ≠ 0 ∧ mo ≠ m ∧ m ≠ 0.
```

Listing 1.10. Interactive LRSW problem

Our tool relies on a generalization of the master theorem introduced by Boneh et al. (2005) and uses SMT solvers for checking constraint satisfiability and computer algebra systems for linear algebra and Gröbner Basis computations.

3 Discussions and conclusions

The primary motivation for our work is to support the construction of independently verifiable proofs of security for cryptographic systems. It is indeed folklore that there is a very significant asymmetry between building and checking a formal proof, and that one can achieve trust in formal proofs by inspecting their statements and the definitions that they use, but without actually inspecting the proofs themselves. On this account, formal proofs provide a pragmatic solution to the unverifiability of cryptographic proofs, and allow the proof reader to shift his focus on checking that definitions and statements are indeed appropriate for the claimed results, which is another important source of mistakes in cryptography. The GenericGroupAnalyzer also makes a step in the direction of validating new security definitions by automatically proving new assumptions are secure with respect to a generic model of computation.

More generally, our experience with the various tools described here tends to show that a careful mix of interactivity and automation is highly desirable when dealing with complex proofs. However, rather than adopting the usual "mostly-automated" approach usually taken in general-purpose program verification, we choose to support a "mostly-interactive" approach to proof building. This allows us to deal with the complex number theoretic and algebraic arguments that appear in cryptography whilst using the automated techniques to discharge the trivial-but-tedious proof obligations generated by the program verification part of the tool. In addition, this mostly-interactive approach encourages the construction of layered tools that may make use of EasyCrypt as a back-end with little to no fear that automation will fail and break the proof. In turn, this allows the development of fully-automated – albeit specialized – tools, or of more intuitive proof construction interfaces that may be used, for example, to teach provable security. [5] Still, EasyCrypt's automation features remain valuable for the construction of proofs that take into account increasing amounts of "real-world" considerations. In particular, the new EasyCrypt prototype was also equipped with an extraction feature, directly producing ML code from formal specifications proved secure in EasyCrypt. It has also been shown that EasyCrypt's proof engine and philosophy, based on program verification techniques, does support direct computational security proofs on C-like code that can be carried through to executable code using certified compilation (Almeida et al., 2013).

Finally, the development of a formal framework to reason about discrete probabilistic programs also allows developments outside of the realm of proofs. Indeed, we have recently developed an EasyCrypt-backed fault attack synthesis algorithm. Given an implementation and a *fault condition* (a set of final states known to yield usable information on the secrets), our algorithm finds variants of the program that follow a chosen fault model and fault policy and guarantee the fault condition, ensuring a successful attack.

About the tools. More information on the tools and projects presented in this chapter, including downloads, documentation and tutorials, can be obtained by contacting its authors at `appa14@projects.easycrypt.info` or by visiting `https://www.easycrypt.info`.

---

[5] For example, ZooCrypt's logic for CPA proofs has been used to construct a proof tutor, accessible through `https://www.easycrypt.info/trac/wiki/ZooCrypt`.

Bibliography

José Bacelar Almeida, Manuel Barbosa, Gilles Barthe, and François Dupressoir. Certified computer-aided cryptography: efficient provably secure machine code from high-level implementations. In *ACM CCS 13: 20th Conference on Computer and Communications Security*, pages 1217–1230. ACM Press, 2013.

Philippe Audebaud and Christine Paulin-Mohring. Proofs of randomized algorithms in Coq. *Science of Computer Programming*, 74(8):568—589, 2009.

Gilles Barthe, Benjamin Grégoire, and Santiago Zanella-Béguelin. Formal certification of code-based cryptographic proofs. In *36th ACM SIGPLAN-SIGACT Symposium on Principles of Programming Languages, POPL 2009*, pages 90–101, New York, 2009. ACM.

Gilles Barthe, Benjamin Grégoire, Sylvain Heraud, and Santiago Zanella-Béguelin. Computer-aided security proofs for the working cryptographer. In *Advances in Cryptology – CRYPTO 2011*, volume 6841 of *Lecture Notes in Computer Science*, pages 71–90, Heidelberg, 2011. Springer.

Gilles Barthe, Juan Manuel Crespo, Benjamin Grégoire, César Kunz, Yassine Lakhnech, Benedikt Schmidt, and Santiago Zanella Béguelin. Fully automated analysis of padding-based encryption in the computational model. In *ACM CCS 13: 20th Conference on Computer and Communications Security*, pages 1247–1260. ACM Press, 2013.

Gilles Barthe, François Dupressoir, Benjamin Grégoire, César Kunz, Benedikt Schmidt, and Pierre-Yves Strub. Easycrypt: A tutorial. In Alessandro Aldini, Javier Lopez, and Fabio Martinelli, editors, *Foundations of Security Analysis and Design VII*, volume 8604 of *Lecture Notes in Computer Science*, pages 146–166. Springer International Publishing, 2014a. ISBN 978-3-319-10081-4. doi: 10.1007/978-3-319-10082-1_6. URL http://dx.doi.org/10.1007/978-3-319-10082-1_6.

Gilles Barthe, Edvard Fagerholm, Dario Fiore, John Mitchell, Andre Scedrov, and Benedikt Schmidt. Automated analysis of cryptographic assumptions in generic group models. In *Advances in Cryptology – CRYPTO 2014*, volume 8616 of *Lecture Notes in Computer Science*, pages 95–112, Heidelberg, 2014b. Springer.

Mihir Bellare and Phillip Rogaway. Optimal asymmetric encryption. In *Advances in Cryptology – EUROCRYPT 1994*, volume 950 of *Lecture Notes in Computer Science*, pages 92–111, Heidelberg, 1994. Springer.

Mihir Bellare and Phillip Rogaway. Code-based game-playing proofs and the security of triple encryption. Cryptology ePrint Archive, Report 2004/331, 2004. http://eprint.iacr.org/.

François Bobot, Jean-Christophe Filliâtre, Claude Marché, Guillaume Melquiond, and Andrei Paskevich. *The Why3 Platform*. Université Paris-Sud, CNRS, INRIA, March 2013. Version 0.81.

Dan Boneh, Xavier Boyen, and Eu-Jin Goh. Hierarchical identity based encryption with constant size ciphertext. In Ronald Cramer, editor, *Advances in*

Cryptology – EUROCRYPT 2005, volume 3494 of *Lecture Notes in Computer Science*, pages 440–456, Aarhus, Denmark, May 22–26, 2005. Springer, Berlin, Germany.

Jean Paul Degabriele, Kenneth G. Paterson, and Gaven J. Watson. Provable security in the real world. *IEEE Security & Privacy*, 9(3):33–41, 2011.

Eiichiro Fujisaki, Tatsuaki Okamoto, David Pointcheval, and Jacques Stern. RSA-OAEP is secure under the RSA assumption. In Joe Kilian, editor, *Advances in Cryptology – CRYPTO 2001*, volume 2139 of *Lecture Notes in Computer Science*, pages 260–274, Santa Barbara, CA, USA, August 19–23, 2001. Springer, Berlin, Germany.

Shai Halevi. A plausible approach to computer-aided cryptographic proofs. Cryptology ePrint Archive, Report 2005/181, 2005.

Anna Lysyanskaya, RonaldL. Rivest, Amit Sahai, and Stefan Wolf. Pseudonym systems. In Howard Heys and Carlisle Adams, editors, *Selected Areas in Cryptography*, volume 1758 of *Lecture Notes in Computer Science*, pages 184–199. Springer Berlin Heidelberg, 2000. ISBN 978-3-540-67185-5. doi: 10.1007/3-540-46513-8_14. URL http://dx.doi.org/10.1007/3-540-46513-8_14.

The Coq development team. *The Coq proof assistant reference manual*. LogiCal Project, 2004. URL http://coq.inria.fr. Version 8.0.

V. I. Nechaev. Complexity of a determinate algorithm for the discrete logarithm. *Mathematical Notes*, 55(2):165–172, 1994.

Victor Shoup. Lower bounds for discrete logarithms and related problems. In Walter Fumy, editor, *Advances in Cryptology – EUROCRYPT'97*, volume 1233 of *Lecture Notes in Computer Science*, pages 256–266, Konstanz, Germany, May 11–15, 1997. Springer, Berlin, Germany.

Victor Shoup. OAEP reconsidered. In Joe Kilian, editor, *Advances in Cryptology – CRYPTO 2001*, volume 2139 of *Lecture Notes in Computer Science*, pages 239–259, Santa Barbara, CA, USA, August 19–23, 2001. Springer, Berlin, Germany.

Victor Shoup. Sequences of games: a tool for taming complexity in security proofs. Cryptology ePrint Archive, Report 2004/332, 2004. http://eprint.iacr.org/.

www.ingramcontent.com/pod-product-compliance
Lightning Source LLC
Chambersburg PA
CBHW071637200326
41519CB00012BA/2334